INTERNET OF THINGS (IoT)

Technologies, Applications, Challenges, and Solutions

INTERNET OF THINGS (IoT)

Technologies, Applications, Challenges, and Solutions

Edited by
B.K. Tripathy
J. Anuradha

CRC Press
Taylor & Francis Group
Boca Raton London New York

CRC Press is an imprint of the
Taylor & Francis Group, an **informa** business

CRC Press
Taylor & Francis Group
6000 Broken Sound Parkway NW, Suite 300
Boca Raton, FL 33487-2742

First issued in paperback 2020

ISBN 13: 978-0-367-57292-1 (pbk)
ISBN 13: 978-1-138-03500-3 (hbk)

Library of Congress Cataloging-in-Publication Data

Names: Tripathy, B. K., 1957- editor. | Anuradha, J., editor.
Title: Internet of things (IoT) : technologies, applications, challenges, and solutions / edited by B.K. Tripathy, J. Anuradha.
Other titles: Internet of things (IoT) (CRC Press)
Description: Boca Raton : Taylor & Francis, CRC Press, 2018. | Includes bibliographical references and index.
Identifiers: LCCN 2017018235| ISBN 9781138035003 (hardback : alk. paper) | ISBN 9781315269849 (ebook)
Subjects: LCSH: Internet of things.
Classification: LCC TK5105.8857 .I57 2018 | DDC 004.67/8--dc23
LC record available at https://lccn.loc.gov/2017018235

Visit the Taylor & Francis Web site at
http://www.taylorandfrancis.com

and the CRC Press Web site at
http://www.crcpress.com

To my wife, Sandhya, for her untiring effort and constant support, which

have helped me in my research endeavors in a big way.

— B.K. Tripathy

To my Teachers, Family, and Friends whose impression

cannot be washed away from my life.

— J. Anuradha

Contents

Preface

Internet of Things (IoT) is the third wave of Internet and is supposed to have a potential to connect about 28 billion items by 2020, ranging from bracelets to cars. The term "IoT," which was first proposed by Kevin Ashton, a British technologist, in 1999, has the potential to impact everything from new product opportunities to shop floor optimization to factory worker efficiency gains that will power top-line and bottom-line gains. It is believed that IoT will improve energy efficiency, remote monitoring, and control of physical assets and productivity through applications as diverse as home security to condition monitoring on the factory floor. Now IoT has been used in markets in the field of health care, home appliances and buildings, retail markets, energy and manufacturing companies, mobility and transportation, logistics companies, and by media.

Equipments are becoming more digitized and more connected, establishing networks between machines, humans, and the Internet, leading to the creation of new ecosystems that enable higher productivity, better energy efficiency, and higher profitability. Sensors help to recognize the state of things, by which they gain the advantage of anticipating human needs based on the information collected per context. These intelligent devices not only gather information from their environment but are also capable of taking decisions without human intervention. IoT technology is being used in our day-to-day life for unlocking the door without a key; in card recognizers, automatic locks, vehicle detecting systems, toll payment system; and for tracking animals, access control, payment systems, contactless smart cards, anti-theft devices, steering column reader, etc. The IoT building blocks will come from those that are web-enabled devices, providing common platforms on which they can communicate, and develop new applications to capture new users.

In this background, this book is an attempt to present updated information on the recent trends on the issues involved, highlighting the challenges involved, and source the solutions for some of these challenges. The book comprises 14 chapters. The chapterwise description of contents in the volume is as follows.

For seamless visual tracking of passengers traveling in a vehicle, the visual light transmission (VLT) and visual light reflectance (VRT) values of the glass windows used in the vehicle should be at a particular value. All vehicle manufacturers follow certain standards. But the vehicle owner/user generally buys window-tinting films from the gray market and pastes the same on the glass windows, which in turn hamper the visibility, thereby preventing the law enforcement authorities from having a clear view of those traveling in the vehicle. In Chapter 1, a proposal for the automatic detection of tint level for vehicles is presented. The system has been designed using ordinary cameras that work in the visible region of the electromagnetic spectrum. The system is capable of identifying a vehicle's registered number using algorithms developed for this purpose. Various filtering techniques are applied to find the level of tint from the captured camera images. Using three different techniques, namely, color segmentation, contour detection, and histogram analysis, the tint level of windshield/window region is estimated. Thus, the IoT system recognizes the numbers on the number plate and can fetch the contact details of the owner from the database. It also communicates the same information to the owner with details on and extent of tint level violation along with documentary evidence.

Supervised and unsupervised learning techniques are reliable tools for the classification and categorization of data. Wearable devices are a relatively recent consumer technology that record large amounts of data by measuring signals around the body. Wearables come in several forms, such as wristbands, armbands, watches, and headbands. Chapter 2 explores the use of supervised and unsupervised learning techniques to identify individuals and activities using a commercially available wearable headband. It also deals with applying machine learning techniques for identifying individuals and their activities that are recorded through various wearable devices available in the market. A deep review on various algorithms applied for sensor data generated by wearable devices is carried out. Supervised and unsupervised learning algorithms are applied on EEG brain signal data for classification. This chapter also discusses challenges involved in handling sensor data and its mining.

IoT is a prototypical example of big data. In order to have viable solutions to that effect, artificial intelligence (AI) techniques are considered to be the right choice. Hence, the IOT system with AI techniques makes us to have smart applications like smart e-health, smart metering, and smart city to name a few. Although IoT is gaining everybody's attention today, security aspects need to be effectively addressed to prevent an intruder from causing disastrous consequences. In Chapter 3, the problem of prediction of energy consumption data generated through the smart grid step to connect all power grids for efficient energy management is discussed. The system applies convolution neural network and addresses the challenges posed by the IoT datasets and is considered to have widespread applications in the future digital world. The system is checked against publicly available smart city, smart metering, and smart health.

Due to enormous growth in the various applications of IoT, it has the potential to replace people as the largest producer and consumer of the Internet. The integration of wireless communication, microelectromechanical devices, and Internet has led to the development of things in the Internet. It is a network of network objects that can be accessed through the Internet, and every object connected will have a unique identifier. The increasing number of smart nodes and constant transfer of data are expected to create concerns about data standardization, interoperability, security, protection and privacy, and other issues. Chapter 4 elaborates on the technical, societal challenges and the impact of IoT applications. With the increasing number of smart nodes and constant transfer of data, it is expected to create concerns about data standardization, interoperability, security, protection, and privacy. This chapter gives a detailed outlook at the issues related to software engineering and security in IoT, from which one can provide solutions based on its understandings.

The social Internet of things (SIoT) is an emerging topic of the digital era with social, economic, and technical significance. The IoT has already proved its dominance in a wide range of sectors such as consumer products, durable goods, transportation, industrial and utility components, and sensors. It is now extended to social media. The evolution of powerful social network data analytic capabilities transforms the social livelihood into a new era of link prediction, community grouping, recommendation systems, sentiment analysis, and more. Chapter 5 covers the evolution of powerful social networking using IoT. The present society is digitally progressing toward an ever connected paradigm. It explains the basics of IoT and its technological evolution. The popularity of social networking describes the emergence of social network analytics in IoT. This chapter further discusses various security issues and research challenges pertaining to IoT analytics. It also provides information and references for further research and developments in applications for those in pro-business and pro-people social IoT services.

It has been established that a habitual typing pattern is a behavioral biometric trait in biometric science that relates to the issues in user identification/authentication systems.

Nevertheless, being nonintrusive and cost-effective, keystroke dynamics is a strong alternative to other biometric modalities. It can be easily integrated with any other existing knowledge-based user authentication system with minor alternation. Obtained accuracy in previous studies is impressive but not acceptable in practice due to problems in intra-class variation or data acquisition techniques. Chapter 6 deals with the prediction of authenticated user depending upon the individual typing pattern. The system aims to identify the user through the data generated based on the typing style. It also identifies the age, gender, ingenuity, and typing skills. A hybrid fuzzy and rough technique is being used and has proved to be effective.

Nanotechnology plays an important role in changing the world through the development of new technologies in various fields such as health care, manufacturing, agriculture, and industrial control systems. Recent developments in the Internet of Nano-Things (IoNT) are instrumental in the production of new nano devices. In Chapter 7, recent trends in nanotechnology are discussed and its roles in various emerging fields are presented. This chapter discusses the challenges and opportunities for the Internet of Forensic Nano-Things (IoNTF) in assisting and supporting reliable digital forensics in the IoNT environment.

Most devices we use today are equipped with sensors that have the capability of communicating and acquiring intelligence. Advancement of such technology supports mankind to assist in daily chores. Ambient assisted living (AAL) has drawn attention from people who are dependent, aged, and are under care. Chapter 8 focuses on this interesting application of IoT, that is, AAL. AAL systems are IoT-enabled that are exclusively designed to assist the aged/senior citizens in their daily activities. This chapter discusses various types of sensors used for this purpose and their strengths and weaknesses. It also describes various approaches adopted to develop an AAL system. Software engineering–based model development in AAL will enhance the functionality of the system, which is crucial in handling emergencies that older people encounter.

IoT is a system of interrelated computing devices, mechanical and digital machines, objects, animals, or people that are provided with unique identifiers and the ability to transfer data over a network without requiring human-to-human-to-computer interaction. Here, we need to know how the devices or things in IoT are communicating and how the name service will translate the meaningful names to machine-understandable form. Chapter 9 discusses the role of the naming service and the high-level requirements of the name service in IoT. It elaborates on the middleware technology to support device naming, application addressing, and profile storage and look-up services in IoT environments. Challenges of naming services in IoT are discussed along with how naming, addressing, and profile server (NAPS) overcomes these challenges. Platform-independent NAPS is discussed in detail with its complete design, including its functionalities, system flows, interfaces, etc.

Generally, an IoT application focuses on timely service and improves the efficiency. With this reason, today many IoT-based healthcare medical devices have been developed to provide proper treatment for patients in a timely manner. In the medical field, such devices have adopted the principle of lightweight protocols to communicate between various IoT-based medical devices. Chapter 10 provides information about recent technologies to mitigate security issues in IoT health care. This chapter discusses various security protocols and standards, concentrating on a detailed study of the security issues in communication among the devices. It offers a simulated approach and provides solutions to security using elliptic curve public key cryptography, which is implemented in a step-by-step manner in Contiki network simulator (Cooja) simulator.

The security issues in many IoT implementations, although already present in the field of information technology, offer a lot of unique challenges. It is a fundamental priority to address these challenges and ensure that the IoT products and services are secure. Chapter 11 deals with the challenges involved in handling security issues on IoT data. IoT devices are vulnerable to attacks as the systems have a very low level of protection and security. In this chapter, several types of intrusions and threats in IoT are discussed. An innovative framework to develop a secured system along with authentication is proposed. The collaborative learning approach has proven to be an ideal model in recognizing the pattern of the intruder, and the system can provide a solution based on it.

Diseases occur because of the interaction of agent, host, and the environment that is altogether called the "epidemiological triad." There are numerous reasons for the failure to prevent the complications of diseases. Physicians address only the biomedical part of the disease and ignore the other important aspects due to lack of experience and time. This need not remain so, provided the science of diseases and the art of healing by physicians are embedded in wearable computational devices. Chapter 12 addresses the electronic health record maintenance by coupling IoT devices with the system where the medical practitioner can view the data on the table. It also emphasizes that IoT is a boon to the medical field that saves doctors' time in handling other necessary calls. Wearable devices like fever watches when used on babies will reduce the risk of diseases by regular monitoring of the sensor data. The goal of this chapter is to make use of IoT to place health in the hands of the patients, which is one of the key principles of primary health care.

Billions of devices are expected to be connected to the Internet in the near future. This will provide interoperability among the things on the IoT which is one of the most fundamental requirements to support object addressing, tracking, and discovery, as well as information representation, storage, and exchange. Defining ontology and using semantic descriptions for data will make it interoperable for users and stakeholders that share and use the same ontology. Chapter 13 provides an overview of semantics in IoT-based analytics. This chapter briefs on semantic technology and its specifications and deployment to develop an application on smart systems.

Autism spectrum disorders (ASDs) show the concepts of instabilities that are disturbing social communication, interaction, and normal behavior in general. Every autistic child is unique in a way. They face severe problems with emotional balance, communal interactions, and communication skills. Such conditions necessitate a high degree of personalization to communicate with the outer world. Today, medical science has not been able to trace the exact reason for autism, but most therapists have proved that it is unpredictable behavior of neurons in human brains. We require a unique set of tools and methodologies to train autistic children. It requires enhancing the intensity of awareness about existing and narrowed competence among autism children to discover the way that unaffected children are performing. In Chapter 14, classification of the opportunities and trends of IoT applications and solutions for autistic children are presented. Also, from an IoT application perspective, diversified smart technologies like wearable environmental sensors mobile apps, home appliances, and others are discussed. The motivation of this chapter is to discuss the prospect of availability of an IoT support and in availing the services for enhancing the lives of autistic children and their supporting families in the society. Also, it helps to teach and train autistic children the basic skills and concepts in their day-to-day requirements.

IoT challenges for the future include ubiquitous data collection, potential for unexpected use of consumer data, and heightened security risks. So, the current technology needs to be improved to enhance privacy and build secure IoT devices by adopting a security-focused approach, reducing the amount of data collected by IoT devices, increasing transparency, and providing consumers with a choice to opt out. In an effort to draw the various issues in IoT, challenges faced, and existing solutions so far, the chapters of this volume have been meticulously selected and studied by knowledgeable reviewers.

It is the wish of the editors of this volume that this effort of theirs in accumulating so many challenges faced in the field of IoT, which have been focused on in the various chapters of this volume, will be helpful for future research in this field. Specifically, the real-world problems and different application areas presented will attract the attention of the researchers in the field and provide them with valuable input.

B.K. Tripathy
J. Anuradha
VIT University

Acknowledgments

The editors are grateful to CRC press for permitting them to edit this volume, *Internet of Things (IOT): Technologies, Applications, Challenges, and Solutions*. Especially, the support, suggestions, and encouragement offered by Dr. Gagandeep Singh are praiseworthy.

We are thankful to the authorities of VIT University for providing a congenial atmosphere and the moral support to carry out our work in a smooth manner.

Also, we are thankful to all the authors for their insightful contributions and the reviewers for their timely support and constructive suggestions, which have improved the quality of the chapters substantially.

Our scholars, R. K. Mohanty and T. R. Sooraj, have assisted during various stages of development of this volume. A special thanks to both of them.

Also, many colleagues and friends have helped in some way or other. We wish to thank them all.

Editors

B.K. Tripathy has received three gold medals for topping the list of candidates at graduation and post-graduation level at Berhampur University. He was a professor and head of the Department of Computer Science at Berhampur University until 2007. Dr. Tripathy is now working as a senior professor in School of Computing Science and Engineering, VIT University, Vellore, India. He has received research/academic fellowships from UGC, DST, SERC, and DOE of Government of India for various academic pursuits. Dr. Tripathy has published more than 400 technical papers in different international journals, proceedings of reputed international conferences, and edited research volumes. He has produced 26 PhDs, 13 MPhils, and 4 MS (by research) under his supervision, and has published two text books on soft computing and computer graphics. He was selected as honorary member of the American Mathematical Society from 1992 to 1994 for his distinguished contribution as a reviewer of the American Mathematical Review. Dr. Tripathy has served as the member of Advisory Board or Technical Program Committee member of several international conferences inside India and abroad. He has edited two research volumes for IGI publications and is editing three more research volumes. Dr. Tripathy is a life/senior member of IEEE, ACM, IRSS, CSI, ACEEE, OMS, and IMS. He is an editorial board member/reviewer of more than 60 journals. He has guest-edited some research journals and has held technical grants for research projects from various funding agencies like UGC, DST, and DRDO. His research interest includes fuzzy sets and systems, rough sets and knowledge engineering, data clustering, social network analysis, soft computing, granular computing, content-based learning, neighborhood systems, soft set theory, social internet of things, big data analytics, theory of multisets, and list theory.

J. Anuradha is an associate professor and head of the Department of Database Systems in School of Computing Science and Engineering, VIT University, Vellore, India. She has more than 13 years of teaching experience and has published at least 30 technical papers in international journals/proceedings of international conferences/edited chapters of reputable publications. She has worked and contributed in the field of data mining, medical diagnosis, and computational intelligence. Her research focuses on machine learning algorithms using soft computing techniques. Dr. Anuradha is currently working on spatial mining and big data analytics. She has coauthored a book, *Soft Computing: Advances and Applications* by Cengage Publishers. Her research interest includes fuzzy set theory and applications; rough set theory; and knowledge engineering and information retrieval.

Contributors

Katpadi Varadarajan Arulalan is a primary care pediatrician. His work follows the principle of primary health care, inter-sectoral coordination, where members of the medical community work in association with other members of the society. This includes the academy council members of the Local Arts and Science College, child welfare organization, education of children of HIV-positive parents, and community-based rehabilitation. He has 32 years of experience in low-cost pediatric health care. His areas of research include prevention of hospital visits, use of mobile phones as personal health records, and healthcare cost reduction through rational pediatric practice.

Kaliyaperumal Ganesan completed his PhD in 1993. Subsequently, he spent 3.5 years at Queen's University of Belfast, United Kingdom as a Post-Doctoral Fellow. He headed the NIIT Computer Centre for about 3 years upon his arrival in India in 1998. He has visited nearly 15 countries. Recently he visited US Universities of repute, including UC Berkeley, California Institute of Technology, Santa Clara University, UC Fresno, University of Southern California, and UC-Irvine. He worked as professor and head of the Department of Information Technology at Arunai Engineering College, Tiruvannamalai, India, for about 3 years. He joined as head of the Department of Computer Science and Engineering in 2002 at VIT University, Vellore. Since 2005, he has been working as the Director of TIFAC-CORE in the Automotive Infotronics Centre. In 2014, he was acting as the dean of the School of Information Technology and Engineering at VIT University for about 14 months. He was instrumental in helping the school obtain ABET accreditation during this period. He has published 100+ international journals cum conference papers. His first PhD student received the "Best PhD Thesis of the Country" award in 2012. The cash award was awarded by the President of India. Dr. K. Ganesan has filed 24 patents, which includes one US patent. He is a member of the Intellectual Property Cell at VIT University. He has conducted nearly 335 training programs for students/faculties/startups in cutting edge technology areas such as embedded systems, sensors, IoT, mobile computing, and big data analytics. He has completed 25 consultancy projects from industries such as Renault-Nissan, Continental, Amaron Batteries, and Delphi TCI. He has received generous grants from various organizations such as IBM and Motorola, in cash Dr. Ganesan has also obtained sponsored research project grants from DRDO, DST, AICTE, TePP, and TIFAC.

Neha Golani is a student pursuing a BTech in computer science and engineering (with specialization in bioinformatics) at VIT University, Vellore. She is currently working on real-time systems and design models based on the internet of things. She has worked on android development that involves efficacious patrolling of crime-prone areas. Her other interests include machine learning and social media mining.

Ezz El-Din Hemdan received his BS and MSc degrees in Computer Science and Engineering from the Faculty of Electronic Engineering, Menofia University, Egypt, in 2009 and 2013, respectively. Currently, he is working toward his PhD degree in the Department of Computer Science, Mangalore University, Mangalore, India. His research areas of interest include image processing, virtualization, cloud computing, networks and information security, digital forensics, cloud forensics, and big data forensics and internet of things/nano forensics.

Pawan Lingras is a graduate of IIT Bombay with post-graduate studies from University of Regina. He is currently a professor and director of Computing and Data Analytics at Saint Mary's University, Halifax. He is internationally active having served as a visiting professor at Munich University of Applied Sciences and IIT Gandhinagar, as a research supervisor at Institut Superieur de Gestion de Tunis, as a Scholar-in-Residence, and as a Shastri Indo-Canadian scholar. He has delivered more than 35 invited talks at various institutions around the world. Dr. Lingras has authored more than 200 research papers in various international journals and conferences. He has coauthored three textbooks, coedited two books, and eight volumes of research papers. His academic collaborations and coauthors include academics from Canada, Chile, China, Germany, India, Poland, Tunisia, UK, and the USA. His areas of interests include artificial intelligence, information retrieval, data mining, web intelligence, and intelligent transportation systems. He has served as the general cochair, program cochair, review committee chair, program committee member, and reviewer for various international conferences on artificial intelligence and data mining. He is also on the editorial boards of a number of international journals. His research has been supported by the Natural Science and Engineering Research Council (NSERC) of Canada for 25 years, as well as other funding agencies, including NRC-IRAP and MITACS. He is also serving on the NSERC's Computer Science peer review committee. Dr. Lingras has been awarded an Alumni Association Excellence in Teaching award, Student union's Faculty of Science Teaching award, and President's Award for Excellence in Research at Saint Mary's University.

Zhixing Liu, Zhicheng Yin, Shuai Zhao, Ziyun Zhong, and Runxing Zhou were part of a research team that graduated from Beijing Normal University, Zhuhai Campus, China with BSc degrees in computer science. They joined Saint Mary's University as students in the Master's program in computing and data analytics at Saint Mary's University.

D.H. Manjaiah is currently a professor in the Computer Science Department at Mangalore University. He earned his BE, MTech, and PhD degrees in computer science and engineering. He has more than 23 years of academic and industry experience. His areas of interest include advanced computer networks, cloud and grid computing, and mobile and wireless communication.

R.K. Mohanty is pursuing his PhD in Computer Science and Engineering under the supervision and guidance of Dr. B.K. Tripathy in VIT University, Vellore, India. His areas of research interest include soft set theory, soft computing, fuzzy set theory, decision-making under uncertainty, and computational intelligence. He is an author in more than 20 research publications.

Manoj Kumar Padhi is associated with Nokia, USA, as an integration professional and the founder of Fan's Global Social NGN LLC, which is involved in research on next generation services in social network. He received his MS in software systems from Birla Institute of Technology and Science, Pilani, India.

G.K. Panda is a senior faculty member in the Department of Computer Science and Engineering, MITS, Biju Patnaik University of Technology, Odisha, India. Presently, he is heading the institute. He obtained his M. Tech. in computer science from Berhampur University and graduated as the head of his class from the University. He received his MPhil from VM University, Tamil Nadu, and his PhD from Berhampur University, India. He has authored many research papers in national/international conferences/journals. Dr. Panda is associated with many professional bodies. His current research interest includes social network analysis, sentiment analysis, anonymization techniques, and rough set theory and applications.

Mrutyunjaya Panda holds a PhD in computer science from Berhampur University. He obtained his MS in Communication System Engineering from the University College of Engineering, Burla, at Sambalpur University; MBA in HRM from IGNOU, New Delhi; Bachelor in Electronics and Tele-Communication Engineering from Utkal University in 2002, 2009, and 1997, respectively. He has 19 years of teaching and research experience. He is presently working as a reader in the PG Department of Computer Science and Applications, Utkal University, Vani Vihar, Bhubaneswar, Odisha, India. He is a member of KES (Australia), IAENG (Hong Kong), ACEEE(I), IETE(I), CSI(I), and ISTE(I). He has published about 70 papers in international and national journals and conferences. Dr. Panda has published five book chapters, edited two books published by Springer, and authored two text books on soft computing techniques and modern approaches of data mining. He is a program committee member of various international conferences. He is acting as a member of an editorial board and active reviewer of various international journals. Dr. Panda's active areas of research include data mining, granular computing, big data analytics, internet of things, intrusion detection and prevention, social networking, wireless sensor networks, image processing, text and opinion mining, and bioinformatics.

P.B. Pankajavalli is currently working as assistant professor, Department of Computer Science, Bharathiar University, Coimbatore, India. She has 13 years of teaching experience. Her areas of specialization include MANETs, wireless sensor networks, and IoT. She has published and presented approximately 30 papers both at national and international conferences. She received the Certificate of Excellence award for teaching from the Lions Club International, Erode in January 2011. She received the Governor Award in January 2013 for the Best Service in Organizing the St. John Ambulance First Aid Training Programme at Kongu Arts and Science College, Erode, Tamilnadu, India. She has acted as a resource person in technical and motivational seminars and workshops conducted by various institutions.

Vijai Shankar Raja is the CEO, HELYXON® Healthcare Solutions Pvt. Ltd., IIT-M Research Park, Chennai, India. He has more than 29 years of experience in developing, maintaining, and installing digital medical equipment. He was one of pioneers to introduce digital x-ray machines on a large-scale across India. Their product Feverwatch® has won Indian Health Care Innovation Award as well as Frost & Sullivan's Best Practices Award in 2017.

Rajkumar Rajasekaran is an associate professor and head for data analytics at Vellore Institute of Technology, since 2002. He received his BE from Madras University in 1999 and his MTech degree from VIT University, Vellore, Tamilnadu. He received his PhD in Computer Science from the Vellore Institute of Technology. His research interests include the internet of things, healthcare data analytics, big data in health care, and mobile cloud health care. He has given numerous invited talks and tutorials, and is a founder of, and consultant to, companies involved in the internet of things. He is involved in teaching and learning new technologies.

Soumen Roy has been with the University of Calcutta for the last 4 years as a research scholar. He has also been working at Bagnan College, Bagnan, Howrah, India, for the past 5 years as a lecturer. He has a year's experience in software development. He has published more than 15 international papers in the form of journal and conference proceedings.

Utpal Roy was with School of Technology (SOT), Assam University, Silchar, India. He is now with Visva-Bharati, Santiniketan, and is the head of the Department of Computer and System Science. He has published more than 64 articles in the form of journal and conference proceedings.

T.R. Sooraj is pursuing his PhD in computer science and engineering under the supervision and guidance of Dr. B.K. Tripathy in VIT University, Vellore, India. His area of research interest includes soft set theory, soft computing, fuzzy set theory, decision-making under uncertainty, and computational intelligence and networking. He is an author of more than 20 research publications.

Shridevi Subramanian completed her MCA from Manonmaniam Sundaranar University, MPhil from Madurai Kamaraj University, and is currently pursuing research at MS University. She has published many papers in international/national conferences and journals. Her areas of interest include semantic web technology, semantic big data, and semantic web services.

Devadatta Sinha has been with the University of Calcutta for the last 30 years. He has 30 years of teaching and research experience. His research interests are in parallel processing, software engineering, and bioinformatics. He acted as chairperson for several national and international conferences.

R. Somasundaram is a research scholar in the School of Computer Science and Engineering at VIT University, Vellore, India. He received his MS degree in computer science and engineering from Arulmigu Meenakshi Amman College of Engineering, Anna University, Chennai. He has 4 years of teaching experience as an assistant professor in the Department of Computer Science and Engineering at Arulmigu Meenakshi Amman College of Engineering, Anna University. His area of specialization is network security. He has presented three papers in national and international conferences.

Mythili Thirugnanam is an associate professor in the School of Computer Science and Engineering at VIT University, Vellore, India. She received a MS degree in Software Engineering from VIT University. She received her PhD in computer science and engineering at VIT University in 2014. Dr. Thirugnanam has 9 years of teaching experience. She has 3 years of research experience in handling sponsored projects funded by the Government of India. Her areas of specialization include image processing, software engineering, and knowledge engineering. She has published more than 20 papers in international journals and presented 7 papers at various national and international conferences.

Matt Triff is a Master's degree candidate in computing and a data analytics at Saint Mary's University. He graduated from the University of Saskatchewan with high honors in computer science, specializing in Software Engineering. He is a veteran of many hackathons and has won awards in the LinkedIn Hackday, 2013 in Toronto and the Data Mining Competition at Joint Rough Set Symposium, 2013 in Halifax. He has software and system development experience with IBM Canada, POS Bio-Sciences, Opencare, and the Government of Canada.

Hrudaya Kumar Tripathy is an associate professor at the School of Computer Engineering, KIIT University, Bhubaneswar, Odisha, India. He earned his MTech degree in CSE from IIT, Guwahati, India, and his PhD in CS from Berhampur University, Odisha, India. He has 16 years of experience in teaching with 6 years of postdoctorate research experience in the fields of soft computing, machine learning, speech processing, mobile robotics, and big data analysis. He has been invited as visiting faculty to Asia Pacific University (APU), Kuala Lumpur, Malaysia, and Universiti Utara Malaysia (UUM), Sintok, Kedah, Malaysia. Dr. Tripathy worked as center head at APTECH Ltd., Bhubaneswar, for 3 years. He is a technical reviewer and member of the technical committee of many international conferences. He has published approximately 60 research papers in reputed international journals and conferences. He received the Young IT Professional Award 2013 at the regional level from the Computer Society of India (CSI). Dr. Tripathy is a member of the International Association of Computer Science and Information Technology (IACSIT), Singapore; member of the International Association of Engineers (IAENG), Hong Kong; senior member of IEEE, India Chapter; member of IET, India; associate member of the Indian Institute of Ceramics, Kolkata; and life member of the Computer Society of India, New Delhi.

Viswanathan Vadivel completed his doctoral degree from Anna University, Chennai, India, by contributing his ideas to the field of semantic web technologies and social media marketing. He has teaching experience of over 20 years in the field of computer applications. His research interests include data mining, semantic web, and social network analysis. He has authored articles in semantic web technologies for renowned publications.

S. Vandhana is a research scholar in the School of Computer Science and Engineering, VIT University, Vellore, Tamil Nadu. She is currently working on spatial mining and big data analytics. Her research focuses on machine learning algorithms using soft computing techniques. Her research interests also include information retrieval and semantic web mining.

Glavin Wiechert is currently a software developer for Lixar IT in Halifax while also attending Saint Mary's University for Honors in Computing Science, Bachelor of Science. He has received first place awards at Volta Lab's 48-hour Hackathon in 2016, IBM/Sobeys Retail Hackathon in 2015, and IBM Bluemix Hackathon in 2015. From 2013 to 2015, he cofounded a Halifax-based startup, Streamlyne Technologies, and developed an online service for predictive analytics using real-time oil and gas data.

1

IoT-Enabled Vision System for Detection of Tint Level

Ganesan Kaliyaperumal

VIT University

Vellore, India

CONTENTS

1.1 Introduction

In a typical automotive system, such as cars, buses, and lorries, we use glass windows at many places. The purpose of the glass window provided at the back side is to provide clear visibility of vehicles that are approaching our vehicle. The glass window provided at the front side of the car (called windshield) gives a clear view of things ahead of our vehicle. The driver moderates the speed of the vehicle according to the objects seen through the windshield. At every row of passenger seats, depending on the type of the vehicle, glass windows are provided on either side. The purpose of these glass windows is to give visibility of outside world to the passengers while traveling. During accidents or emergency situations, one can even break these windows and escape out of the vehicle. Apart from this, a mirror has been provided near the driver seat. Using it, a driver can look at the approaching vehicle(s) and also roughly estimate their speed which in turn will help him to take proper decisions on how to maneuver the steering wheel and avoid accidents.

The law enforcement authorities such as police personnel can look through these glass windows from outside and find out who are all traveling inside the vehicle. This kind of tracking is normally done by police personnel whenever some untoward incident such as accident or bomb explosion happens. In many such untoward incidents, typically the criminals/extremists escape in the vehicles to remote locations. Then, the police personnel are forced to track all vehicles at many strategic locations to identify or find where the criminals are. This kind of tracking is possible only if the police personnel are able to see the driver/passengers who are traveling in a vehicle from outside.

For seamless visual tracking of passengers traveling in a vehicle, the visual light transmission (VLT) and visual light reflectance (VLR) values of the glass windows used in the vehicle should be at a particular value. All vehicle manufacturers follow certain standards. But the vehicle owner/user generally buys window tinting films from grey market and pastes the same on the glass windows which in turn will hamper the visibility, thereby preventing the law enforcement authorities from having the clear vision of those traveling in the vehicle. In many cases, the tinting films are such that the passengers traveling inside the vehicle can clearly see the outside world through the glass window while the outsiders may not be able to see who are all inside the vehicle. To regulate this, law enforcement authorities in each country/state have prescribed certain threshold values on the VLT/VRT value to be used after fixing the tinted films on the glass window.

In this chapter, we propose a video-based automatic technique that will not only identify the vehicle (say, car) but also estimate the VLT percentage. If the VLT level is below the prescribed limit (say, 35%) then the envisaged system will automatically identify the vehicle number plate. Then the identified vehicle registration number will be searched in the appropriate database to locate the violator's contact details. Subsequently, an automatic message will be transmitted to the concerned person along with the date, time, and place of violation. Thanks to IoT (Internet of Things), the proposed system not only identifies the tint level violation but also provides documentary evidence and communicates the same to the relevant person.

1.2 Literature Survey

"Window tinting" refers to the methods used to block certain levels of light from passing through the glass window of vehicles. Most of the window glasses in the vehicles are coated or treated in order to filter the harmful ultra-violet (UV) rays entering the vehicle. The window tinting has various other effects. (1) Dark tinting will reduce driver's vision, particularly while driving at night. (2) Tinting windows will distort the vision of aged people and people with problems with eye sight. (3) If anyone wears polarized sunglasses, it will produce some visual patterns on the window. (4) Over a period of time, tinting can even lead to bubbles, peel, or develop cracks that may bring down the value of a vehicle and it general look. (5) Many vehicles come with window glasses that are engineered in such a way that in the event of an accident, they will "break away." At times, the tinting material will prevent windows from breaking during accidents. This in turn can cause or even worsen injuries during accidents.

Computing systems are generally introduced wherever speed and accuracy is needed. A typical computing system works on the principle of input–process–output. In the conventional computing system, the input is keyed in by human beings (as we see in many reservation counters, banks, etc.) and the software algorithm running at the computing system processes the inputted data and produces an output. In general, these outputs are displayed on a display device such as a monitor or printed using a printer. If people responsible are lethargic or lazy, often the speed and accuracy of the computing system is lost. However, IoT systems help us to resolve this issue.

In the case of an IoT system, the inputs are replaced by relevant sensor(s). Hence, the possibility of human errors can be eliminated to a great extent. Whenever we use a vision-based IoT system, the camera acts as an input sensor. Unlike a computing system, the processing application is located at a central server (or cloud system) in the case of an IoT system. In many IoT systems, the sensors forward their data to a local computing system via short-range wireless communication protocols such as Bluetooth, Zigbee, and WiFi. The local computing system preprocesses it and sends it to a central server (cloud) at regular intervals using long-range wireless communication protocols such as SMS, 2G, 3G, 4G, and WiMax. The central server (cloud) software collates the data and takes some informed decision (generally using analytics) and sends it back to the local server/end user. But in the case of an IoT system, one uses an actuator (controlling a valve) or an event driver notification system (such as an SMS/e-mail alert in a mobile phone). Thus, an IoT system uses sensor–processor–actuator principle. This can bring in next level of automation in many industries. The important challenge in IoT is designing proper sensors, actuators, and interfacing with wireless communication protocols. In general, the data handled by the servers are unstructured.

Thus, a typical IoT-enabled system consists of sensors, embedded system, (wireless) communication, data storage via Internet into a cloud server, an analytics software system to take informed decisions at the cloud platform, and an optional receiver end device (mobile). In few cases, the receiver end system will be an actuator that will execute certain actions on the targeted hardware. We can choose appropriate object(s) (such as lockers, curtains, consumer electronics devices such as cookers) and interface it (them) with an appropriate sensor(s). The embedded system attached with the object can sense the required data and monitor it continuously. At periodic, pre-defined time intervals, the necessary data are transfer to a remote location for storage using wired or wireless communication protocols (such as UDP, TCP/IP, Wi-FI, Bluetooth, ZigBee). The data aggregated from various objects are stored in the server, preferably a cloud server. At periodic intervals, the aggregated data are analyzed/diagnosed using appropriate analytics software running in the server (cloud). The decisions are informed accordingly to the end user/system. In most cases, it turns out to be the notification in the mobile device or actuation of some subsystems of the end device (such as rotating a fan, opening the valve system, switching ON/OFF).

Currently, there are many devices to measure the VLT/VRT values, called tint meters. There are two popular types of tint meters available in the market, namely one-piece tint meter and two-piece tint meter. The one-piece tint meter is used to check the tint level of side (glass) windows. If we use the one-piece tint meter, then one has to request the driver of the vehicle to be inspected to stop the vehicle and ask the driver/passenger to roll down the (side) windows partially. The tint meter has a slot built in to it (as shown in Figure 1.1). One has to insert the partially opened window into this slot and by pressing a button the light waves of specific wavelength are transmitted from one side of the tint meter and a receiver at the other side of the tint meter will analyze the received light waves. Depending on the transmitivity/opacity of the glass along with the black film pasted on it, we get values ranging from 0% (opaque material such as card board) to 100% (for pure air). Generally, the lighter glasses have high transmittance (65% to 80%) and the darker glasses have low transmittance (20% to 35%). In many states/countries, the allowed transmittance level is 35%. If the vehicle's glass tint level goes below this level, then one can issue a ticket and initiate legal action.

In the case of two-piece tint meters, we have two pieces, one back meter and one front meter (as shown in Figure 1.2). These kind of tint meters also are used to measure the

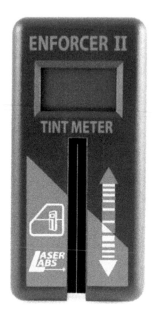

FIGURE 1.1
One-piece hand-held tint meter.

FIGURE 1.2
Two-piece hand-held tint meter.

tint level of glass windows used at the front side (wind shield) and rear (back) side of the vehicle. One can also use them for measuring the tint level of side windows of vehicles. The back meter is attached at the back side of the glass. It has a suction cup so that it can firmly stick to the glass window. The front meter is attached at the front side of the glass to be tested. The magnet helps both the pieces to be held together. By pressing a button, we can get the exact light transmittance value.

The main problem in both these types of tint meters is that one has to stop the suspected vehicle and carry out the experiment. We would like to design an IoT-enabled vision system that can resolve this problem. That means with our proposed system, we can measure the light transmittance value of glass windows of vehicles when it is moving. Our approach is based on image/video processing techniques. In our proposed method, we first extract vehicle windshield region using Gaussian kernel–based background subtraction, histogram equalization, optimal edge detection, and extreme point detection techniques. Then, in the detected region, we estimate the tint and transparency level. If the vehicle violates the government norms, then we crop the number plate and locate the owner's communication address and generate a ticket automatically by attaching the evidence.

1.3 Overview of the Proposed Methodology

Figure 1.3 shows the overview of the proposed windshield tint level detection system based on real-time vision. The input to our system is a live video captured through surveillance camera. From the captured video, two frames are extracted; one is the background image (taken when no vehicle is moving) and another one is the foreground image (taken when the vehicle is moving). Both the images have certain common properties such

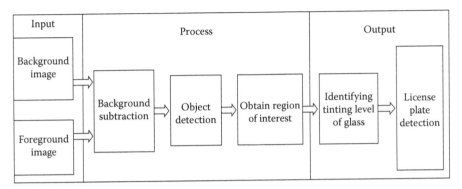

FIGURE 1.3
Overview of the proposed windshield tint level detection system based on real-time vision.

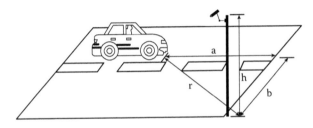

FIGURE 1.4
The operational environment of the proposed system.

as image color, image position, image size, image type, and origin. The images are initially preprocessed. The preprocessing includes filters, brightness and contrast equivalization, and image conversion. Using adaptive background subtraction method, the foreground object (moving vehicle) is only extracted. From this processed image, the region of interest (ROI), namely windshield/window region, is extracted, and then this ROI is cropped from the original color image and is passed on to the tint level calculation module. Three different techniques of tint detection were employed in the present work. If the tint level exceeds the prescribed limits, then the process switches over to license plate detection to uniquely identify the vehicle, and a ticket is issued, if necessary.

Figure 1.4 shows the operational environment of the proposed system in which the camera is placed at a certain height h from the ground level. The height is generally about 8–10 feet above the ground level so that the windshield/window can be easily visible with desirable coverage area that can also easily locate the number plate of the vehicle, if necessary. While capturing the image, the vehicle is in a position such that "r" is the shortest distance from the base of the pole on which the camera is mounted on the roadside and the front end of the vehicle. Here a and b are the horizontal and vertical components of this distance r, respectively. The camera inclination position should generally be not more than 30° with respect to the windshield/window level so that the internal view of the captured image (vehicle internal view) will be good.

Figure 1.5 shows the methodology used for finding the distance of the vehicle from the camera which in turn can be used for classifying the vehicle. For this purpose, the ground region is segmented into various imaginary grid lines both along x- and y-axes (x_1, x_2,... and y_1, y_2, y_3,...) and is overlaid on the captured image. Assume that the vehicle is present

FIGURE 1.5
Distance calculation using overlaid coordinates on the acquired image.

at the segmented graph location denoted as (x_{25}, y_{21}). Let the length and breadth of each block in the segmented graph be fixed. Then, the horizontal component gives the value "25 (from x component) * length of block" and the vertical component gives the value of "21 (from y component) * breadth of black." From these horizontal lengths and vertical breadths, one can calculate the aspect ratio, to approximately classify the vehicle type. Identification of vehicle types is very important in our case. The number of windows and location of windows depend on the type of vehicle.

1.4 Implementation Details

Figure 1.6 shows the detailed adaptive background subtraction (ABS) algorithm implementation. Here, two images, that is, background (Bg) and foreground (Fg) images are given as input to the system, and an appropriate filtering algorithm is applied to both the images for removing the noises (if any) from the images. Then, the images are converted into gray-scale images and their corresponding histograms are equalized for the desired background subtraction. The whole process is performed by pixel-by-pixel comparison of foreground and background images. If the absolute difference value of the foreground image pixel and its corresponding background image pixel is greater than a specified threshold value, then it is considered as foreground image and hence it is substituted with a high-intensity value (say, 255), else it is substituted with a low-intensity value (say, 0). Along with this subtraction process, the updation of background image is also done by comparing the foreground and background image pixels. This updation is performed to change the background image according to the change in environment. In this comparison, if the background image and foreground image intensity of specific location is the same, then the background remains same (no updation). Otherwise, if the value of the background pixel intensity of a specific location is greater than the value of the foreground pixel intensity of the corresponding location, then the background pixel intensity value is increased by one, if not the intensity value is decreased by one.

In Figure 1.6, an image "I" consists of a finite set of pixels and our mapping assigns to each pixel $p = (Px, Py)$, a pixel value $I(p)$. Consider that the mapping is done in Z^2 plane. So, in this algorithm, we have two images; one is background image which is represented as "Bg" and its intensity value $Bg(p)$ which is mapped in Z^2 plane at $p = (Px, Py)$, and the

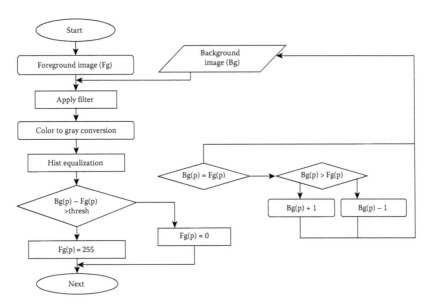

FIGURE 1.6
Data flow diagram explaining the adaptive background subtraction algorithm.

other is foreground image which is represented as "Fg" and its intensity value Fg(p) is also mapped in Z^2 plane at p = (Px, Py). The mapping of both images is the same because they have the same property.

1.4.1 Noise Removal

Most of the dynamic scenes exhibit persistent motion characteristics. Therefore, in order to reduce noise, image must be preprocessed with an appropriate filter. In our case, we used Gaussian kernel filter. Gaussian filtering is done by convolving each point in the input array with a Gaussian kernel and then summing them all to produce the output array. Gaussian kernel for N (N = 1, 2, 3,...) dimension is given by

$$G_{ND}(\bar{x};\sigma) = \frac{1}{\left(\sqrt{2\pi}\sigma\right)^N} e^{-\frac{|x|^2}{2\sigma^2}} \tag{1.1}$$

Here, σ determines the width of the Gaussian kernel. In statistics, when we consider the Gaussian probability density function, it is called the standard deviation, and the square of it namely, σ^2, is called the variance. The normalized Gaussian kernel has an area under the curve of unity, that is, as a filter it does not multiply the operand with an accidental multiplication factor.

The output of Gaussian filter is converted to a format (image format) which is easy to process. For this purpose, we use gray-scale range of shades wherein the darkest possible shade is black, which is the total absence of transmitted or reflected light and the lightest possible shade is white, which has the total transmission or reflection of light at all visible wavelengths. Intermediate shades of gray are represented by equal brightness levels of the three primary colors (red, green, and blue) for transmitted light, or equal amounts of the three primary pigments (cyan, magenta, and yellow) for reflected light.

1.4.2 Adaptive Background Subtraction

After preprocessing, the main task of background subtraction is to find d which is defined as $d(p) = |Fg(p) - Bg(p)|$. Here, Fg refers to foreground image, and Bg refers to back ground image. "p" refers to the position of various pixels on the image. Thus, d(p) refers to the distance between foreground and background image pixels. This difference must maintain the value less than the threshold value which is determined by the user. We store the $d(p)$ value in a visual pattern which is done by creating a matrix of size similar to foreground image which we call as "abs image." If the $d(p)$ is below a definite threshold value, then the "abs" image is padded with lowest intensity value and if $d(p)$ goes beyond the definite threshold value, then the "abs" image is padded with the highest intensity value. The abs image gives the subtraction of background and foreground image. Now we need to change the background image according to the change in the environmental or dynamic road condition such as a vehicle. It checks the background frame and verifies whether there is any change in the background image with respect to the foreground image. If $Bg(p) > Fg(p)$ for the location p, then the background intensity $Bg(p)$ is added with some arbitrary value, that is, "$Bg(p)$ + value" and if $Bg(p) < Fg(p)$, then $Bg(p)$ is subtracted with the same arbitrary value, that is, "$Bg(p)$ – value," and if $Bg(p) = Fg(p)$, then no change is done on the background image.

Our next intention is to increase the global contrast of foreground and background gray-scale images, especially when the usable data of the image is represented by the nearest contrast values. Through this adjustment, the intensities can be better distributed on the histogram. This allows for areas of lower local contrast to gain a higher contrast. Histogram equalization accomplishes this by effectively spreading out the most frequent intensity values.

1.4.3 Object Detection

The processed image ("abs" image) now contains high-intensity levels (which corresponds to the foreground and background intensity difference) indicating the presence of an object (i.e., vehicle). Then, we apply morphological operations on this processed image. In a morphological operation, the value of each pixel in the output image is based on a comparison of the corresponding pixel in the input image with its neighbors. The shape and size of the structural element is used to remove imperfections added during segmentation.

Figure 1.7 shows the auto-cropping algorithm used for obtaining the ROI from the chosen input foreground image. The algorithm is performed on the black and white images obtained from the background subtraction algorithm, and from this image one can identify the pixel location values for cropping the vehicle from the given foreground image. Initially, the size of image is stored in two variables, in which one is used to define the row size (say, row) and the other is used to define the column size (say, col). Now the variables m and n are set such that they store the starting location, namely top left corner of the image [say, $p(1,1)$] and bottom right corner of the image [say, p(row, col)]. We look for the first transition from low to high-intensity value of pixels from all sides (top, bottom, left, right) of the image. By scanning the binary image

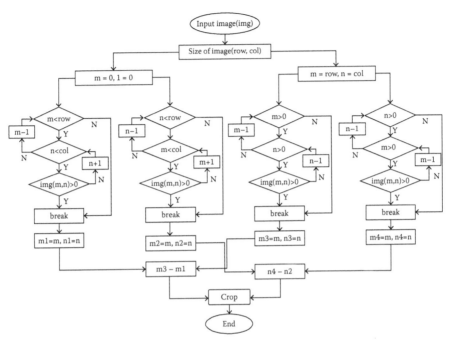

FIGURE 1.7
Data flow diagram for finding regions of interest and cropping the objects of interest.

from top to bottom and from left to right, we locate the extreme points (pixel locations) [say, top left pixel (m1, n1), top right pixel (m2, n2), bottom left pixel (m3, n3), and bottom right pixel (m4, n4)]. From these four extreme points, we can find the height and breadth of the vehicle which can fit into the rectangular box. Using these parameters (pixel extreme locations, length and breadth) on the original color image frame, we can crop the vehicle portion alone.

From the ROI, we can find the ratio of height and width of the vehicle which is directly related with the width and breadth of image rectangle. From this obtained ratio, the type of vehicle is determined such as hatchback or notchback (sedan class). The windscreen/ window area generally changes according to the type of vehicle. For example, in the sedan class vehicles, the windscreen location is nearly at the middle whereas in hatchback vehicles, it is slightly at the rear side of the vehicle. Thus, the vehicle tint location can be easily obtained for different class of vehicles. Once the vehicle class/type is extracted, then we can extract the windshield/window tint area of the vehicle and on this cropped area we apply our three basic techniques of tint detection, namely contour detection, histogram analysis, and color segmentation.

1.4.4 Tint Level Detection Algorithm

The first technique is color segmentation which identifies different color percentages present in the windshield/window area of the extracted image. It is also beneficial to identify approximately which color tint is applied on the screen on the basis of color percentage of different channels. In the present technique, the three channels of extracted area are separated out (i.e., RGB channel). In the separated channel, a specific point is defined for finding the intensity level in the image. For better accuracy, two or more points are identified within the window region and the average intensity of the pixels is used. Then, this intensity value is represented by their intensity percentage. According to the database available for VTL percentage for different environmental conditions, the approximate tinting level is determined. Table 1.1 shows the various tinting level on green/blue channels and the corresponding RGB level of region in the image.

The second technique used is known as contour detection which improves the tint detection of windscreen/window of vehicle. Contours can be explained simply as a curve joining all the continuous points (along the boundary), having the same color or intensity. In this technique, first we find the edges of the image by using relevant edge detection technique. After edge detection, the image becomes a function of two variables which are curves joining all continuous points, and these curves are called contours. These contours are in different numbers depending on the reflection of light through windshield/window glass. The number of contours is counted in the given image. From the available database (calculated manually) of contour for various intensity levels, the threshold (here is the number of contours) is defined and is used for determining the presence or absence of tint. If the number of contours is below the specified threshold limit, then the tested windshield/window tinting is not allowed according to the norms, and the process will switch to the next module, namely number plate detection.

The third technique is called "Histogram analysis" and is used for visually judging whether the image is in an appropriate range of gray level. Ideally, digital image should use all available gray-scale range, from minimum to maximum. From histogram of tint image (final extracted image), we can judge the presence of tinting on windshield/window. If the histogram of the extracted image has only two peaks of gray level, then it has undesired tinting level. If the histogram of the extracted image shows multiple peaks in the histogram, then we can conclude that there is desirable level of tinting. These three methods collectively can determine the tinting level of window/windshield region. Figure 1.8 shows the flow chart of various tint level detection algorithms used in our present work.

After identifying the tinting level of vehicle window/windshield using these three techniques, the process is switched to number plate identification module. If tinting level is more than the desired level (prescribed norms by state/country), then it switches to number plate identification module; otherwise, the test ends for the current vehicle. After identifying the license plate registration number of a given vehicle, the system

TABLE 1.1

Various Tinting Level on Green/Blue Channels and Corresponding RGB% Level Present in the Image

Tint VTL%	Green(grass) Value			Blue(sky) Value		
	R%	G%	B%	R%	G%	B%
No tint	49	55	20	67	75	68
50%	30	34	12	57	60	67
35%	16	22	8	40	40	40
30%	14	14	6	25	25	29
20%	9	9	3	19	19	21
15%	6	6	2	12	12	13
5%	0.4	0.4	0.4	0.4	0.4	0.4

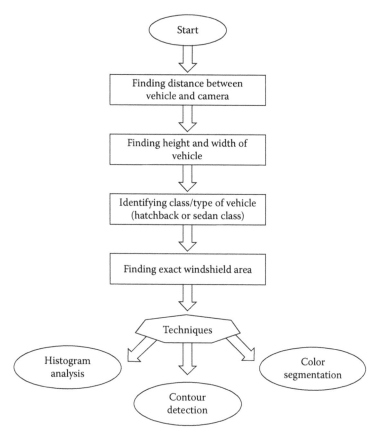

FIGURE 1.8
Various window tint level detection techniques used in the proposed system.

captures the image of the vehicle for the purpose of proof and saves it in a controller (or server). The controller is interfaced with a GPS device which provides the location details of the place where the images were captured and appends the system date and time. All these parameters are sent to the server system which stores all the received data. The vehicle registration number is extracted from the captured image and is searched in the available database, and the vehicle owner's contact details are traced for sending evidence to issue a ticket.

1.4.5 License Plate Detection

This is an important stage in vehicle license plate detection in the location of the license plate. The license plate area contains a good amount of edge and texture information. The license plate of the vehicle consists of several characters, and hence contains rich edge information. At times, the background of the vehicle image also contains much edge information. The interesting fact is that the background areas around the license plate mainly include some horizontal edges whereas the edges in the background mainly contain long curves and random noises. Also, the edges in the plate area cluster together and produce intense texture feature. If only the vertical edges are extracted from the vehicle image and most of the background edges are removed, then the plate

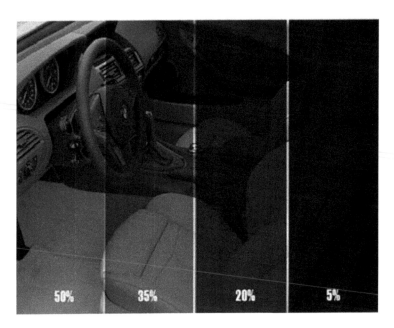

FIGURE 1.9
Effect on vision due to various tinting levels applied on windows.

area will be isolated out distinctly. The algorithm proposed in the present work contains four parts: image enhancement, vertical edge extraction, background curve and noise removal.

Figure 1.9 shows the effect on vision due to various tinting levels applied on windows and the corresponding darkness value. From Figure 1.9, we could clearly see that when the VTL value is less than 35%, it is very difficult to see the objects/persons present inside the vehicle. Due to this reason, many law enforcement authorities in various states/countries insist that the threshold value of VTL should be 35% or more.

Figure 1.10 shows the design of controller module for further processing. The vehicle under surveillance is tested for windshield/window tinting level. If the tinting level is up to the desired level, then the testing is continued for the next vehicle; otherwise, for the same vehicle, we switch to number plate identification module. After identifying the "vehicle number" of the vehicle, the GPS data or location ID, time, and date along with the proof of tinting level (processed image) are sent from the client system (surveillance camera system) to the server. The server checks the vehicle registration number in its database and locates the owner's contact details and accordingly issues a ticket to the owner of the vehicle.

1.5 Results and Conclusions

Figure 1.11 shows the various input images which are preprocessed and subtracted from background image. Then, the output is given to find the object(s) present in the image. Finding its extreme points gives us extracted or segmented image from which our main object of interest, namely the window is extracted. For the first set of input images, the RGB percentages are 45%, 43.16%, and 44%, respectively, which are greater

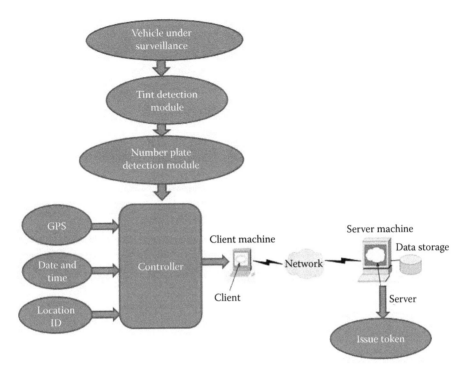

FIGURE 1.10
The controller modules for issuing ticket with evidence.

than 35% VTL color percentages. For the second set of input images, the RGB percentages are 14%, 16%, and 16%, respectively, which are less than 35% of VTL, and hence one can conclude that the dark tinting is present in the windows of those vehicles. The various input images given clearly identify and classify the vehicles based on tinting levels.

In conclusion, in the present work, we extract the relevant, meaningful images from the captured video frames. From the captured images, through motion detection techniques, the presence or absence of a vehicle is determined. If a vehicle is present, it is classified using appropriate imaging techniques. Depending on the type of vehicle, the ROI is identified and vehicle image alone is cropped for further processing. Depending on the vehicle type, the windshield/window region is identified and that portion of the image is used for further processing. Using three different techniques, namely color segmentation, contour detection, and histogram analysis, the tint level of windshield/window region is estimated. The tint level on the detected vehicle is verified against the database of government permissible limits. If the tint level exceeds government norms, the imaging system extracts the license plate of the vehicle. If the level exceeds the limits of regulation bodies, the controller/processor interfaced with the surveillance system extracts the GPS details from the GPS receiver and appends it with date and time. A messaging system (such as SMS/MMS/e-mail) interfaced with the surveillance system automatically generates an evidence consisting of latitude, longitude, date, time, license/plate image, vehicle with tinted window/windshield image for issuing necessary tickets. The messaging system stores the evidence at the central server. At the server side, using the number

(a)   (b)   (c)   (d)   (e)   (f)

FIGURE 1.11
Experimental results. (a) Input foreground image, (b) input background image, (c) result of background subtraction where car (object) location is represented as white pixel, (d) detected portion from foreground image, (e) from cropped car, window position is identified, and (f) drawn contour on windshield portion and the available contour in image is identified.

plate information, the contact details of the owner is identified and ticket is issued with necessary proof or evidence.

1.6 Scope for Future Studies

The proposed tint level detection system has been designed using ordinary cameras which work in the visible region of the electromagnetic spectrum. In reality, we drive our vehicles during early mornings, late evenings, and night times as well. If one wants to use the proposed system, then one has to use infrared cameras during those poor lighting conditions. Then, the proposed algorithm needs to be modified. The designed algorithm has been tested while the vehicles were moving at low speeds. One has to test the algorithm while vehicles are moving at high speeds.

Acknowledgments

This work forms part of the Research and Development activities of TIFAC-CORE in Automotive Infotronics at VIT University, Vellore, India. The author would like to thank the Department of Science and Technology, Government of India, for providing the necessary software and hardware facilities for successfully carrying out the present work.

Bibliography

1. H. Zhiwei, L. Jilin and L. Peihong, New Method of Background Update for Video-based Vehicle Detection, in 2004 IEEE Intelligent Transportation Systems Conference, Washington, DC, pp. 582–584 (2004).
2. P. Arbeláez, M. Maire, C. Fowlkes and J. Malik, Contour Detection and Hierarchical Image Segmentation, *IEEE TPAMI*, Vol. 33(5), pp. 898–916 (2011).
3. T. Ko, S. Soatto and D. Estrin, Background Subtraction on Distributions, Computer Vision-ECCV 2008, *Lect. Notes Comput. Sci.*, Vol. 5304, pp. 276–289 (2008).
4. S.Y. Elhabian, K.M. El-Sayed and S.H. Ahmed, Moving Object Detection in Spatial Domain using Background Removal Techniques—State-of-Art, *Recent Patents Comput. Sci.*, Vol. 1(1), pp. 32–54 (2008).
5. A. Elgammal, D. Harwood and L. Davis, Non-parametric Model for Background Subtraction, ECCV-2000, *Lect. Notes Comput. Sci.*, Vol. 1843, pp. 751–767 (2000).
6. P.W. Power and J.A. Schoonees, Understanding Background Mixture Models for Foreground Segmentation, in Proceedings of Image and Vision Computing, University of Auckland, Auckland, New Zealand, pp. 267–271 (2002).
7. C. Stauffer and W.E.L. Grimson, Adaptive Background Mixture Models for Real-Time Tracking, *Comput. Vision Pattern Recogn.*, Vol. 2, pp. 252–258 (1999).
8. A. Mittal and N. Paragios, Motion-Based Background Subtraction Using Adaptive Kernel Density Estimation, in Proceedings of 2004 IEEE Computer Society Conference on Computer Vision and Pattern Recognition, Washington, DC, pp. 302–209 (2004).
9. X. Hao, H. Chen, Y. Yang, C. Yao, H. Yang and N. Yang, Occupant Detection through Near-Infrared Imaging, *Tamkang J. Sci. Eng.*, Vol. 14(3), pp. 275–283 (2011).
10. M.R.J. Baldock, A.J. McLean and C.N. Kloeden, *Front Side Window Tinting Visual Light Transmittance Requirements*, CASR Report Series, CASR002, Centre for Automotive Safety Research, The University of Adelaide, Adelaide, Australia (2004).
11. M.-Y. Ku, C.-C. Chiu, H.-T. Chen and S.-H. Hong, Visual Motorcycle Detection and Tracking Algorithms, *Wseas Trans. Electron.*, Vol. 5(4), pp. 121–131 (2008).
12. K. Ganesan, C. Kavitha, K. Tandon and R. Lakshmipriya, Traffic Surveillance Video Management System, *Int. J. Multimedia Appl.*, Vol. 2(4), pp. 28–36 (2010).
13. C.-C. Chiu, M.-R. Ku and C.-Y. Wang, Automatic Traffic Surveillance System for Vision Based Vehicle Recognition and Tracking, *J. Inform. Sci. Eng.*, Vol. 26(2), pp. 611–629 (2010).
14. J. Lee and M. Park, An Adaptive Background Subtraction Method Based on Kernel Density Estimation, *Sensors*, Vol. 12, pp. 12279–12300 (2012).
15. A. Raju, G.S. Dwarakish and D.V. Reddy, A comparative Analysis of Histogram Equalization Based Techniques for Contrast Enhancement and Brightness Preserving, *Int. J. Signal Proc. Image Proc. Pattern Recogn.*, Vol. 6(5), pp. 353–366 (2013).
16. L. Kabbai, A. Sghaier, D. Ali and M. Machhout, FPGA Implementation of Filtered Image Using 2D Gaussian Filter, *Int. J. Adv. Comp. Sci. Apps.*, Vol. 7(7), pp. 514–520 (2016).
17. M.K. Kaushik and R. Kasyap, A Review Paper on Denosing Filter Using 2D Gaussian Smooth Filter for Multimedia Application, *Int. Res. J. Comp. Sci.*, Vol. 5(3), pp. 21–26 (2016).
18. A. Aslam, E. Khan and M.M. Sufyan Beg, Improved Edge Detection Algorithm for Brain Tumor Segmentation, *Procedia Comp. Sci.*, Vol. 58, pp. 430–437 (2015).
19. C. Jin, T. Chen and L. Ji, License Plate Recognition Based on Edge Detection Algorithm, in *2013 Ninth Int. Conf. on Intelligent Information Hiding and Multimedia Signal Processing*, Beijing, China, pp. 395–398 (2013).
20. M.M. Shidore and S.P. Narote, Number Plate Recognition for Indian Vehicles, *Int. J. Comp. Sci. Netw. Secur.*, Vol. 11(2), pp. 143–146 (2011).
21. Traffic Surveillance Video Management System, *Int. J. Multimedia Appl.*, Vol. 2(4), (2010).

2

Supervised and Semi-Supervised Identification of Users and Activities from Wearable Device Brain Signal Recordings

Glavin Wiechert, Matt Triff, Zhixing Liu, Zhicheng Yin, Shuai Zhao, Ziyun Zhong, Runxing Zhou, and Pawan Lingras

Saint Mary's University Halifax

Nova Scotia, Canada

CONTENTS

2.1 Introduction

Supervised and unsupervised learning techniques are reliable tools for the classification and categorization of data. Wearable devices are a relatively recent consumer technology that record large amounts of data by measuring signals around the body. Wearables come in many different forms, such as wristbands, armbands, watches, and headbands. This chapter explores the use of supervised and unsupervised learning techniques to identify individuals and activities using a commercially available wearable headband.

Wearable headbands typically measure the electroencephelogram (EEG) signals generated by the user's brain from many different locations around the head. Depending on the duration of measurement, these signals generate a large stream of values of variable length on multiple channels, from multiple locations. It is not easy to use the raw representation generated by such devices for meaningful data mining activities. This chapter illustrates a method of creating a compact representation of the data streams from multiple channels without losing the essence of the patterns within the data.

The data were collected from a commercially available wearable headband, called Muse. The Muse headband records EEG signals from four different locations around the head, as well as acceleration and some facial movement data. The dataset was created based on the recordings of five individuals performing five tasks each; reading, playing computer games, relaxing, listening to music, and watching movies. The activities were repeated a number of times for each participant. The raw data were converted to the proposed knowledge representation for use with the various supervised and unsupervised learning techniques.

The viability of the proposed knowledge representation is demonstrated through the usage of a number of well-known supervised classification [1] techniques, including decision trees [2], support vector machines (SVM) [3], neural networks [4], and random forests [5]. The usage of a variety of classification techniques show the summarized frequency distribution is effective for representing the time series of signals, independent of the classification techniques that are used. The classification techniques successfully predicted the persons as well as the activities based on the data.

This chapter also explores the usage of unsupervised and semi-supervised learning techniques with the proposed data representation. Clustering using the K-means algorithm is one of the most popular unsupervised learning techniques. K-medoids is an alternative of K-means that finds an object that is most similar to all the other objects in the cluster, as opposed to determining the centroid of the cluster. These methods focus on optimizing within cluster scatter and separation between the clusters. Both K-means and K-medoids do not provide the capability of incorporating additional optimization criteria. The fact that K-medoids offers a discrete search space, limited by the total number of objects in the dataset, can provide advantages for evolutionary searching. By combining K-medoids with an evolutionary algorithm, it is possible to perform multi-objective clustering. Peters [6] first proposed the use of evolutionary computing in the context of rough set theory, Lingras [7,8] subsequently explored both K-means and K-medoids based on rough set theory.

The K-means and the proposed semi-supervised crisp and rough K-medoid algorithms are compared using the proposed knowledge representation structure. By extending the evolutionary rough K-medoid algorithm to optimize the precision of the known categorization of signals in the dataset, it is shown that this approach may be effective for improving the precision of known category information in some cases.

Through comprehensive testing with supervised, unsupervised, and semi-supervised techniques, this chapter shows the viability of the proposed data representation to effectively capture the essence of the large amount of recorded raw data. It is shown that the various classification techniques are effective in predicting persons and activities, and that various clustering techniques also provide reasonable results.

2.2 Review of Wearables

Wearable technologies are evolving quickly and companies are discovering innovative ways to utilize the enormous amount of data they now have access to. There are many different categories of wearables. Each category collects its own unique type of data that can be used for meaningful data mining activities. This chapter will specifically investigate the collection of EEG brain signals [9] from a commercially available wearable, the Muse headband.

2.2.1 Wristbands and Watches

Smart watches such as the Pebble or the Apple Watch provide an interface to notifications, such as messages, calendar events, or breaking news, that are typically viewed on a smart phone. Additionally, as a wearable, they have their own sensors such as GPS, accelerometers, magnetometers, and heart rate sensors. These sensors provide additional capabilities for users, who can now track activity levels or health information.

Wristbands, such as Fitbit devices or the Nymi band, provide more specialized functionality and typically do not rely on an ongoing connection to a smart phone. The Nymi band, for example, measures an individual's ECG to create a unique biometric identifier that can be used in place of a traditional password. Fitness trackers such as the Jawbone UP or Fitbit devices are wearable wristbands that can measure health and activity information. They can record information such as how long and how far the individual has performed a specific exercise, the duration and quality of their sleep, and heart rate levels. Fitness tracker wearables such as the Fitbit devices aim to motivate users and provide feedback through the measurements they gather. Fitness trackers often tie in to smart phone or online applications to better analyze the data they collect, and to provide feedback to users.

2.2.2 Armbands

Armband wearables, such as the Myo, use sensors to monitor the user's gestures and movements. The Myo, for example, uses electromyography sensors to detect the electrical activity changes as the user activates different muscles and combines the EMG input with gyroscope, accelerometer, and magnetometer sensors.

2.2.3 Headbands, Headsets, and Smartglasses

Head mounted wearables in general either augment the wearer's capabilities by displaying additional information through a heads up display, such as in the case of Google Glass or the Microsoft HoloLens, or use sensors to measure the wearer's brain activity, such as the Emotiv and Muse headbands.

The Google Glass headset provides a display and sensors to show the information found on a smart watch or smartphone. Additionally, a camera in the device can be used not only for taking pictures but also for computer vision tasks such as object recognition or, in combination with the display, augmented reality where computer-generated graphics are displayed as an overlay to highlight objects in the real world. Other headsets, such as the Microsoft HoloLens, are expected to provide a similar augmented reality experience, serving as a way to improve both creation and consumption of multimedia. The HoloLens also promises to improve interactions with real-world objects, such as providing step-by-step instructions on repairing a light switch.

Headbands such as Muse or Emotiv provide brain–computer interfaces (BCI). The Muse headband uses four EEG sensors, two on the forehead and two behind the ears, and three additional reference sensors. Muse can also record blinking and jaw clenching. These sensors are used to detect and measure electrical activity in the brain. Currently, the Muse headband is targeted as a mindfulness training device, helping users calm their minds. However, developers have used the device as a BCI to control robots and perform research. Similarly, the Emotiv makes use of even more EEG channels, accelerometer sensors, magnetometers, and gyroscope sensors to detect a wide variety of facial expressions, emotional states, and mental commands, such as push, pull, left, or right.

2.2.4 e-Textiles: Smart Clothing, Smart Textiles, Smart Fabrics

e-Textiles comprise various clothing and fabrics with integrated electronics that allow them to communicate, measure, and transform. Smart clothing, such as the smart shirts, developed by OMsignal and Hexoskin, provides in-depth biometric measurements, including heart rate, breathing rate, step counts, calorie counts, and more. The smart clothing is used by athletes to improve training and athletic performance, and by researchers to research sleep patterns, stress levels, respiratory ability, and air pollution.

Smart fabrics such as Smart Skin, while currently is not wearable by humans, is used for products on packaging lines to measure various factors such as pressure and orientation to provide higher levels of quality assurance.

2.3 Review of Supervised Learning

In this chapter, we compare the results of many classification and clustering algorithms. A brief overview of the algorithms used in this study is provided below.

2.3.1 Decision Tree

The objective of a decision tree algorithm is to determine the rules that can be used to classify an instance based on the value of the instance's attributes [2]. The possible combination of rules includes all partitions that can be obtained from the process of recursively splitting the data [10], which may include multiple splits on the same attribute [10].

To determine the optimal attribute for splitting and the corresponding cut-point value, impurity reduction is used. The impurity in each node is calculated with entropy measures, such as the Gini Index or the Shannon Entropy [10]. Impurity reduction is measured as the difference between the impurity value for the parent node and the average impurity value for the child nodes.

The stopping criteria for the recursive decision tree algorithm is commonly a threshold for the minimum number of remaining instances in a node's partition or a threshold for the minimum difference in impurity calculated between the parent and child nodes [10]. Overfitting occurs when a classification model performs well when tested against training data but poorly when applied to unseen data. Tree pruning techniques [10,11] and statistical stopping criteria [10] have been used to mitigate overfitting.

2.3.2 Random Forest

The random forest technique, proposed by Breiman (2001), leverages multiple decision trees to predict an outcome [12, 34]. Its output is determined by the prediction that appears the most often in each of the individual decision trees [10,12]. Multiple trees, or an ensemble of trees, can be used to mitigate the instability of a single decision tree [10]. An ensemble of trees is created with random samples picked from the input training data [10,11]. The instances excluded with each random sample can be considered "out-of-bag" and used as test samples for measuring out-of-bag prediction accuracy.

Tree ensembles minimize overfitting with a set of diverse trees that tend to converge when the set is sufficiently large [12]. By randomly restricting the attributes used to generate the trees, attributes that would otherwise not have been chosen in a single decision tree can result in the discovery of cross-attribute correlations and patterns that otherwise would have been missed [10]. This has the potential to improve global prediction and accuracy.

2.3.3 Support Vector Machine

Support vector machine is used for binary classification [3,13–15]. The attributes of input training data are referred to as features. SVM works by first mapping the input data into a higher dimensional feature space. The SVM model then works to produce an optimal hyperplane in the new high dimensional feature space. The hyperplane separates the data into two groups, representing the two classes of the input data. In a two-dimensional space, we can separate instances into two groups with a line. In a higher dimensional space, we use hyperplanes. An optimal hyperplane maximizes the margin, or separation, between the two groups. The dataset used in this chapter has more than two classes. There are five classes of activities and four classes representing the persons. Since SVM are explicitly designed to classify into two groups, a specialized approach is required to handle multiple classes in the classification dataset, referred to as multiclass classification.

The implementation of SVM used in this chapter used the "one-against one" or "one-versus-one" approach for multiclass classification [16]. For k classes, there are $\frac{k(k-1)}{2}$ binary classifiers trained, and then a voting scheme decides the appropriate single class predicted [13,16]. Each binary classifier is given a newly constructed training dataset, such that one class is considered the positive class and another class is considered the negative class [13]. Finally, each of the binary classifiers can vote on the single class that they predict is correct, and the class with the most votes is considered the combined prediction.

2.3.4 Neural Network

Artificial neural networks, used for information processing, are inspired by the interactions within the biological nervous system of the human brain [17]. Neurons are connected and communicate to each other through synapses, dendrites, and axons. The dendrites of a neuron act as a set of inputs while, in contrast, the axon acts as the neuron's output. The synapse is a junction that bridges the communication between one neuron's output, the axon, and another neuron's input, the dendrites.

Artificial neural networks work by mimicking these biological processes, using a network of artificial neurons. The neurons are based on their input variables and their assigned weights. The weight corresponds to the strength of the synapse in the network. If the input values meet the required level to fire, the neuron will pass on an output value to the next set of neurons. This process repeats until the final neuron is fired upon and outputs the classification. Both biological and artificial neural networks learn by changing their weight values, which is the strength of the synapses in biological networks [17]. For supervised learning, the back propagation algorithm is used to quickly train artificial neural networks and adjust the weights to minimize the error between predicted and desired outputs [17,18].

2.4 Semi-Supervised Learning with Genetic Algorithms and Rough Set Theory

2.4.1 K-means Clustering

The K-means algorithm is provided with the number of clusters k to find, which also determines the number of centroids used [19]. A centroid is the average position of all the points within the cluster. The K-means algorithm starts with random centroids as initial guesses, and then determines the optimal and stable centroids for the clusters through subsequent iterations.

The k initial centroids can be selected randomly from the dataset, or created as random locations, within the input space. Each of the input data instances are then assigned to the closest centroid. The closest centroid is determined by a distance metric, such as Euclidian distance. After each instance has been assigned to a centroid, a new centroid for each cluster is calculated as the average of all of the instances assigned to it.

This process is repeated and the entire dataset is reassigned to the new centroids. The algorithm will repeat until the cluster centroids do not change with subsequent iterations [19].

2.4.2 Adaptation of Rough Set Theory for Clustering

Rough set theory represents a set of both lower and upper approximations instead of the traditional nonoverlapping sets. In conventional nonoverlapping sets, the boundaries of the sets or clusters are not always clearly defined, and objects may be equidistant from the center of multiple clusters. In contrast, rough sets are more flexible because they allow overlapping clusters and are also less descriptive (specific) than fuzzy sets. The lower approximation of a rough set is a set comprising only the elements that definitely belong to the subset. Objects in this set are located in the positive region. The upper approximation of a rough set comprises elements that both definitely belong, and those that possibly belong to the subset. The lower approximation is a subset of the upper approximation. Elements that are outside the upper approximation are located in the negative region, whereas elements in the outer approximation, but not in the lower approximation, are located in the boundary region. Let U be a set of objects. Rough sets were originally proposed using equivalence relations on U. However, it is possible to define a pair of lower and upper bounds $\left(\underline{A}(C), \overline{A}(C)\right)$ or a rough set for every set $C \subseteq U$ as long as the properties specified by Pawlak [20,21] are satisfied. Yao [22] described various generalizations of rough sets by relaxing the assumptions of an underlying equivalence relation. Such a trend toward generalization is also evident in rough mereology proposed by Polkowski [23], and the use of information granules in a distributed environment by Skowron [24]. The present study uses a generalized view of rough sets. If one adopts a more restrictive view of rough set theory, the rough sets developed in this chapter may have to be looked upon as interval sets [25]. Let us consider a hypothetical clustering scheme

$$U / P = \{C_1, C_2, ..., C_k\} \tag{2.1}$$

that partitions the set U based on an equivalence relation P. Let us assume that due to insufficient knowledge, it is not possible to precisely describe the sets, $C_i, 1 \leq i \leq k$, in the partition. However, it is possible to define each set $C_i \in U / P$ using its lower $\underline{A}(C_i)$ and upper $\overline{A}(C_j)$ bounds based on the available information. We will use vector representations $\vec{u}, \vec{v}$ for objects and $\vec{c}_i$ for cluster C_i

We are considering the upper and lower bounds of only a few subsets of U. Therefore, it is not possible to verify all the properties of the rough sets [20,21]. However, the family of upper and lower bounds of $\vec{c}_i \in U / P$ are required to follow some of the basic rough set properties such as:

(P1) An object $\vec{v}$ can be part of at most one lower bound

(P2) $\vec{v} \in \underline{A}\left(\vec{c}_i\right) \Rightarrow \vec{v} \in \overline{A}\left(\vec{c}_i\right)$

(P3) An object $\vec{v}$ is not part of any lower bound m

$\Updownarrow$

$\vec{v}$ belongs to two or more upper bounds.

Property (P1) emphasizes the fact that a lower bound is included in a set. If two sets are mutually exclusive, their lower bounds should not overlap. Property (P2) confirms the fact that the lower bound is contained in the upper bound. Property (P3) is applicable to the objects in the boundary regions, which are defined as the differences between upper and lower bounds. The exact membership of objects in the boundary region is ambiguous. Therefore, property (P3) states that cannot belong to only a single boundary region. Note that (P1) – (P3) are not necessarily independent or complete. However, enumerating them will be helpful in understanding the rough set adaptation of evolutionary, neural, and statistical clustering methods.

2.4.3 Genetic Algorithms

A genetic algorithm is a search process that follows the principles of evolution through natural selection. The domain knowledge is represented using a candidate solution

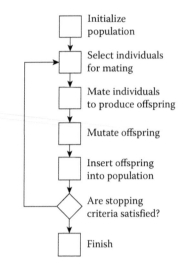

FIGURE 2.1
Flowchart of a generational genetic algorithm [33].

called an organism. Typically, an organism is a single genome represented as a vector of length n:

$$c = (c_i \leq i \leq n), \qquad (2.2)$$

where c_i is called a gene.

An abstract view of a generational genetic algorithm (GA) is given in Figure 2.1. A group of organisms is called a population. Successive populations are called generations. A generational GA starts from initial generation $G(0)$, and for each generation $G(t)$, generates a new generation $G(t + 1)$ using genetic operators such as mutation and crossover. The mutation operator creates new genomes by changing values of one or more genes at random. The crossover operator joins segments of two or more genomes to generate a new genome. The process is repeated until a specified stopping criterion has been met, as described in Figure 2.1.

2.4.4 Genetic Algorithms for Rough K-Medoid Clustering

This section describes a variation of the evolutionary rough K-means approach proposed by Lingras [7]. The proposal replaces K-means with K-medoids. A medoid is the most centrally located object in a given cluster. For k clusters, we will have K-medoids. A genetic algorithm can be used to search for the most appropriate K-medoids. The genome will contain k genes, each corresponding to a medoid. This reduces the size of a genome from km by a factor of m to k. The smaller genomes will reduce the space requirements and facilitate faster convergence. The genes in the rough K-means algorithm were continuous real variables with no restriction on their values. The values of genes for the medoids will be discrete and limited to the number of objects in the dataset. If we number the objects from $1,...,n$, then each gene can take an integer value in the range $1,...,n$. This restriction on the values of genes will further reduce the search space allowing for even faster convergence.

The next step in the development of a rough K-medoid algorithm is to assign an object to lower and/or upper bound of one of the clusters. The process is similar to the rough

K-means algorithm. The major difference is that we use medoids instead of centroids of the clusters.

Let us assume that the clusters are represented by K-medoids: $\vec{c}_1, \vec{c}_2, ..., \vec{c}_k$. For each object vector, $\vec{v}$, let $d(\vec{v}, \vec{c}_j)$ be the distance between itself and the medoid of cluster $\vec{c}_j$

Let $d(\vec{v}, \vec{c}_i) = \min_{i \leq j \leq k} d(\vec{v}, \vec{c}_j)$. The ratios $d(\vec{v}, \vec{c}_i) / d(\vec{v}, \vec{c}_j)$, $1 \leq i, j \leq k$, are used to determine

the membership of $\vec{v}$. Let $T = \left\{ j : \dfrac{d(\vec{v}, \vec{c}_i)}{d(\vec{v}, \vec{c}_j)} \leq threshold \text{ and } i \neq j \right\}$.

1. If $T \neq \phi$, $\vec{v} \in \overline{A}(\vec{c}_i)$ and $\vec{v} \in \overline{A}(\vec{c}_j)$, $\forall j \in T$. Furthermore, $\sim v$ is not part of any lower bound. The above criterion guarantees that property (P3) is satisfied.

2. Otherwise, if $T \neq \phi$, $\vec{v} \in \underline{A}(\vec{c}_i)$. In addition, by property (P2) $\vec{v} \in \overline{A}(\vec{c}_i)$

It should be emphasized that the approximation space A is not defined based on any predefined relation on the set of objects. The lower and upper bounds are constructed based on the criteria described above.

The next step in calculating the fitness of a genome is to measure the validity of a clustering scheme. We will use one of the most intuitive distance-based validity measures. The measure will accumulate the distances of the objects assigned to a cluster and its medoid as determined by the GAs:

$$\Delta = \sum_{i=1}^{k} \sum_{\vec{u} \in \vec{c}_i} d(\vec{u}, \vec{c}_i) \tag{2.3}$$

where the function d provides the distance between two vectors. The distance $d(\vec{u}, \vec{c}_i)$ is given by:

$$d(\vec{u}, \vec{v}_i) = \sqrt{\dfrac{\sum_{j=1}^{m}(u_j - v_j)}{m}} \tag{2.4}$$

We need to adapt the above measure for the rough set theory by creating lower and upper versions of the error as:

$$\underline{\Delta} = \sum_{i=1}^{k} \sum_{\vec{u} \in \underline{A}(\vec{c}_i)} d(\vec{u}, \vec{c}_i) \text{ and} \tag{2.5}$$

$$\overline{\Delta} = \sum_{t=1}^{k} \sum_{\vec{u} \in \overline{A}(\vec{c}_i) - \underline{A}(\vec{c}_i)} d(\vec{u}, \vec{c}_i) \tag{2.6}$$

The rough error is then calculated as a combination of the lower and upper error:

$$\Delta_{\text{rough}} = w_l \times \underline{\Delta} + w_u \times \overline{\Delta} \tag{2.7}$$

The rough error described above is based on the distances between patterns. However, we know the categorization of the patterns based on activity and the person performing the activity. In this experiment, we will focus on the categorization of the data based on the person. We first need to make a correspondence between a cluster and the most predominant class in the upper bound of that cluster. We then count the number of correctly classified patterns. The error in classification will be the number of incorrectly classified patterns, *wrongClasses*. We then take a weighted combination of Δ_{rough} and *wrongClasses*:

$$objective = w_d \times \Delta_{rough} + w_c \times wrongClasses, \tag{2.8}$$

where w_d is the weight attached to the rough error and w_c is the weight attached to the classification error. Our GAs will minimize the objective functions given by Equation 2.8.

2.5 Study Data

The Muse headband is a commercially available wearable product for consumer use. The usage of a commercial product in this study leaves the decisions for sensor selection and positioning to the device manufacturer. The Muse headband uses four sensors, two located on the forehead, and two behind the ears. Three additional sensors on the forehead are used as reference sensors by the device. Although the Muse API also provides access to the signals for muscle movement and accelerometer data, our research focused on the EEG data. Muse provides this data in a variety of formats. At the lowest level, the analog microvolt signals are recorded by the four sensors which, by default, are sampled at a rate of 220 Hz. These EEG signals are then compressed in order to stream the data over Bluetooth. The signals are compressed via Golomb Encoding and further quantized to reduce their size. Full details on Muse's compression algorithm are available via their online manual [26]. Muse offers both absolute band powers and relative band powers. The absolute band power is computed as the logarithm of the sum of the power spectral data of the EEG over the specified frequency range (alpha, beta, delta, gamma, theta). The power spectral density is computed via fast fourier transform on the device. The relative band powers normalize the absolute band powers as a percentage of the total absolute band powers [27], resulting in values between 0 and 1. This is calculated by:

$$s_r = \frac{10^{S_{abs}}}{10^{\alpha_{abs}} + 10^{\beta_{abs}} + 10^{\delta_{abs}} + 10^{\gamma_{abs}} + 10^{\theta_{abs}}} \tag{2.9}$$

where s_r is the relative frequency range (alpha, beta, delta, gamma, or theta) being calculated and S_{abs} is the frequency range's absolute value.

It was assumed that the signals will be able to help us extract signature patterns for individual users as well as various activities. Five individuals participated in the original data collection programs. These individuals worked very closely with each other and used similar setups for data collection. Five activities that represent day-to-day functions performed by most people were identified as a proof of concept. These activities were as follows:

1. Reading: A user read a magazine for 1–3 minutes.
2. Doing nothing: A user sat quietly for 1–3 minutes.

3. Watching video: A user watched a video for 1–3 minutes.

4. Game playing: A user played a computer game for 1–3 minutes.

5. Listening to music: A user listened to music for 1–3 minutes.

The above activities were repeated 10 times for five individuals resulting in a total of 50 datasets.

2.6 Knowledge Representation

One of the key aspects of any data mining activity is representing real-world entities using the pertinent numeric data available for them. In our case, we want to capture individuals and their activities using the signals emanating from their brain. This chapter focuses on summarizing, and not manipulating, the signals collected from the wearable into a fixed-length representation. As mentioned before, the Muse headband collects data from four positions around the head. Each position provides five types of waves: alpha, beta, gamma, delta, and theta. That means at any given point in time, we receive a record with 20 values. There will be a stream of these 20-valued records that will be recorded at a discrete time interval of 0.1 seconds. That means if we record activity for one person for one minute we will have a total of 600 records. Realistically, if we are recording an activity for a person, we cannot put an exact time limit on each person. In our experiment, the recording time per activity ranged anywhere from 60 to 180 seconds.

Table 2.1 shows the summary statistics for all the five waves: alpha, beta, gamma, delta, and theta for each of the four locations for all the participants. It can be seen that the values for each type of wave have similar ranges in all locations. However, the frequency range for different waves varies considerably. For example, the absolute frequency range of alpha values is from 7.5 Hz to 13 Hz, while beta values range from 13 Hz to 30 Hz [27]. In our dataset, we used Muse's relative band powers, which normalize the absolute band powers as a percentage of the total absolute band powers. This resulted in alpha values ranging from 0.00 to 0.98, while beta values range from 0.00 to 0.95, as shown in Table 2.1. Using the relative band powers decreased the variability of the value ranges; however, our collected data still had slight variations in the range for each wave. Therefore, we normalized the values for each wave using 90% of the maximum value for that wave, as shown in Equation 2.10. That made sure that the values for all the waves were in the same range. For instance, the value of $wave_{maximum}$ would be 0:98 for an alpha value, denoted by $value_{relative}$, determined using relative band powers.

$$value_{normalized} = \frac{value_{relative}}{0.90 * wave_{maximum}} \qquad (2.10)$$

Another issue with the data collection was the variable length of time for different activities as well as the length of each record. The length of the record could vary anywhere from 60 to 180 seconds. Figure 2.2 shows an example of one of the waves, alpha, from position 3 while player 1 was playing a game for 60 seconds. In order to fix the length of the record to a fixed and more manageable value, we studied the frequency distribution of the records. After experimenting with different number of bins, we decided

TABLE 2.1

Statistical Summary of Signals (Relative Band Powers) from all the Channels

Location	Wave	Min	25%	Median	75%	Max	Mean	Std. Dev.
Front right	Alpha	0.01	0.15	0.22	0.30	0.80	0.23	0.11
Front right	Beta	0.01	0.09	0.14	0.20	0.68	0.15	0.09
Front right	Delta	0.00	0.11	0.17	0.23	0.68	0.18	0.09
Front right	Gamma	0.01	0.14	0.21	0.29	0.77	0.23	0.12
Front right	Theta	0.00	0.09	0.15	0.21	0.59	0.16	0.08
Front left	Alpha	0.01	0.09	0.16	0.24	0.62	0.17	0.10
Front left	Beta	0.00	0.08	0.15	0.21	0.60	0.16	0.10
Front left	Delta	0.00	0.09	0.15	0.22	0.62	0.16	0.10
Front left	Gamma	0.01	0.21	0.32	0.47	0.95	0.35	0.17
Front left	Theta	0.02	0.23	0.38	0.56	0.96	0.40	0.21
Back right	Alpha	0.01	0.25	0.39	0.54	0.98	0.41	0.20
Back right	Beta	0.00	0.20	0.32	0.47	0.95	0.34	0.18
Back right	Delta	0.00	0.05	0.08	0.12	0.52	0.09	0.06
Back right	Gamma	0.00	0.07	0.13	0.19	0.68	0.14	0.09
Back right	Theta	0.00	0.05	0.09	0.15	0.69	0.11	0.08
Back left	Alpha	0.00	0.04	0.08	0.13	0.62	0.10	0.08
Back left	Beta	0.01	0.12	0.17	0.22	0.59	0.17	0.07
Back left	Delta	0.01	0.09	0.13	0.18	0.58	0.14	0.07
Back left	Gamma	0.01	0.10	0.15	0.20	0.52	0.15	0.07
Back left	Theta	0.01	0.11	0.16	0.22	0.62	0.17	0.08

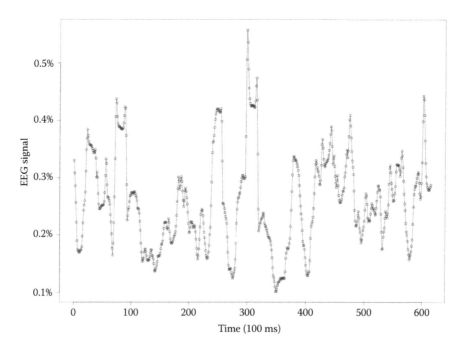

FIGURE 2.2

The alpha wave from position 3 for person 1 playing a game.

to use 5%, 15%, 25%, 50%, 75%, 85%, and 95% values to represent the histogram. Here if x is 5% value, it means that 5% of the values will be less than or equal to x. Figure 2.3 shows the histogram of values for the wave shown in Figure 2.2. As we can see, the representation consists of only eight values and captures the essence of each wave, position, and person in a concise and consistent manner. Since we have a total of four positions, five waves, and eight histogram bins, we have a total of 4*5*8 = 160 values to represent each record of an activity. For five persons and five activities, this process was planned to be repeated 10 times, giving a total of 5*5*10 = 250 records. For reasons detailed below, the results presented in this chapter only used 178 of the 250 records. We will use these records to train classification models to predict the person as well as the activity based on a recording from the Muse headband.

2.7 Experimental Design

2.7.1 Data

There were five individuals involved in the data collection. There were 5 records collected for one of the individuals, 23 records for another individual, and the remaining individuals had 50 or more records. The individual with five records tended to get neglected by the classifiers in favor of individuals with a larger number of records. Therefore, in the initial experiments reported in this chapter, we have used data from four individuals. Individual 0 had 23 records, individuals 1 and 2 had 50 records each, and individual 3 had 55 records, resulting in a total of 178 records.

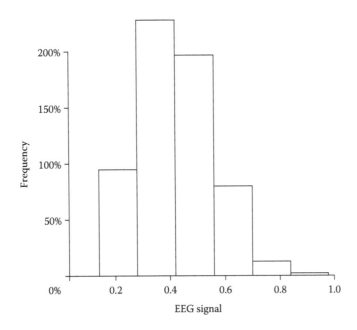

FIGURE 2.3
The histogram of the alpha wave from position 3 for person 1 playing a game.

2.7.2 Classification

The dataset used to develop the classification models therefore consists of 178 records. Each record contains 160 attributes. We used the following four well-known classifiers:

1. Decision tree
2. Random forest
3. Support vector machines
4. Neural networks

The classifiers were trained using default parameters and options. Each classifier was applied to three predictions:

1. Predicting a person—Five classes corresponding to five persons
2. Predicting an activity—Five classes corresponding to five activities
   - Doing nothing
   - Playing a game
   - Listening to music
   - Watching a video
   - Reading
3. Predicting a person as well as the activity—Twenty-five classes corresponding to the cross-product of the set of persons and the set of activities.

To test how well our knowledge representation and classifiers describe the activities, we first applied the classifiers to the entire dataset. This allowed us to study the importance of different attributes in the classification process.

In Tables 2.4(a), 2.4(b), and 2.5, you can see the most significant independent variables obtained by the random forest classifier when classifying by person, activity, or person and activity, respectively.

Using a complete dataset for training can lead to overtraining of the classifiers. The classifier may not work for new datasets. In order to see if the models were general enough to predict new recordings, we applied ten fold cross-validation.

In ten fold cross-validation [28], 10% of randomly selected data from the dataset is set aside for testing the model. The remaining 90% of the dataset is used for training the model. This process is repeated 10 times and the results are summarized.

From our tests we learned that our knowledge representation paired with SVM, random forest, or neural networks could predict a class for person or activity with fairly high precision. However, using a decision tree classifier proved to be ineffective.

The primary objective of the study was to explore the effectiveness of a number of well-known classification techniques combined with a data representation that reduces variable length signals to more manageable and uniformly fixed-length distributions. Using this representation and the classification techniques, the hypothesis was that the EEG brain signals will be identifiable for both persons and activities with a reasonable degree of precision.

2.7.3 Semi-Supervised Evolutionary Learning

We used the two-point crossover technique for the genetic algorithm. The two-point crossover technique uses two points to divide both of the parents' genomes into three partial genomes. The children genomes are generated by swapping the middle partial genome, between the two points, of the parents' genomes [29].

The parameters for rough clustering were set as follows: threshold = 1.1, $w_u = 0.7$; $w_l = 0.3$. For multi-objective optimization, the weight for distance-based optimization w_d was set at 0.75, and the weight for classification w_c was equal to 0.25. Since the assignment of patterns to clusters is based on distance measure, the distance tended to influence the optimization more than the wrong classification. Therefore, the evolutionary semi-supervised crisp and rough clustering was run multiple times and solutions that provided the best classifications were chosen. These multiple runs also allowed us to steer out of locally optimal solutions.

The primary objective of this experiment was to explore the effectiveness of Euclidean distance between brain signals to identify an individual person. The hypothesis was that signals from an individual will be similar and belong to the same cluster. The hypothesis can be tested with the well-known K-means clustering algorithm. The K-means clustering algorithm was selected because of its simplicity. Furthermore, the study explored the possibility of influencing the clustering with the known categorization using evolutionary crisp and rough K-medoid algorithms. We used confusion matrices for detailed analysis and the precision of clustering in identifying the individuals as two evaluation measures.

2.8 Classification Results

Tables 2.2 and 2.3 show the precision [30,31] of prediction for two of the classifiers, SVM and random forest, respectively. For each classifier, Tables 2.2 and 2.3 report the accuracy for predicting the activity, person, or both. The results include training for the entire dataset as well as by ten fold cross-validation.

TABLE 2.2

Prediction Accuracy of Support Vector Machine Classifiers

	Classifier	Prediction Variable	Precision	
			Entire Dataset	Ten-fold Cross-Validation
25	Support vector machine (Linear)	Person	100	95.39
26	Support vector machine (Linear)	Activity	100	77.51
27	Support vector machine (Linear)	Person+activity	100	84.97

TABLE 2.3

Prediction Accuracy of Random Forest Classifiers

	Classifier	Prediction Variable	Precision	
			Entire Dataset	Ten-fold Cross-Validation
1	Random forest	Person	100	92.85
2	Random forest	Activity	100	75.78
3	Random forest	Person+activity	100	77.71

Our tests from training on the entire dataset showed SVM and random forest were able to successfully predict a class for person, activity, or person and activity with 100% precision. Furthermore, neural networks were also able to predict person with 100% and activity with 95% accuracy, although it did not perform well for predicting person and activity together. Overfitting was most prominent for predicting activity; however, the precision values remained fairly high after switching to using ten fold cross-validation.

Precision tells us the likelihood of the classifier being correct when predicting a class. Precision can be calculated by dividing the diagonal value in a row by the sum of the row of a confusion matrix [32]. The precision values for ten fold cross-validation are all above 92% for predicting the person with SVM, neural networks, or random forest. In comparison, the precision values for the decision tree classifiers showed that they do not perform well using our knowledge representation of the EEG data and at best only achieve a precision of 65%.

Tables 2.4(a), 2.4(b), and 2.5 show the top five significant variables obtained by the random forest classifier when classifying by person, activity, or person and activity, respectively. The variable names include the bin number, wave, and location. For example, bin 7 (85% to 95%) of the theta wave at location 4 would be named "B7.Theta.4."

The most significant independent variable for predicting a person, activity, or person and activity was bin 1 (0% to 5%) of beta wave at location 1 with a significance score of 25%, bin 1 (0% to 5%) of theta wave at location 2 with a significance score of 6%, and bin 7 (85% to 95%) of theta wave at location 4 with a significance score of 6%, respectively. The theta wave at location 2 is significant for predicting person or activity with a significance score of 8% and 6%, respectively.

Figure 2.4 shows a graphical representation of the decision tree classifier that helps us understand the logical process of the person classification. Notice individual 0, with only 23 records instead of 50 records, is not included in the decision tree. This demonstrates that individuals with fewer records were neglected.

Tables 2.6(a), 2.6(b), 2.7(a), 2.7(b), 2.8(a), and 2.8(b) show the confusion matrices for the most accurate classifiers using SVM, neural networks, and random forest for classifying

TABLE 2.4

Most Significant Independent Variables Using
Random Forest Classifiers

Rank	Variable	Significance Score
(a) Person		
1	B1.Beta.1	25.32
2	B3.Theta.2	8.29
3	B1.Delta.3	7.04
4	B8.Gamma.3	6.43
5	B4.Delta.1	4.04
(b) Activity		
1	B1.Theta.2	5.59
2	B6.Beta.3	4.66
3	B3.Delta.4	4.33
4	B1.Delta.3	4.10
5	B7.Delta.4	4.05

TABLE 2.5

Person+Activity: Most Significant Independent
Variables Using Random Forest Classifiers

Rank	Variable	Significance Score
1	B7.Theta.4	6.43
2	B7.Theta.3	5.02
3	B6.Beta.2	4.33
4	B1.Delta.3	4.30
5	B4.Delta.1	4.26

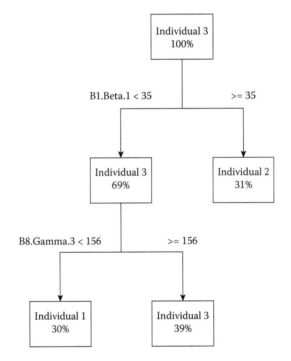

FIGURE 2.4
The decision tree for predicting a person.

by person or activity, respectively. For the ten fold cross-validation confusion matrices, the
entries are percentages of table totals.

Tables 2.6(a), 2.7(a), and 2.8(a) show correctly predicting individual 0 is not as likely as the
other persons, which can be expected since individual 0 has only 23 instead of 50 records.

2.9 Semi-Supervised Evolutionary Learning Results

Table 2.9 shows the confusion matrix resulting from K-means clustering. The rows represent clusters and columns represent classes. The correspondence between clusters and

TABLE 2.6

Confusion Matrix: Person–SVM Ten fold Cross-Validation

(a) Person				
	0	**1**	**2**	**3**
0	11.13	0.00	0.56	0.56
1	0.00	27.55	0.00	0.59
2	1.14	0.59	27.58	0.59
3	0.59	0.00	0.00	29.13

(b) Activity					
	Game	**Music**	**None**	**Reading**	**Video**
Game	21.40	1.15	0.00	2.89	3.41
Music	0.00	13.50	1.67	0.00	0.00
None	0.00	2.75	15.70	0.53	0.00
Reading	0.59	0.00	0.59	13.45	2.26
Video	1.14	2.20	1.11	2.20	13.47

TABLE 2.7

Confusion Matrix: Neural Network (1,000 iterations) Ten fold Cross-Validation

(a) Person				
	0	**1**	**2**	**3**
0	10.08	0.00	0.53	0.00
1	0.00	27.61	1.11	0.00
2	1.64	0.53	25.91	0.00
3	1.14	0.00	0.59	30.86

(b) Activity					
	Game	**Music**	**None**	**Reading**	**Video**
Game	19.18	0.56	1.18	1.18	4.52
Music	0.53	14.07	3.90	1.05	2.16
None	0.56	3.31	12.92	1.08	2.26
Reading	0.00	0.53	0.56	14.69	1.21
Video	2.85	1.15	0.56	1.08	8.92

classes was based on the most dominant class in a given cluster. It is clear that cluster 1 matches class 1 reasonably well. Similarly, cluster 3 matches class 3 reasonably well. Most members of cluster 2 are from class 2. Cluster 0 is a little difficult to match with a class. Most of the members from class 0 belong to cluster 0. However, the most dominant class in cluster 0 is class 2. Since we had already assigned another cluster to class 2, we associated cluster 0 with class 0. This assignment leads to the precision, recall, and F-measure values are shown in Figure 2.5.

With K-means clustering, our precision is quite high for persons 1 and 3. However, the value is very low for person 0 and middling for person 2. The likelihood of us classifying

TABLE 2.8

Confusion Matrix: Random Forest Ten fold
Cross-Validation

	(a) Person			
	0	1	2	3
0	11.20	0.00	0.59	0.53
1	0.00	25.96	0.00	0.59
2	0.00	1.05	25.96	0.00
3	1.64	1.14	1.61	29.73

	(b) Activity				
	Game	Music	None	Reading	Video
Game	19.12	1.08	0.53	2.78	3.93
Music	0.59	16.34	2.88	0.00	0.59
None	0.00	0.56	14.60	0.00	0.53
Reading	0.56	0.56	0.00	14.02	2.32
Video	2.82	1.14	1.11	2.25	11.70

TABLE 2.9

Confusion Matrix: K-Means

Cluster/Class	0	1	2	3
0	14	1	29	7
1	0	49	0	11
2	3	0	20	12
3	6	0	1	25

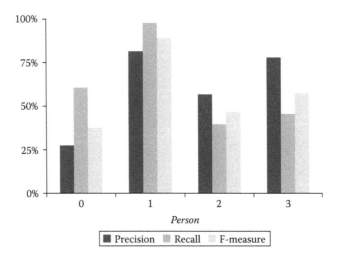

FIGURE 2.5
Precision and recall and F-measure: K-means.

a given person in the correct cluster is given by recall. Recall is calculated by the diagonal value for a column and the sum of the column. Recall of K-means for person 1 is reasonably high, fair for person 0, but low for person 2 and 3. F-measure is the harmonic mean of precision and recall. F-measure is calculated as:

$$F = 2 \times \frac{\text{precision} \times \text{recall}}{\text{precision} + \text{recall}} \tag{2.11}$$

Table 2.10 shows the confusion matrix resulting from the evolutionary semi-supervised K-medoids clustering. As before, we matched the clusters and classes based on the most dominant class in a given cluster. This assignment leads to the precision, recall, and F-measure values shown in Figure 2.6. In comparison to the precision, recall, and F-measure values from K-means shown in Figure 2.5, the evolutionary semi-supervised K-medoids provide more reasonable values. All the precision, recall, and F-measure values are above 65%, as opposed to the 40% and 45% values seen for recall of person 2 and 3, respectively, with K-means clustering.

Since rough clustering provides upper and lower bounds of clustering, we analyze them separately. It should be noted that the lower bounds are exclusive. A pattern belongs to a lower bound when we are almost certain of its membership. The upper bound, on the contrary, is inclusive. If there is a reasonable chance that a pattern may belong to a class, we assign it to the upper bound of the class.

TABLE 2.10

Confusion Matrix: Semi-Supervised K-Medoids

Cluster/Class	0	1	2	3
0	15	0	4	0
1	0	48	2	14
2	0	0	36	2
3	8	2	8	39

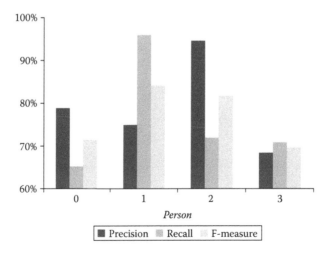

FIGURE 2.6
Precision and recall and F-measure: K-medoids.

TABLE 2.11

Confusion Matrix: Semi-Supervised Rough K-Medoids

Cluster/Class	0	1	2	3
(a) Upper Bound				
0	3	10	24	11
1	1	50	33	40
2	19	14	41	44
3	11	20	46	51
(b) Lower Bound				
0	10	0	0	0
1	0	30	0	4
2	2	0	4	0
3	2	0	0	0

Table 2.11(a) shows the confusion matrix of upper bounds resulting from the evolutionary semi-supervised rough K-medoids clustering. As before, we matched the clusters and classes based on the most dominant class in a given cluster. The cluster and class matching from the lower bounds was used to create the confusion matrix shown in Table 2.11(b). The precision values for the lower and upper bounds are shown in Figure 2.7. Since the boundary region represents ambivalence and has a higher likelihood of containing the wrong classes, the precision of the upper bounds is seen to be lower than the corresponding K-medoids clustering. On the contrary, the precision of the lower bounds reached as high as 100% for person 0, and 88% for person 1, higher than previous methods. However, the precision for person 3 was the worst of all methods, with no correct matches. The higher precision of the lower bound is due to its exclusive nature, while the inclusive nature of the upper bound leads to generally lower precision.

The calculations of recall for the rough clustering shown in Figure 2.7 cannot use the sum of the columns because of the overlap between the upper bound clusters. Instead, we

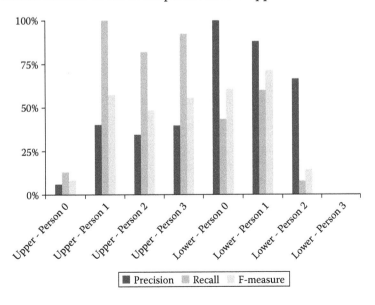

FIGURE 2.7

Precision and recall and F-measure: Semi-supervised rough K-medoids.

divide by the cardinality of the known classification. The inclusive nature of upper bound provides us with higher recall values. On the contrary, the exclusivity of lower bounds leads to lower recall values.

2.10 Conclusion

Wearable technology provides exciting opportunities for data mining. This study demonstrates the viability of supervised, unsupervised, and semi-supervised learning techniques for identifying individuals and activities based on EEG data collected from a commercially available wearable headband. In a real-world scenario, the signals will have even more variation in length than what has been evaluated here. Most data mining techniques are based on fixed-length object representation. This chapter shows that by using histograms of EEG brain signals, we can apply a wide variety of data mining techniques. Histograms are very advantageous as they reduce the raw variable length data to fixed-length representations. No matter the length of the data recorded, the representation will be the same.

All four supervised learning techniques explored were successful in identifying persons based on their collected EEG brain signals. The ten fold cross-validation results varied from 95% to 93% for SVM, neural networks, and random forest. Decision tree provided the least precision with 65%. SVM's precision of 78% for activities and 85% for person and activity showed that the proposed representation is quite credible for predicting activities from brain signals as well. Optimizing and tuning the classifier parameters may yield increased accuracy and precision; however, that was not explored in this chapter.

The traditional clustering methods had precision values ranging from 27% to 82% for K-means, and improved values of 68% to 95% for K-medoids. The proposed rough K-medoids evolutionary clustering method provided varying results. The precision of the lower bounds had a very large range, from 0% to 100%. Recall values were lower overall for K-means, with a range of 40% to 98%, as compared to K-medoids, with a range of 65% to 96%. For the proposed rough K-medoids algorithm, the recall for the upper bound was overall higher, but with a wide range from 13% to 100%.

The proposed rough K-medoids algorithm provided widely varying results depending on the person. This is demonstrated by the 0% recall and precision values for person 3 and the 100% precision value for person 0 in the lower bound. Correspondingly, in the upper bound, the variation is demonstrated by the 6% precision and 13% recall for person 0 and the 100% recall for person 1. With rough clustering, the upper bounds are inclusive in nature. If a pattern has reasonable chance of belonging to a cluster, it goes into its upper bound. This leads to generally higher recall values. The lower bounds, on the contrary, are exclusive. A pattern goes into the lower bound of a cluster only if there is a very high chance that the pattern belongs to the cluster. This leads to lower recall and higher precision values for the lower bounds of clustering. An analyst can use the upper bounds of a cluster when higher recall is desired. Lower bounds of the clusters will be useful when the precision is an overwhelming criterion.

All clustering methods were reasonably successful with matching clusters to persons. K-medoids provided the most reliable results, while the proposed rough K-medoids algorithm had both very good and very poor results. This suggests that further optimization and tuning of the algorithm may be necessary.

References

1. T. G. Dietterich, Approximate statistical tests for comparing supervised classification learning algorithms, *Neural Computation* 10 (7) (1998), 1895–1923.
2. J. R. Quinlan, Induction of decision trees, *Machine Learning* 1 (1) (1986), 81–106.
3. C. Cortes and V. Vapnik, Support-vector networks, *Machine Learning* 20 (3) (1995), 273–297.
4. K. Hornik, M. Stinchcombe, H. White, Multilayer feed forward networks are universal approximators, *Neural Networks* 2 (5) (1989), 359–366.
5. V. Svetnik, A. Liaw, C. Tong, J. C. Culberson, R. P. Sheridan, B. P. Feuston, Random forest: A classification and regression tool for compound classification and QSAR modeling, *Journal of Chemical Information and Computer Sciences* 43 (6) (2003), 1947–1958.
6. G. Peters, Evolutionary rough k-medoid clustering, *Transactions on Rough Sets VIII* (2008), Springer, Berlin, pp. 289–306.
7. P. Lingras, Evolutionary rough k-means algorithm, *Proceedings of Rough Set and Knowledge Technologies 2009, Lecture Notes in Computer Science 5589*, Springer Verlag (2009), Gold Coast, QLD, Australia, pp. 68–75.
8. P. Lingras, Rough k-medoid clustering using gas, *Proceedings of ICCI 2009*, Hong Kong, pp. 315–319.
9. W. Dement and N. Kleitman, Cyclic variations in EEG during sleep and their relation to eye movements, body motility, and dreaming, *Electroencephalography and Clinical Neurophysiology* 9 (4) (1957), 673–690.
10. G. T. Carolin Strobl and J. Malley, An introduction to recursive partitioning: Rationale, application, and characteristics of classification and regression trees, bagging, and random forests, *Psychological Methods* 14 (4) (2009), 323–348.
11. A. J. Myles, et al., An introduction to decision tree modeling, *Journal of Chemometrics* 18 (6), 275–285.
12. E. Fernandez-Blanco, et al., Random forest classification based on star graph topological indices for antioxidant proteins, *Journal of Theoretical Biology* 317 (2013), 331–337.
13. Z.-X. Yang, et al., Multiple birth support vector machine for multi-class classification, *Neural Computing & Applications* 22 (1) (2013), S153–S161.
14. P. L. Zhibin Li, et al., Using support vector machine models for crash injury severity analysis, *Accident Analysis and Prevention* 45 (2012), 478–486.
15. M. Sun, A multi-class support vector machine: Theory and model, *International Journal of Information Technology & Decision Making* 12 (6) (2013), 1175–1199.
16. A. Zeileis. K. Hornik, A. Smola and A. Karatzoglou, kernlab-an S4 package for kernel methods in R, *Journal of statistical software* 11 (9) (2004), 1–20.
17. C. M. Bishop, Neural networks and their applications, *Review of Scientific Instruments* 65 (6) (1994) 1803–1832.
18. M. Gevrey, et al., Review and comparison of methods to study the contribution of variables in artificial neural network models, *Ecological Modelling* 160 (2003), 249–264.
19. R. A. McIndoe, et al., Parakmeans: Implementation of a parallelized k-means algorithm suitable for general laboratory use, *BMC Bioinformatics* 9 (200) (2008).
20. Z. Pawlak, Rough sets, *International Journal of Information and Computer Sciences* 11 (1982), 145–172.
21. Z. Pawlak, *Rough sets: Theoretical aspects of reasoning about data*, Kluwer Academic. Boston, MA. 1991.
22. Y. Yao, Constructive and algebraic methods of the theory of rough sets, *Information Sciences* 109 (1998), 21–47.
23. A. S. L. Polkowski, Rough mereology: A new paradigm for approximate reasoning, *International Journal of Approximate Reasoning* 15 (4) (1996), 333–365.

24. J. S. A. Skowron, Information granules in distributed environment, in Zhong, N., Skowron, A., and Ohsuga, S. (eds.), *New Directions in Rough Sets, Data Mining, and Granular-Soft Computing, Lecture notes in Artificial Intelligence 1711*, Springer, Berlin (1999), pp. 357–365.

25. Y. Yao, Interval-set algebra for qualitative knowledge representation, *Proceedings of the 5th International Conference on Computing and Information*, in Chang, C. K., and Koczkodaj, W.W. (eds.), IEEE Computer Society Press, Sudbury, Ontario, Canada. (1993), pp. 370–375.

26. Interaxon. Protocols: Compressed EEG packets. http://developer.choosemuse.com/protocols/bluetooth-packet-structure/compressed-eeg-packets (2016).

27. Interaxon. Research tools: Available data, relative power bands. http://developer.choosemuse.com/research-tools/available-data#Relative_Band_Powers (2016).

28. R. Kohavi, et al., A study of cross-validation and bootstrap for accuracy estimation and model selection, *International Joint Conference on Artificial Intelligence*, Montreal, Quebec, Canada. Vol. 14 (1995), pp. 1137–1145.

29. S. N. Sivanandam and S. N. Deepa, Introduction to genetic algorithms. Springer Science & Business Media, (2007).

30. J. Davis and J. M. Goodrich, The relationship between precision-recall and roc curves, in *Proceedings of the 23rd International Conference on Machine Learning*, ACM (2006), pp. 233–240.

31. J. T. Townsend, Theoretical analysis of an alphabetic confusion matrix, *Perception & Psychophysics*, 9 (1) (1971), 40–50.

32. D. M. Powers, Evaluation: From precision, recall and f-measure to roc, informedness, markedness and correlation, *Journal of Machine Learning Technologies* 2 (1) (2011), 37–63.

33. M. Wall, *GAlib: A C++ Library of Genetic Algorithm Components.* Mechanical Engineering Department, Massachusetts Institute of Technology. https://www.cs.montana.edu/~bwall/cs536b/galibdoc.pdf.

34. B. Leo, Random forests. *Machine Learning* 45 (1) (2001), 5–32.

3

Internet of Things and Artificial Intelligence: A New Road to the Future Digital World

Mrutyunjaya Panda

Utkal University

Bhubaneswar, India

B.K. Tripathy

VIT University

Vellore, India

CONTENTS

3.1 Introduction

With the rapid development of Internet of Things (IoT), it has now become a buzzword for everyone who works in this area of research. Further, it is seen that with the rapid development of sensors and devices with their connection to IoT become a treasure trove for big data analytics. It has found numerous applications in developing smart cities where predictions of accidents and traffic flow in the cities can be effectively monitored; smart health care where the doctor is able to get useful information from the implant sensor chip in the patient's body; industrial production can also be enhanced manifolds by efficient prediction of the working of machinery and smart metering in helping the electric distribution company to understand the individual household energy expenses and making smart homes with connected appliances to name a few.

The 21st century is for IoT, where it is viewed as a network of physical devices coming together from electronics, sensors, and software. It is envisioned that the network of approximately 27 billion of physical devices on IoT are presently available and the list grows. These devices (Cars, Refrigerators, TVs etc.) can be uniquely identifiable through embedded computing system and can be connected from anywhere through suitable information and communication technology, to achieve greater service and value.

The "THING" in IoT means everything and anything around us that includes machines, buildings, devices, animals, human beings, etc. Today's, smart health care, smart homes, smart traffic, and smart household devices use this technology for a better digital world.

3.1.1 Working Principles of IoT

IoT has a unique identification that is embedded relying on the RFID connections which does not need any human or human–computer intervention for its working (Ashton, 2009). The IoT devices use IPv6 addressing for a huge address space, which makes it operational with active monitoring by computers with network connectivity and controlled by sensors attached to that devices. This monitoring and control can be interestingly seen in smart home applications where one can turn on the air conditioner while returning from office on the way home.

3.1.2 The Internet of Insecure Things

IoT technology is undergoing incredibly fast development and is a heterogeneous network of small yet lightweight devices; security aspects are to be addressed for IoT deployment so that it can suitably address the issues that may come in the way while used for both personal and commercial purposes (Eijndhoven, 2016). To that effect, the networking capability is incorporated into the IoT devices with proper encryption; firewalls and anti-virus methods are applied to counter the security concerns. This way, the confidentiality of the data and privacy of the users are guaranteed. But, this poses some challenges also. For example, in our houses, we usually have a physical means of access control to decide on who can access our house and similarly we have physical means of authentication in deciding who can enter into our house and can use the available artifacts inside the house. Suppose we want to turn on the air conditioner in our house to make the room cool. Here, two possibilities arise: one is to do it by physically being present in the house and following the procedures to switch on the AC. The other is, if we are away from home, to switch on the AC before we reach home so that the room is already cool by the time we reach home. In the second case, one is physically not present to switch on the AC. In order to switch on the AC, even though one is physically not present in the house, the AC should be connected as networked objects to have access control, which needs an extra layer of security in comparison to the early nonnetworked objects concepts, for preventing any unauthorized or malicious access to such devices.

The rest of the chapter is organized as follows. Section 3.2 discusses artificial Intelligence (AI) basics followed by IoT basics in Section 3.3. The fusion of IoT and AI is discussed in Section 3.4. Section 3.5 discusses the implication of deep learning in IoT scenario. While Section 3.6 discusses the proposed methodology, Section 3.7 presents the results and discussion. Finally, we conclude in Section 3.8.

3.2 Artificial intelligence

As we move toward the highly connected digital world, everything or anything goes smart with the use of small lightweight sensors with distributed intelligence; a huge amount of data is being collected from such networked devices, which poses real-time challenges in dealing with them for better insights and making corrective action thereto. The data is so BIG a data that even if one takes a sample of it for processing, time and accuracy becomes a challenge. For example, in the case of wearable computing, where sensors are implanted in the human body and are interconnected and connected to the Internet, any health-related issues arising from the patient's body can immediately be send to the concerned doctor for taking the necessary action. This type of real-time processing poses challenges to IoT for its effective and efficient implementation. Here, IoT combined with AI may be thought of as a viable solution to address the issue at large and help to uncover the hidden information from the data, for intelligent decision-making (Aadhityan, 2015).

3.2.1 Machine Learning

Machine learning introduced in 1950 is considered to be a technique for AI, initially aimed at robust and viable algorithms for numerous applications such as bioinformatics, intrusion detection, spam detection, forecasting, and the smart grid to name a few. Machine learning is a powerful tool for analyzing IoT datasets (Xu, 2015).

3.3 IoT Basics

IoT is considered to be the next step in the Internet evolution. As per European Commission, in the coming years, the integration of Internet with wireless communications and embedded wireless sensor networks will provide a paradigm shift in transforming our everyday devices into intelligent and context-aware ones (EU_Commission, 2009). Due to its technological structures, market shares, values, and earnings, it has found its place in almost all facets of human life, which can be unavoidable (Bandyopadhyay and Sen, 2011; Jain et al., 2011).

It is also envisaged that, in the near future, Internet will be integrated into a multitude of things such as clothes, toothbrush, and food packaging (Liu and Tong, 2010), with context awareness capability, pseudo-intelligence on processing capability among the connected things; also, efficient consumption of limited available power demands new forms of communication between things and people as well as between things themselves (Castellani et al., 2010; Mao et al., 2010).

The knowledge hierarchy showing how raw data are transformed into actionable intelligence and finally help in the decision-making process in the context of IoT is shown in Figure 3.1.

The raw sensory data can be thought of as the lowest layer in the knowledge hierarchy process, where a large amount of data are being collected from many IoT devices in terms of Exabyte (EB) or even more than that as time progresses. The next layer preprocesses the raw data to obtain a structured, filtered, and machine understandable data ready for processing to get the information. The third layer provides us the knowledge by uncovering the hidden information from the structured data for taking intelligent action at the end.

3.3.1 Technology Challenges in IoT

As IoT tries to connect the things in a single network and generate a large amount of data for actionable intelligence, it poses several challenges to be addressed. Some of them are discussed below.

3.3.1.1 Data Integration from Multiple Sources

The data generated from multiple sources such as sensors, social networking feeds, and mobile devices are all in different contexts; hence, the integration of all types of data is a challenge and if done efficiently, should be of a huge value addition for decision-making.

3.3.1.2 Scalability

As IoT generates a huge amount of data, dealing with data volume, variety, velocity, and veracity poses a challenge for real-time operation to efficiently handle the data with meaningful analysis.

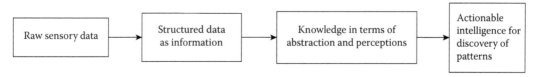

FIGURE 3.1
Knowledge hierarchy in IoT context.

3.3.1.3 Device Management

In an IoT scenario, even though a large number of devices are interconnected (not all) among each other with servers for a wide region and can share the data effectively, still managing a device not connected to network but somehow involved in data communication process may pose some kind of data linking issues.

3.3.1.4 Device Diversity

As many companies develop their products in a different way with different standards, making such devices in IoT to perform together is really a challenge.

3.3.1.5 Flexibility

An IoT scenario shall be developed in such a way that the new device and technology improvements may suitably be taken care without much hindrance.

3.4 IoT and AI

Machine learning as a part of AI along with IoT finds lots of application in both research and industry (Poniszewska-Maranda and Kaczmarek, 2015).

3.4.1 IoT Challenges and Capabilities

As IoT is gaining everybody's attention, there is an emergent need to understand the role of AI methods to gain insights into the market scenario and the readiness of the competitors to address the situation by then. In an article published by *Harvard Business Review*, it has been well said that IoT should be capable of addressing the following four basic issues such as monitoring, control, optimization, and autonomy for making it more sensible for the use of the customer in a smart connected environment. While monitoring is needed for the effective operation of sensor nodes in the working environment with utmost control, optimization is needed for improving the performance based on the feedback received from first two steps. Finally, autonomy makes the IoT work independently with self-diagnosis and repair.

3.4.2 IoT Won't Work without AI

As IoT produces big data where city traffic data can be used to predict the accidents and crime, it helps in building smart homes by digitally connected household appliances and much more. The information extraction from the sheer volume of data that are being collected from such IoT scenarios is a real challenge in order to see that IoT meets our expectations. It is also envisaged that dealing with such a huge amount of data (even if a sample of it) with traditional methods is too much of a time-consuming process. Hence, one needs to use AI methods incorporated to IoT data for improving the speed and accuracy. The negative consequences like, in home applications, all connected devices not working together will certainly annoy the customer; similarly, traffic can be mishandled with hundreds of cars in line or may be disastrous in health with malfunction of the pacemaker, and the lists goes on.

3.5 Deep Learning and its Role for IoT

In this section, an introduction to deep learning and its suitability to IoT datasets are discussed. In deep learning, sometimes referred to as fog clouding as shown in Figure 3.2, edge nodes are more capable of local computation among them through the network. In the more technical sense, deep learning is a single or set of algorithms that learn in the layer and mimics the brain as well, so that a hierarchy of complex concepts out of simple ones may be built by a computer.

3.5.1 How Does a Computer Learn? and How Do Deep Learning Algorithms Learn?

To understand deep learning with ease, it is required to have knowledge of how thinking and learning are done by computer with a top-down approach with all possible rules for its operations. But, now, it has a paradigm shift to have a bottom-up approach where the computer can learn from labeled data and the system is suitably trained based on responses. A notable example of this scenario is playing chess with 32 chess piece as primitives and 64 actions squares, but in real-life scenario, deep earning presents huge problem space with infinite alternatives. Such a huge problem suffers from the curse of dimensionality, making the computer difficult to learn. In the case of text mining, such as sentiment analysis or recognizing words or face recognition, the data available are intuitive (or sparse) in nature and the problem domain is not finite; hence, it is very difficult to represent possible combinations to have meaningful analysis.

Deep learning is a machine learning algorithm that is best suited to address these intuitive problems that are not only hard to learn but also have no rules with high dimensionality (Najafabadi et al., 2015), in order to deal with hidden circumstances without knowing the rules a priori.

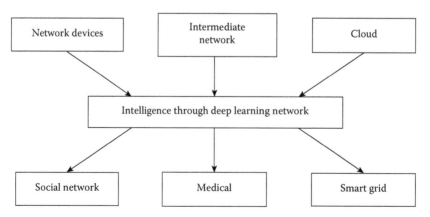

FIGURE 3.2
Deep learning and IoT.

Deep learning is inspired from the working of the human brain. This mimics the concepts of identifying a "cat" by a child by identifying the cat's behavior, shape, the tail, etc. and then putting them all together for a bigger idea generation as the "cat" itself.

Following the example cited above, deep learning progresses through multiple layers and divides the intuitive problem into parts, with each part mapped to a specific layer. The first layer is the input or visible layer where the input is being provided, then a series of hidden layers selected randomly for specific mapping with input data. In the image processing example, layer-wise information progress is made like: from input, pixels to edge identification at the first hidden layer, then corners and contours by second hidden layer, then parts of objects are identified in the third hidden layer, and finally the whole object is identified at the last and final hidden layer. This is shown in Figure 3.3.

In this chapter, our focus is to answer the following in a IoT scenario based on deep learning:

- The intuitive deep learning applications in smart city datasets
- The performance metrics used for better prediction that carry an intuitive component

3.5.2 Complementing Deep Learning Algorithms with IoT datasets

Although extensive research is being carried out in the area of energy load forecasting and its suitability in using neural network (Bhattacharyya and Thanh, 2004; Rodrigues et al., 2014), deep neural architecture is the most promising in this application scenario (Marino et al., 2016). Following are some of the emerging strategies/techniques that may be useful for complementing deep learning algorithms with IoT datasets.

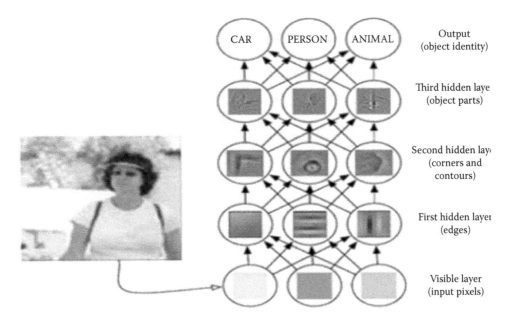

FIGURE 3.3
Deep learning example. (From Goodfellow, L. et al., *Deep Learning*, MIT Press, 2016.)

3.5.2.1 Deep Learning Algorithms and Time Series Data

Considering that most sensor data collected from an IoT environment are time series in nature, applying deep learning techniques for energy forecasting is gaining the attention of researchers for developing a sustainable smart grid as a part of the future digital world (Busseti et al., 2012; Connor et al., 1994; Hong, 2010). This high-dimensional dataset when implemented with a deep neural network is found to perform better in comparison with other existing approaches such as linear and kernelized regression techniques and more importantly do not overfit.

3.5.2.2 IoT Implications

Since 2015, IoT has been emerging but the real impact is yet to be realized as the year progresses with the development and deployment of a wide area network with 5G 2020 and beyond. Considering the availability of such a Tsunami of mostly time series data, it will lead to an exponential demand for predictive analytic and for the development of a viable model and address the challenges thereto.

Our prime focus in this chapter shall be on using deep learning techniques, as a practical solution to IoT data analytic.

3.5.2.3 Implications for Smart Cities

As one can see, smart cities are an application domain for IoT, where digital technologies are used to enhance the performance of the IoT system with a reduction in cost and resource usage with an aim of active engagement of citizens for effective implementation to find its benefit at large for well-being of self. The applications include but are not limited to energy sector, health sector, transport sector, to name a few.

3.5.2.4 Deep Learning as a Solution to IoT

Deep learning basically concerns with neural network structure where many layers are used to solve the complex situations built from the simpler ones. The ability of deep learning to solve the curse of dimensionality problems with no rules make it an attractive one in IoT–big data scenario (Marino et al., 2016). The concept takes birth from the question of learning to identify a cat to recognizing it from its behavior, shape, etc.

As IoT mostly depends on the resource-constraint computing devices, our aim shall be to check whether deep learning can provide any interesting predictive analytic to obtain meaningful observations.

3.6 Proposed Methodology and Datasets Used

The proposed methodology of using deep learning in IoT dataset shall address some of the following questions:

- The applicability of deep learning in smart cities, smart health care, etc.
- What is the performance metrics for prediction?

The proposed methodology adopted in this chapter is shown in Figure 3.4.

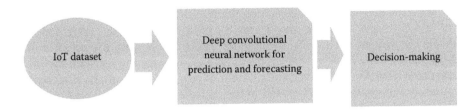

FIGURE 3.4
Proposed methodology.

3.6.1 Deep Learning with CNN

Even though Deep CNN is somehow similar to linear neural network, the main difference lies in using a "convolution" operator as a filter that can perform some complex operation with the help of convolution kernel. It is to be noted that Gaussian kernel is used for smoothing an image; Canny kernel is used for obtaining the edges of an image, and, for gradient features; Gabor kernel as a filter is widely used in most image processing applications. At the same, while comparing with autoencoder and restricted Boltzmann machines, it is pointed out that DCNN is intended to find a set of locally connected neurons while the others learn from a single global weight matrix between two layers.

The central idea of using DCNN is not to be serious on using predefined kernels rather learn data-specific kernels where low-level features can be translated to the high-level ones and learning is done from the spatially close neurons.

3.6.2 Training DCNN

The training process of DCNN consists of two phases: feed-forward phase and back-propagation phase. In the first phase, all the tasks are passed through the input layer to the output layer and the error is computed. Based on the error obtained, the back-propagation phase starts with bias and weight updates for minimization of the error obtained in the first phase. Several additional parameters named as hyperparameters, such as learning rate and momentum, are set properly in the range of 0 to 1. Further, a number of iterations (epochs) that are required for training is also to be mentioned, for an efficient learning of the DCNN model so that the error gradient shall be below the minimum acceptable threshold. The learning rate should be chosen in such a way that it should not overfit or overtrain the model. The momentum value may be chosen with trial-and-error method for its best adaptability to the situations (Panda, 2016). It should be noted that if we chose high learning rate and high momentum value (close to 1), then there shall be always a chance that we may skip the minima.

3.6.3 DCNN Layers and Functions

Initially, the concepts of Deep CNN were coined by LeCun et al. (2010). There are three layers in deep convolutional neural network such as input layer, one or more hidden layers, fully connected layers, and an output layer. The preprocessing starts at the input layer by applying the whole dataset to it. The middle layer is the hidden layer which is the heart of the DCNN layer and the number of hidden layers to be used in this stage depends largely on the input data. Convolution function is the process that involves in scanning a data filter over the dataset used. The filter is thought of as an array of 2-D data of the size 3×3, $5 \times 5, 7 \times 7$, etc. The set of nodes obtained from the convolution process is termed as feature maps. While the dataset used is the input to the first hidden layer, the number of feature maps obtained from the previous layers becomes the input for the subsequent hidden layers. At the same time, for each layer, the number of output feature maps shall be the same as the number of filters that is to be used in the convolution process. Based on the type of convolution function used in the process, the size of the output feature maps can be determined. As the feature extraction process depends on the size and number of filters used in each hidden layer, processing the large dataset with huge parameters demand a high-end computing environment with large memory capabilities for providing optimum performance (Panda, 2016). In this, use of the random filter is a good choice that is represented in the range of [–1 to +1]. The output from the convolution process is applied to an activation function for linear or nonlinear transformation across different layers so that the output falls in the range of [–1,+1] or [0,1]. Although several activation functions are available, we use rectified linear unit activation function for our analysis which says that for input x and output y, $f(y) = 1$, if $x>0$, and $f(y) = 0$, otherwise. Finally, pooling function is used (here, we use max-pooling) for down sample and to downsize the input features' map to half of its original size. Dimensionality reduction on the datasets shall be achieved by the end of this stage.

The fully connected layer is the last but one to the DCNN layers where the result is obtained as a single vector after consolidating the previous layers single-node output feature maps. As this stage leads to classification of the dataset, proper weights and bias are applied to this layer so that the cost of misclassification falls below a certain threshold or else the back-propagation process gets initiated for better weight and bias updates across layers for error minimization. Finally, the output layer presents the desired output. Different hyperparameters such as learning rate and momentum are then specified to have a faster convergence.

3.6.4 Smart Grid

Cloud computing also has many useful applications in smart grid, and data mining is one of its most significant attributes. A smart grid generates huge volumes of data through the weather conditions, solar or wind characteristics, network intrusions, people's electricity consumption, and how much electrical power people add from their own rooftop systems continuously to the grid. Utilities will need to store that data, and cloud computing is the solution for that. Once smart grid data are stored in the cloud, utilities will need data mining techniques to develop knowledge from the raw data. They will use the data to determine the relationship between demand and supply or to explain customer impressions or opinions about their smart grid services. So cloud computing, data mining, and smart grids are all very closely related.

It is worth noting that the quality of the intelligence we gain from data mining will be influenced by the quality and quantity of data we have available to us. Smart grid is creating a wonderful database that is a resource for monitoring the system and even solving operational problems. But the database needs to be comprehensive and demands a good database for every section of the smart grid. This way, it is envisaged that "Internet of Things" will help implement smart grids

The IoT generally connects to any object, whether the object is an element in the smart grid, a human being, or any other physical entity. It is capable enough to transfer the data automatically over a network without any personal or human-to-computer interaction. It uses a combination of wireless technologies, micro-electro-mechanical systems (MEMS), and the Internet.

The IoT will play a significant role in smart grid (Al-Ali and Aburukba, 2015). It is intended to perform network safety management, network operations, and maintenance, and further can be used to monitor end-user interactions and finally more importantly to make the smart grid secure. The IoT is not only a big world of connected devices but also a tool that we can use to implement a smart grid.

In order to meet the growing demand for electricity, we cannot be afraid of these new techniques and systems. The technologies and applications needed for smart grid are a 50–50 combination of modern information and communication technologies (ICT) and traditional grid technologies. We have these technologies. We just need to integrate existing systems.

While smart grid technologies are available, people from both ICT and utility engineering communities do need to work together to make the smart grid dream come true.

3.6.5 Metrics Used

Smart grid data and analytics will revolutionize the way power is managed, delivered, and sold. While collecting accurate, timely, and relevant data is of prime importance for any data analytic program, the data need to be put into an appropriate context to obtain useful information. The following are some of the fundamental analytical data transformations which have immediate relevance to smart grid applications: aggregations, correlations, trending, exception analysis, and forecasting. Many high-value analytical processes combine several of these techniques as part of an overall analytical process (Ranganatham, 2011). We use forecast method for our implementation in this chapter.

3.6.5.1 Forecasts

Forecasts are predictions of future events or values using historical data (Aung et al., 2012). For instance, a forecast of power consumption for a new residential subdivision can be created using history from similar homes. Forecasts can also be built using correlation data. For example, a forecast could be a simple prediction to understand the effect of one incremental megawatt of power consumption for each one-degree rise in summer temperature above 40°C, or it could be a more finer and sophisticated set of algorithms that forecast maintenance expenses based on the age of equipment, utilization trends, and past service trends for similar equipment.

3.6.6 Datasets Used

This section discusses the datasets used in our experiments.

3.6.6.1 UMass Smart* Microgrid Data Set

A microgrid data (Barker et al., 2012) is a new paradigm developed by the inclusion of distributed generation of smart grids. This dataset includes electrical data from over 400 homes. In the course of the Smart* project, investigation is made based on design, deployment, and efficient use of smart homes for sustainability.

3.6.6.2 *Residential Energy Consumption Survey Data*

The Residential Energy Consumption Survey (RECS) data (Kavousian et al., 2012) were collected in 2005 through a national survey on residential energy-related data, where 4,381 households in housing units were randomly selected to represent the 111.1 million housing units in the United States. These data were obtained from residential energy suppliers with the consumption and expenditures per unit sample. The consumption and expenditures and intensities data are divided into two parts: In the first part, the data provide energy consumption and expenditure by census region, population density, climate zone, type of housing unit, year of construction, and ownership status, whereas in the second part, the same data are provided according to household size, income category, race, and age. The next update to the RECS survey (2009 data) made available in 2011, summarized in Table 3.2, has been used in this study.

3.6.6.3 *The Reference Energy Disaggregation Data*

The Reference Energy Disaggregation Data set (REDD) (Kolter and Johnson, 2011) contains home electricity data: one is with high-frequency current/voltage waveform data of the two power mains and the other is lower-frequency power data with mains and individual labeled circuits. The low_freq directory contains average power readings for both the power mains and the individual circuits of the house with plug loads and plug monitors. The data are logged at a frequency of about once a second for mains and once every three seconds for the circuits. AC waveform data for power mains and a single-phase voltage for household purposes are present in the high-frequency directory. In order to reduce the data to a manageable size, the waveform may be compressed using lossy compression. This is mainly because the voltage signal in most homes is approximately sinusoidal (unlike the current signals, which can vary substantially from a sinusoidal wave), and zero-crossings of the voltage signal to isolate a single cycle of the AC power are found. For the time spanned by this single cycle, both the current and voltage signals are recorded, and the entire waveform is then reported. However, because the waveform remains approximately constant for long periods of time, the current and voltage waveform at "change points" in the signal are only reported. The high-frequency raw directory finally contains raw voltage and current waveform without alignment and compression, for a small number of sample points throughout the data as the entire dataset consists of more than a terabyte of data.

3.7 Experimental Results and Discussion

All the experiments are conducted in an Intel Pentium 2.8 GHz CPU with 200 GB HDD and 2 GB RAM and Microsoft XP professional. At first, we use RECS data (Residential Energy Consumption Survey-2010, PART-1 and 2) as per the statistics provided below: with household size in Figure 3.5 and Table 3.1, with census region and division in Table 3.2, and then with a most populated state in Table 3.3.

3.7.1 Experiment 1: Prediction and Forecasting Using Deep Convolutional Neural Network with RECS Dataset

All the experiments are conducted with the full training set to predict and forecast the model with deep convolutional neural network, with 95% confidence interval and 10 years

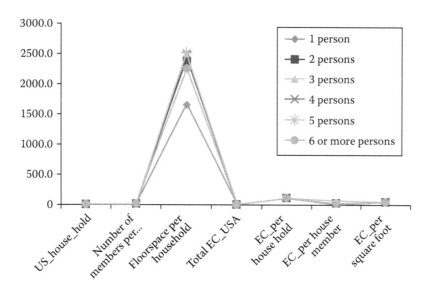

FIGURE 3.5
Energy consumption based on household size.

TABLE 3.1

Energy Consumption Based on Household Size

Household Size	US_House_Hold	Number of Members per Household	Floor Space per Household	Total EC_USA	EC_per Household	EC_per House Member	EC_per Square Foot
1 Person	30.0	1.00	1,671	2.12	70.7	70.7	42.3
2 Persons	34.8	2.00	2,297	3.36	96.4	48.2	42.0
3 Persons	18.4	3.00	2,324	1.91	104.1	34.7	44.8
4 Persons	15.9	4.00	2,460	1.72	108.4	27.1	44.1
5 Persons	7.9	5.00	2,539	0.92	117.1	23.4	46.1
6 or more persons	4.1	6.75	2,246	0.51	123.8	18.3	55.1

ahead, for getting insights for effective decision-making process. The results obtained for RECS data with all attributes and with energy consumption attribute only are provided in Figures 3.6 and 3.7.

3.7.2 Experiment 2: Prediction and forecasting using Deep Convolutional Neural Network with Microgrid Dataset

In the second experiment, we use microgrid dataset for predicting and forecasting the consumer behavior for a step further taken toward developing a sustainable smart grid environment. This is shown in Figure 3.8.

TABLE 3.2

Energy Consumption Based on Census Region and Division

Census Region and Division	US_House_Hold	Number of Members per Household	Floor Space per Household	Total EC_USA	EC_per Household	EC_per House Member	EC_per Square Foot
Northeast	20.6	2.56	2,334	2.52	122.2	47.7	52.4
New England	5.5	2.34	2,472	0.71	129.3	55.3	52.3
Middle Atlantic	15.1	2.64	2,284	1.81	119.7	45.3	52.4
Midwest	25.6	2.47	2,421	2.91	113.5	46.0	46.9
East North Central	17.7	2.49	2,483	2.09	117.7	47.3	47.4
West North Central	7.9	2.43	2,281	0.82	104.1	42.9	45.7
South	40.7	2.52	2,161	3.25	79.8	31.6	37.0
South Atlantic	21.7	2.50	2,243	1.65	76.1	30.4	33.9
East South Central	6.9	2.42	2,137	0.60	87.3	36.1	40.9
West South Central	12.1	2.62	2,028	1.00	82.4	31.4	40.6
West	24.2	2.76	1,784	1.87	77.4	28.1	43.4
Mountain	7.6	2.67	1,951	0.68	89.8	33.7	46.0
Pacific	16.6	2.80	1,708	1.19	71.8	25.7	42.0

TABLE 3.3

Energy Consumption Based on Populated State of United States

Most Populated State	US_House_Hold	Number of Members per Household	Floor Space per Household	Total EC_USA	EC_per Household	EC_per House Member	EC_per Square Foot
New York	7.1	2.72	1,961	0.84	118.2	43.5	60.3
Florida	7.0	2.51	1,869	0.42	60.0	23.9	32.1
Texas	8.0	2.76	2,168	0.65	81.5	29.5	37.6
California	12.1	2.75	1,607	0.81	67.1	24.4	41.7
All other states	76.9	2.51	2,307	7.82	101.8	40.5	44.1

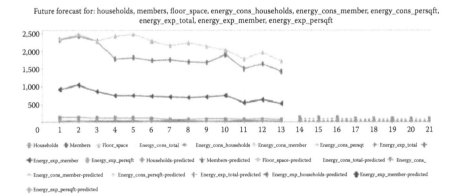

FIGURE 3.6
RECS data future predictions for all attributes.

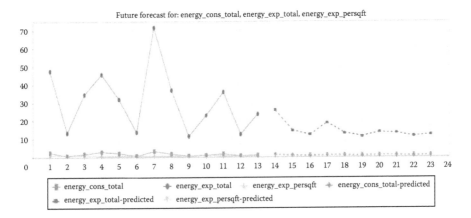

FIGURE 3.7
RECS data future prediction for some attributes.

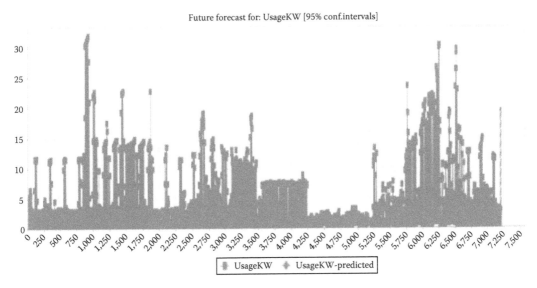

FIGURE 3.8
Future predictions for microgrid data.

3.7.3 Experiment 3: Prediction and forecasting using Deep Convolutional Neural Network with REDD Dataset

Finally, the third experiment is conducted in the REDD dataset for more knowledge on IoT scenario with an understanding of the energy consumption in an hourly to yearly basis for smart applications. The results obtained are presented in Figures 3.9 and 3.10.

From all the results obtained above, one can envisage the behavior of the consumer in using energy for their daily life at a suitable hour, daily, monthly and yearly basis for making effective decision-making process.

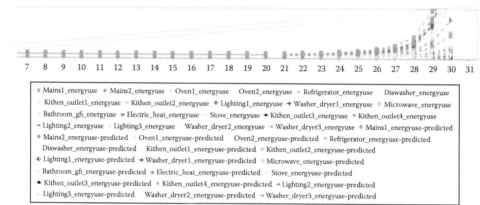

FIGURE 3.9
Energy usage forecasting on a monthly basis.

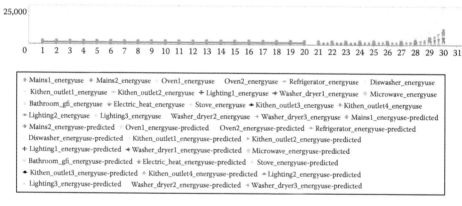

FIGURE 3.10
Forecasting energy consumption for 10 steps ahead in yearly basis.

3.8 Conclusions and Future Scope

Deep learning is an emerging field in AI which can address the challenges posed by the IoT datasets and is considered to have widespread applications in the future digital world. Due to the data unavailability, many real-world applications are not yet explored in this promising area of research. We have proposed to use the smart grid as a step to connect all power grids for efficient energy management. In future, several experiments are proposed to be conducted in smart traffic, smartphones, smart parking system, etc., in order to obtain a viable solution for the several potential IoT applications of AI into tomorrow's digital world.

References

Aadhityan, A. 2015. A novel method for implementing artificial intelligence, cloud and Internet of Things in robots. In *IEEE Sponsored 2nd International Conference on Innovations in Information Embedded and Communication Systems ICIIECS'15*, March 19–20, Coimbatore, India.

Al-Ali, A. R. and Aburukba, R. 2015. Role of Internet of Things in the smart grid technology, *Journal of Computer and Communications*, 3, 229–233.

Ashton, K. 2009. That "Internet of Things" thing, *RFiD Journal*, http://www.rfidjournal.com/articles/pdf?4986 (accessed February 16, 2017).

Aung, Z., Toukhy, M., Williams, J. R., Sanchez, A., and Herrero, S. 2012. Towards accurate electricity load forecasting in smart grids. In *DBKDA 2012: The Fourth International Conference on Advances in Databases, Knowledge, and Data Applications*, IARIA, February 29–March 5, Sainet Gilles, Reunion Island, pp. 51–57.

Bandyopadhyay, D. and Sen, J. 2011. Internet of things: Applications and challenges in technology and standardization, *Wireless Personal Communications*, 58(1), 49–69.

Barker, S., Mishra, A., Irwin, D., Cecchet, E., Shenoy, P. and Albrecht, J. 2012. Smart*: An open data set and tools for enabling research in sustainable homes. In *Proceedings of the 2012 Workshop on Data Mining Applications in Sustainability (SustKDD 2012)*, Beijing, China, August 2012, pp. 1–6.

Bhattacharyya, S. C. and Thanh, L. T. 2004. Short-term electric load forecasting using an artificial neural network: Case of Northern Vietnam, *International Journal of Energy Research*, 28(5), 463–472.

Busseti, E., Osband, I. and Wong, S. 2012. *Deep Learning for Time Series Modeling*, Technical Report, Stanford University, https://pdfs.semanticscholar.org/a241/a7e26d6baf2c068601813216d3c-c09e845ff.pdf (accessed June 24, 2017).

Castellani, A. P., Bui, N., Casari, P., Rossi, M., Shelby, Z. and Zorzi, M. 2010. Architecture and protocols for the Internet of Things: A case study. In *IEEE International Conference on Pervasive Computing and Communications Workshops (PERCOM Workshops)*, http://www.dei.unipd.it/~rossi/papers/WOT_camera_2010.pdf

Connor, J., Martin, R. D. and Atlas, L. E. 1994. Recurring neural networks and robust time series prediction, *IEEE Transactions on Neural Networks*, 5(2), 240–254.

Eijndhoven, D. 2016. *The Internet of (Insecure) Things: Cyber Security Goofs in IoT Devices*. http://www.altran.nl/fileadmin/medias/NL.altran.nl/documents/Don_Eijndhoven_-_The_Internet_of_Insecure_Things.pdf (accessed February 16, 2017).

EU_Commission. 2009. *Internet of Things—An Action Plan for Europe*. http://ec.europa.eu/information_society/policy/rfid/library/index_en.html (accessed June 24, 2017).

Goodfellow, I., Bengio, Y. and Courville, A. 2016. *Deep Learning*, MIT Press, Cambridge, MA.

Hong, T. 2010. *Short Term Electric Load Forecasting*, https://repository.lib.ncsu.edu/bitstream/handle/1840.16/6457/etd.pdf?sequence=2&isAllowed=y (accessed June 24, 2017).

Jain, P., Noor, A. and Sharma, V., 2011. Internet of Things an introduction. In *Proceedings of ASCNT C-DAC India*, March 2, Noida, India, pp. 1–9.

Kavousian, A., Rajagopal, R. and Fischer, M. 2012. *A Method to Analyze Large Data Sets of Residential Electricity Consumption to Inform Data -Driven Energy Efficiency*, CIFE Working Paper #WP130, Stanford University, Stanford, CA.

Kolter, J. Z. and Johnson, M. J. 2011. REDD: A public data set for energy dis-aggregation research. In *Proceedings of the SustKDD Workshop on Data Mining Applications in Sustainability*, http://redd.csail.mit.edu/

LeCun, Y., Kavukcuoglu, K. and Farabet, C. 2010. Convolutional networks and applications in vision, In *Proceedings of the International Symposium on Circuits and Systems (ISCAS'10)*, May 30–June 2, Paris, France, pp. 253–256, IEEE.

Liu, J. and Tong, W., 2010. Dynamic services model based on context resources in the Internet of Things. In *Procededings of 2010 6th International Conference on Wireless Communications Networking and Mobile Computing (WiCOM)*, September 23–25, Chengdu City, China, pp. 1–4.

Mao, M. et al., 2010. Solution to intelligent management and control of digital home. In *3rd International Conference on Biomedical Engineering and Informatics (BMEI)*. IEEE Conference Publications, October 16–18, Yantai, China.

Marino, D. L., Amarasinghe, K. and Manic, M., 2016, *Building Energy Load Forecasting Using Deep Neural Networks*. https://arxiv.org/abs/1610.09460 (accessed September 16, 2016).

Najafabadi, M. M., Villanustre, F., Khoshgoftaar, T. M., Seliya, N., Wald, R. and Muharemagic, E. 2015. Deep learning applications and challenges in big data analytic, *Journal of Big Data*, 2(1), 1–21. DOI: 10.1186/s40537-014-0007-7.

Panda, M. 2016. Towards the effectiveness of the deep convolutional neural network based fast random forest classifier. https://arxiv.org/ftp/arxiv/papers/1609/1609.08864.pdf (accessed June 24, 2017).

Poniszewska-Maranda, A. and Kaczmarek, D. 2015. Selected methods of artificial intelligence for Internet of Things conception. In *Proceedings of the Federated Conference on Computer Science and Information Systems*, September 13–16, Technical university of Lodz, Poland, pp. 1343–1348.

Ranganatham, P. 2011. Smart Grid data analytics for decision support. In *IEEE 2011 Electrical Power and Energy Conference*, October 3–5, Winnipeg, MB, Canada. pp. 1–6.

Rodrigues, F., Cardeira, C. and Calado, J. M. F. 2014. The daily and hourly energy consumption and load forecasting using artificial neural network method: A case study using a set of 93 households in Portugal. In *6th International Conference on Sustainability in Energy and Buildings*, SEB-14, Energy Procedia, June 25–27. Cardiff, Wales, UK. pp. 220–229.

Xu, Y. 2015. Recent machine learning applications to Internet of Things (IoT), http://www.cs.wustl.edu/~jain/cse570-15/ftp/iot_ml.pdf (accessed January 22, 2017).

4

Technical and Societal Challenges: Examining the Profound Impacts of IoT

P.B. Pankajavalli

Bharathiar University

Coimbatore, India

CONTENTS

4.1 Introduction

In the digital era, the rapidly growing Internet of Things (IoT) comes with enormous solutions for relating and communicating among smart objects. Right now there are 14 billion IoT devices installed, and IoT acts as an umbrella covering different fields like manufacturing and logistics, agriculture, social networks, transport, health care, and hospitals. In 2014, Gartner estimated that by 2020 there would be 25 billion connected devices. Cisco predicted the number would be closer to 50 billion [1]. The importance of IoT in future can be easily depicted through Figure 4.1 [2].

The impact of IoT revolution assures to transform many aspects of our life. IoT technology offers the possibility to reach out in all market segments and transform all fields by increasing the availability of information using networked sensors. However, IoT raises many issues and challenges that need to be considered and addressed in order to realize the potential benefits of IoT [3]. IoT researches have to be carried out based on the needs, including hardware and software technologies, as shown in Table 4.1 [4].

IoT can be related to five steps of maturity. Industries need capabilities to achieve each step process. Maturity continuum is shown in Figure 4.2.

4.2 Standards and Interoperability

With the expansion of IoT, it may be necessary to review various regulatory constraints and take steps to ensure sufficient capacity for expansion like seeking additional radio spectrum allocation as it becomes available. Interoperability is the capability of more than one system to exchange data and use information. Interoperability challenges can be taken in consideration of different dimensions like technical interoperability, syntactical interoperability,

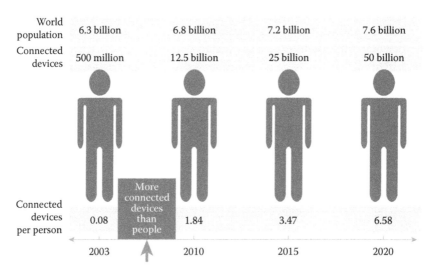

FIGURE 4.1
Comparison of connected devices with human population. (From Ondemir, O., et al., *IEEE Trans. Ind. Inform.*, 8(3), 719–728, 2012.)

TABLE 4.1

Internet of Things Scenario

IoT Research Needs	2011–2015	2015–2020	Beyond 2020
Identification technology	• Combination of IP, IDs, and addressing scheme • Unique ID • Multiple IDs for specific cases • Enhancing the IDs • Electromagnetic identification (EMID)	Beyond EMID	Multimethods—one ID
IoT architecture	• Extranet of things • Partner to partner applications, basic interoperability, billions-of-things	• IoT with global perspectives	
SOA software services for IoT	• Composed of IoT services	• Process IoT services	
IoT architecture technology	• Variation of symmetric encryption and public key algorithms • Worldwide certification of objects • Graceful recovery of tags following power loss • Increased memory • Reduces utilization of energy consumption • Location tracing embedded systems • IoT governance scheme	• Code in tags to be executed in the tag or in trusted readers • Global applications • Adaptive coverage • Context awareness • Object intelligence	Intelligent and collaborative functions

Source: Vermesan, O. and Friess, P., *Internet of Things—Converging Technologies for Smart Environments and Integrated Ecosystems*, River, Alaborg, 2013.

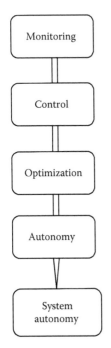

FIGURE 4.2
IoT Maturity Model.

and semantic and organizational interoperability [5]. Table 4.2 depicts the challenges in IoT technical interoperability.

A specific challenge in this regard is ensuring global interoperability, especially for things and devices which make use of radio spectrum. Without interoperability, devices cannot be networked around them. It is important to connect between two or more random devices which exchange messages anytime anywhere. Users don't wish to get committed to a single vendor and they expect their devices to be connected among different vendors. The approaches to be addressed are (1) Do all devices speak the same language? (2) Are intermediates required to translate their communication? The first issue arises due to the heterogeneity of the IoT, and this issue can be solved if we can agree on standards instead of constantly aiming to disrupt, define our own aspects of this growing ecosystem. The second issue is translating between many protocols and systems automatically, which means we are moving from concrete data models to an abstract information model. The open-source project Eclipse Smart Home is an example of an intermediate-based approach.

The standard can be defined as a model or a specific template to be followed. Standards should be designed to support a range of applications, and it should address the requirements of industry,

TABLE 4.2
IoT Technical Interoperability Challenges

Requirements	Rationale and Remarks
Best Practice Awareness Avoid spreading effort in addressing interoperability for worldwide protocols	• Supports interoperability market worldwide to support specifications and protocols • Develop market acceptance roadmap • Use clear specification development and methodologies, leading to improved quality while reducing time costs in a full chain optimized development cycle • Define profiles, if needed, to improve interoperability
Validations of Specifications Reduce ambiguities in specifications and development time	• Specifications development time would be too long • Ambiguities in development could lead to major noninteroperability issues • Quality, time, and cost factors lead to the needs of models and automation
Tests Specification Provide market accepted test specifications ensuring minimum accepted level of interoperability	• Lack of test specifications lead inevitably to different specification, implementation, and interoperability issues • Development of test specification is too expensive for the limited standard of stakeholders and effort should be collectively shared • Tools processing and automation are the only way to reduce time and to market
Tools and Validation programs Develop market accepted and affordable tools used in the market accepted validation programs	• Development of test tools is expensive • Available test tools developed spontaneously by market forces can have test scopes overlapping and may not even answer all test needs • Full chain of specification for tool development is not considered • Providing final confident to end users with consistent tests is not considered

Source: Vermesan, O. and Friess, P., *Internet of Things—Converging Technologies for Smart Environments and Integrated Ecosystems*, River, Alaborg, 2013.

environment, society, and individuals. Standards are required for two-way communication and for exchange of information between things and environment. The design of IoT standards should focus on efficient use of energy and network capacity [6]. It also has to take into account other regulations which restrict frequency bands and power levels for radio frequency communication. For IoT, European Telecommunications Standards Institute (ETSI) produces globally applicable standards related to information and communication technology (ICT), and their goal is to strengthen the standardization efforts of Machine to Machine (M2M). Internet Engineering Task Force (IETF) has come out with the new group 6LoWPAN (IPV6 over Low Power Wireless Personal Area Networks) to integrate sensor nodes with IPV6 networks. Another group named ROLL (Routing Over Low power and Lossy) has come out with the routing protocol draft. It aims to provide basis for routing over low-power and lossy networks.

IETF along with other relevant organizations is working to come out with more standards. Many organizations are putting a lot of effort to standardize protocol to lower the cost of the IoT product. Although many protocols are emerging, at the application layer, the contribution toward data formats and encoding of independent standardization of information (i.e., information model) is lacking. It is obvious that the emerging idea is to consider IoT standardization as an integral part of the future standardization process. Various issues faced in standardization are discussed in the following section.

4.2.1 Integration Issues

IoT is challenged in using mixture of devices interacting low-power devices with low capabilities high end machines. This requires standards to enable a platform to communicate and operate, and program devices regardless of their model, manufacture, and industry application.

Devices with sensing capabilities are in the form of smart objects, where wireless sensor in IoT is expected to play a significant role. Integrating IoT with sensor networks raises challenges in terms of communication and collaboration.

The following section will explore ways and challenges faced in integrating IoT with sensor networks.

4.2.1.1 Network-Based Integration

In network-based integration topology, the sensors communicate to the Internet through their gateway as shown in Figure 4.3. In the case of a multi-hop mesh wireless topology, the sensors rely on a base node, which may have its own gateways to communicate directly or it might have connections indirectly through another gateway. Wireless sensor network (WSN) base node is used to connect indirectly through another gateway.

4.2.1.2 Independent Integration

In independent integration topology, the sensors can connect directly to the Internet without the intervention of base point. Figure 4.4 shows the independent integration topology. However, providing independent IP address for each small sensor which has smaller packet size is the challenge behind this topology in assigning individual IP address for every IoT device.

This leads to communication and processing overhead associated with the use of TCP protocol.

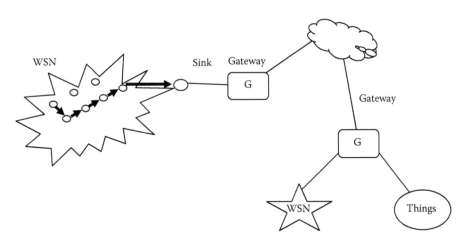

FIGURE 4.3
Network-based Integration.

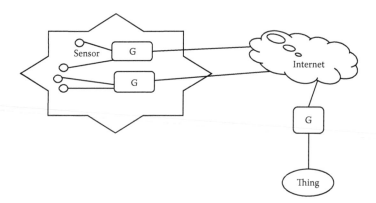

FIGURE 4.4
Independent integration.

4.2.1.3 Approaches to Overcome Challenges

To overcome this challenge, we need a platform-independent Application Programming Interface (API) for application developers and a way for different platforms to discover how to interoperate with each another. A further approach is the need to enable platforms to share the same meaning when data exchange happens.

The below case study discusses about the challenges in opting wireless standard for kidney stone formation monitoring system.

Kidney stone formation monitoring system is an innovative, technical, IoT-based solution that has been developed to minimize the possibility of kidney stone formation in human beings. Biologically, stones are formed due to unbalanced levels of pH (acidity and alkalinity) of urine. Therefore, by using the proposed system, doctors can monitor the pH level of urine and prescribe diet changes in order to avoid stone formation. The system senses the pH levels in urine using pH sensor directly when a person urinates in the closet and sends the pH level data over the Internet to the doctor using wireless standard. The major challenge behind this system is selecting the appropriate wireless standard among the different standards available. Standards play a major role in creating a worthy market for a new technology and increasing the interoperability [1].

There are different types of wireless standards significant for IoT. These standards can be operated with distance as their main constraint, which ranges from a few centimeters to many kilometers. For short- to medium-range communications, wireless personal and local area network (WPAN/LAN) standards are recommended. For long range communications, wireless wide area network standards (WWAN) are recommended. Among these types of wireless standards, short- to medium-range communications, WPAN/LAN standards are highly compatible for the proposed system because data are sensed by a pH sensor and sent to cloud platform for remote accessing. WPAN/WLAN offers many different standards for communications, such as Wi-Fi, ZigBee, and 6LoWPAN [2].

LOW-POWER WI-FI MODULES

The Wi-Fi module has an advantage in reducing the power consumption of battery used in sensor-based applications and also eradicates the installation of extra gateways to communicate with existing Wi-Fi modules. It supports IEEE 802.11 protocol, which provides transferring of data throughput in the range of hundreds of megabits per second. Wi-Fi devices can be interoperable with other devices using certain gateway mechanisms.

ZIGBEE AND 6LOWPAN MODULES

ZigBee and 6LoWPAN supports IEEE 802.15.4 protocol for low-power consumption applications such as low-power wireless sensor networks. Hence, they are suitable in IoT applications in terms of low power consumption. The data rate of ZigBee and 6LoWPAN is in the range of 250 kilobits per second. ZigBee devices can interoperate with other ZigBee devices using the Bluetooth profile. 6LoWPAN also offers interoperability with other wireless 802.15.4 devices as well as with devices of any other standard using a simple bridge device.

Based on the study and implementation, Wi-Fi outperforms the other standards. The transmission time of data on any network depends on the data rate, message size, and the distance between two devices. When comparing Wi-Fi with ZigBee and 6LoWPAN, Wi-Fi acquires less time for data transmission because its data rate is higher than the other two standards. Wi-Fi standard operating at higher data rates also requires less power consumption and also sends or receives large amounts of data. While considering the distance coverage range, Wi-Fi covers the maximum 70 m^2 at indoor systems and 225 m^2 at outdoor systems, whereas others cover a minimum range of distance and do not fit the outdoor system. Therefore, the Wi-Fi module is preferred for this proposed system of data transmission because of its high data rate, short-range connectivity, interoperability, and less power consumption as compared with other protocols.

REFERENCES

1. Mahmoud Shuker Mahmoud, Auday A. H. Mohamad, "A Study of Efficient Power Consumption Wireless Communication Techniques/Modules for Internet of Things (IoT) Applications." Advances in Internet of Things Vol.06 No.02(2016).
2. 6 Key Challenges of the Internet of Things—A Guide to Building IoT Ready Devices. eBook Tektronix.

4.2.2 Open Interconnect Consortium

Many standard bodies, consortiums, and alliances are currently working on IoT standard issues. Few of the standards are shown in Table 4.3 [7].

Overcoming the challenge of interoperability and standardization may be the single most important hurdle for IoT for mass usage, as it enables the boundless connections in a connected world.

TABLE 4.3

Open Interconnect Consortium

Standard	Launched	Scope
IoTivity	2014	It provides common communication framework based on industry standard technologies regardless of form factor, OS, or service provider.
AllJoyn	2013	Open-source software environment intended to enable compatible smart devices to communicate irrespective of OS and network protocols.
Thread		The protocol gives each device an IPv6 address and utilizes mesh networks that scale to hundreds of devices without a single point of failure.
ZigBee	2002	It is a common wireless language where everyday devices connect to one another.
AVB/TSN	2009	It creates interoperable and low latency requirements of various applications applying open standards through certification.
OneM2M	2012	Provides a common M2M service layer that can be embedded within various hardware and software to connect IoT devices.
Wi-Fi HaLow	2016	It doubles the distance and cuts the power consumption of traditional Wi-Fi.
ITU-T SG20	2015	An international standard to enable the coordinated development of IoT technologies, including M2M communications and ubiquitous sensor networks.
Brillo & Weave	2015	Brillo is an IoT OS, which consists of an android-based OS, core platform services, and a developer kit which links IoT devices with other devices in the cloud. Brillo uses Google's communication protocol, Weave, the standard that Google uses to promote, as the default standard for all IoT devices.

Source: Sharron, S.L. and Tuckett, N.A., *The Internet of Things: Interoperability*, Industry Standards & Related IP Licensing, Socially Aware, 2016.

4.3 Security

Every device involved in IoT represents a risk, and it is a major threat to an organization about the confidentiality of the data collected and the integrity of the dataset. Connecting low-cost IoT devices with minimal security mechanism will face ever-increasing potential security threats. IoT applications pose a variety of security challenges. For instance, in factory floor automation, the deeply embedded programmable logic controllers (PLCs) which operate robotic systems are typically integrated with the enterprise IT infrastructure. How can those PLCs be shielded from human interference at the same time protecting the investment in the IT infrastructure and leveraging the available security controls [8].

4.3.1 IoT Security Requirements

IoT involves sharing of data among users and objects or among objects themselves. In this environment, certain security requirements should be implemented. Basic IoT framework is shown in Figure 4.5 and the security at each layer is mandatory.

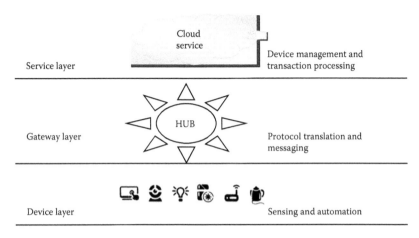

Cloud service

Service layer

Device management and transaction processing

HUB

Gateway layer

Protocol translation and messaging

Device layer

Sensing and automation

FIGURE 4.5
Basic IoT framework.

4.3.1.1 Device Layer Security

The device layer includes objects, places, and things can be simple devices like light bulb or complex devices like medical equipment. IoT security should be considered during design itself, and appropriate encryption should be made to maintain integrity and privacy. Devices should be designed to be tamperproof and the required software updates should be done constantly.

4.3.1.2 Gateway Layer Security

Gateway layer security indicates the messaging between the Internet enabled devices and other services. Gateway layer has to consider the security when communicating over protocols and has to ensure confidentiality and integrity.

4.3.1.3 Service Layer Security

The service layer security represents the IoT management activities like policies and rules, and automation of devices [9]. It has to focus on role-based access control and audit trail of changes done by devices or the users. Data monitoring should be done to identify compromised devices during abnormal behavior.

A wide security risk has been uncovered in the IoT device. Open Web Application Security Project (OWASP) has identified the following top 10 issues in IoT devices.

- Insecure web interface
- Insufficient authentication/authorization
- Insecure network services
- Lack of transport encryption
- Privacy concerns
- Insecure mobile interface
- Insecure cloud interface

- Insufficient security configurability
- Insecure software/firmware
- Poor physical security

Figure 4.6 shows the recent research done by Hp Fortify, which shows about IoT security vulnerabilities.

4.3.2 Security Challenges

IoT should be built in such a way that it assures consumers that their communication is done in a safe environment. To prevent an unauthorized user from accessing private data, IoT faces many security challenges and they are shown in Figure 4.7.

IoT research should address the following issues:

- Event-driven agents should be available to enable an intelligent/self-aware behavior of networked devices.
- Privacy-preserving technology for heterogeneous sets of devices.
- Models for decentralized authentication and trust.
- Energy-efficient encryption and data protection technologies.
- Security and trust for cloud computing.
- Data ownership.
- Legal and liability issues.
- Repository data management.
- Access and user rights, rules to share added value.
- Artificial immune systems solutions for IoT.
- Secure low-cost devices.
- Privacy policies management.

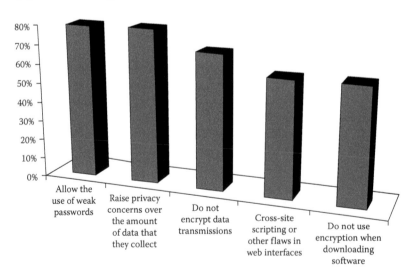

FIGURE 4.6
Device-level IoT security vulnerabilities.

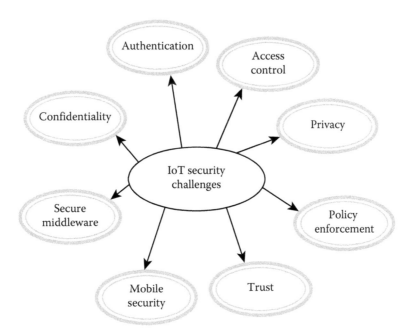

FIGURE 4.7
Security challenges in IoT.

4.3.3 Approaches to Solve Security Challenges

To solve various security challenges, many approaches have to be taken, and security should be considered during design stage itself. Building security at the design stage starts at the operating system level and then at the design level and should be extended till deployment in particular application. Apart from designing, the organizations should provide appropriate access rights to the right people in the workplace and should also apply efficient data protection policies. Continuous monitoring of devices and its performance measure should be maintained.

In the broad perceptive, IoT requires customized security and many researches have to be carried out in the IoT security field.

4.4 Data Collection, Protection, and Privacy

Data collection and protection of privacy in an IoT environment is the most significant thing to be focused on.

4.4.1 Privacy Concept

In IoT communications, privacy should be enforced over personal information and constrained should be placed over the storage and disclosure of information to others. The notion of privacy varies across different cultural perspectives. In one perspective, it is viewed as a fundamental human right and in another view, it has less importance. The notion of privacy has changed and shifted over time. Initially, the installation of surveillances camera was an invasive tool but now it has become common. IoT captures and communicates with any device that may involve using the existing infrastructure and Internet.

Today, Internet users, afraid of revealing personal information online, fill registration forms using false address details and name. In IoT, the background communication will happen between people, toys, washing machine and other appliances.

Deployment of sensors and a certain connection mechanism is required for applications or to render services in an organization. IoT is a combination of both public and private sectors. Traffic control systems and many other applications involve mass communication, with high degree of data getting transferred, connectivity being established, and interoperability happening. Everyday household items could be exploited by cybercriminals to gain access.

The question is who will have the control over the data which is surrounded in the environment. To influence users to participate in the IoT, an efficient mechanism for privacy and protection should be established. Users should be assured about the data and should have trust in communicating. To hold the trust and to protect the data, the following domains should be taken care of. The domains include technical, industrial, regulatory, and sociological.

4.4.2 Privacy Exploitation

IoT comes with a lot of beneficial features for the users to enjoy but also poses challenges for the stakeholders. Security of the private data and control over the data are the main factors to be considered. One of the recent studies by ITU has found that 80% of mobile users surveyed in South East Asia are receiving one to five unsolicited messages from their mobile network operator per month.

Privacy challenges need to be solved before technology becomes a part of our daily lives. The issues involved in sacrificing private data would be the following:

1. Obtaining personal data without the knowledge of the concerned person is done through object communication.
2. Issue of incomplete and asymmetric information between data subjects and the person or objects that gather data.

Sensors and radio frequency identification (RFID) are the potential threats to the security of a user's personal information [10].

4.4.3 Challenges

The IoT presents new challenges and needs response in the following areas which will have an impact on market, society, consumer rights, and policy making (Figure 4.8) [11].

The various challenges faced are:

- To ensure continuity and availability of data to provide the services.
- Protection should be incorporated at the design stage.
- To ensure protection rights for individuals and law for organizations.
- Multiple layers of security should be enforced.

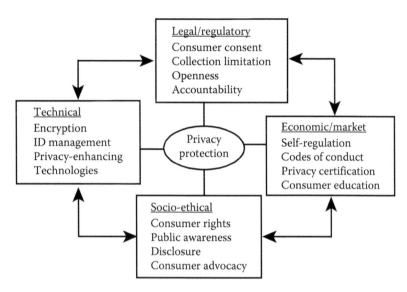

FIGURE 4.8
Privacy protection. (From ITU, *International Telecommunication Report 2005*, ITU Publication.)

- To ensure data privacy and location privacy.
- Certain law enforcement is required for enhancing technologies.
- Trust should be established for communication among objects.

Major issue in IoT privacy and confidentiality is about the data and the user involved. To ensure confidentiality, numerous standard encryption technologies exist. The challenge behind is (1) the speed at which the encryption algorithm works and (2) the amount of energy it consumes. In IoT, the heterogeneity and mobility of things will create more complexity in privacy-preserving technology. Network and data anonymity can provide the basis for privacy but certain technologies are supported with high bandwidth and computing power. A similar question arises for authentication of devices and for establishing trust.

Organizations have to identify the risk levels which are exposed to IoT and should think about the privacy and security requirements associated with volume of data. Trust in data privacy and protection should be taken into consideration for further research in IoT.

4.5 Socio-Ethical Considerations

Technological developments in this digital era will have an impact on the society either directly or indirectly. Although technologies like sensors, nano, and RFID have a role in threatening the privacy of data, these are also factors to be considered in socio-ethical issues. The issue at stake is that although recorded video, navigation, and biometric records are used for the purpose of observation, they still create a fear about the usage of data. Communication between people is influenced by technology, as people prefer to text rather than talk. Although nanotechnology increases safety and security, it still raises issues involving privacy, environmental hazards, and in the field of biomedical engineering.

Due to the pace of current technological innovations, it is important to ensure that the required and unrequired needs are categorized. Although IoT offers significant benefits to the users as well the industries, the complexities and its impact on the societies should be considered. The socio-ethical issues can be considered only if there is increased awareness about the new technology innovations and their impact.

4.6 Scale

Scalability is the ability of the product to perform well besides the size of the device or volume of the data. The next wave of Internet is how things get done without intervention of humans. There are more exciting applications like intelligent car and gateways, and brilliant hospitals built on connected devices. To achieve success and overall adoption, the crucial concept must be the ability to scale. Data distribution service (DDS) is a data-centric technology and many DDS IoT real-time applications like air traffic system and connecting complex medical devices are complex tasks to be carried out. These sort of systems require good performance, have to achieve reliability, and support scale.

Messaging design allows distributed systems to scale to larger sizes. IoT systems will be built on thousands of different applications like hospitals comprising tens of thousands of different devices [12].

4.7 Radio Spectrum

IoT is set to enable a large number of unconnected devices to communicate and share data with one another and the wireless world communication leads to the need for more spectrums. In the next decade, connecting 50 million devices to the Internet would require more spectrum than is available today. Almost half of 389 MHz is involved in the pilot program of spectrum which is shared by the government and the commercial world. Although shared programs come with a solution, it is only a small initiative. Larry Strickling, the head of the National Telecommunications and Information Administration, insisted the requirement of the government's goal of unleashing 500 MHz of new spectrum by 2020.

4.8 Other Barriers

The deployment of IoT and its development can slow down its pace due to the following barriers:

- Deployment of IPV6: IoT connects many tiny devices where billions of devices require unique IP address. This will prevent or will slow down the IoT's progress and more security features are to be offered.

- Sensor Energy: Sensors involved in IoT need to be self-sustaining. Changing batteries among billions of devices deployed across the planet is quite a tedious task. Sensor energy is another challenging issue to consider in IoT.

- Legal, Regulatory Rights: Usage of IoT devices raises many legal questions and magnify many legal issues concerning the Internet. One issue concerns the cross-border data flows transmitting data from one jurisdiction to another. Legal and regulatory challenges have to be taken into consideration for evolving IoT laws.

- Flexibility and Evolution of Applications: Everyday sensors and devices evolve with new capabilities. This leads to the development of new techniques and algorithms to analyze data. The development of applications with minimal effort and a comfortable platform is another challenge in IoT.

- Integration of Data from Multiple Sources: In IoT applications, data are streamed from different places like sensors, data from mobile devices, social network feeds, and from other web resources. Semantics of data should be part of data itself, and it should not be based on the application logic.

A survey conducted by Accenture in January 2016 was about barriers of purchasing IoT devices from the group of people aged greater than 14 years. The result of the survey is shown in Figure 4.9 [13].

- IoT Threats: Other security threats include (1) common worms jumping from ICT to IoT, (2) unprotected webcams, stealing content, (3) organized crime, and (4) cyber terrorism.

- Business adoption cycle: New technology adoption always goes through the business adoption life cycle. IoT is fundamentally the core model of the business in various fields. IoT is most disruptive technology and has to transform radically in all fields like medical, retail, transportation, manufacturing, and so on. Adoption of IoT depends on the removal of the barrier faced.

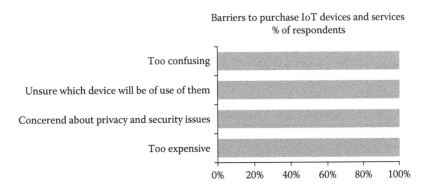

FIGURE 4.9
Accenture Survey. (From Accenture, *Igniting Growth in Consumer Technology*, Communication, Media & Technology, 2016.)

4.9 Green IoT

Impact of IoT in the next few years will create a mass environment surrounded by sensors, devices, and objects. By connecting physical objects and sharing information on the wireless medium, IoT plays a remarkable role in the growing world economy. Green IoT can be placed in two perspectives. On the one hand, it encompasses approaches that can be taken to cut carbon emissions and pollutions to enhance energy efficiency and on the other hand it is a way to design energy-efficient computing devices, communication protocols, and networking approaches for interconnecting devices [14].

4.9.1 Green ICT toward Green IoT

ICT is a platform which relates any application, technology related to information, and communication which involves transmission. Green RFID is an approach to reduce the size of the RFID tags which decreases the amount of nondegradable material used in manufacturing. Energy-efficient algorithm should be applied to optimize tag estimation and adjust transmission power. Green WSNs can be achieved by making sensor nodes sleep during idle time. Boosting energy can be done through the environment sources like harvesting, kinetic energy, through solar panels, and through vibrations. Energy-efficient routing techniques can be applied using clustering approach and multipath routing [15].

4.9.2 Future Work

The Green IoT can be incorporated with technology-based solar panels. More future work has to be carried out in Green IoT. The IoT solutions should give a feedback on energy usage and has to direct the user to efficiently allocate energy. Energy-harvesting approaches should be carried out in the environment. Energy issues such as energy harvesting and low-power chipsets are core to the development of IoT. Current technology support is inadequate for processing power and energy limitations in the near future.

Although many efforts have been placed in the advancement of IoT technologies, Green IoT still faces many significant challenges in different aspects. More research has to be carried out in this area.

4.10 Conclusion

The vision of IoT is an omnipresent range of devices leap to the internet in shaping human life. The prospects and benefits of IoT on the global scenario places various challenges related to technical and societal factors. Number of potential challenges, specifically in the area of standardization, interoperability, legal rights, protection, and privacy, have to be addressed for the benefit of people, society, and economy. Overcoming these challenges in industry and academia will be a prevailing approach for network and communication research to ensure that IoT is a powerful driving force in the future.

References

1. Evas, D., *The Internet of Things How the Next Evolution of the Internet Is Changing Everything*, Cisco IBSG, San Jose, CA, 2011.
2. Ondemir, O., Ilgin, M.A. and Gupta, S.M., Optimal end-of life management in closed-loop supply chains using RFID and sensors, *IEEE Trans. Ind. Inform.*, 8(3), 719–728 (2012).
3. Mukhopadhya, S.C., *Internet of Things Challenges and Opportunities*, Springer, Switzerland, 2014.
4. Vermesan, O. and Friess, P., *Internet of Things—Converging Technologies for Smart Environments and Integrated Ecosystems*, River, Alaborg, 2013.
5. Elkhodr, M., Shahrestani, S. and Cheung, H., The Internet of Things: New interoperability, management and security challenges, *Int. J. Netw. Secur. Appl.*, 8(2), 85–102 (2016).
6. Chen, S., Xu, H., Liu, D., Hu, B. and Wang, H., A vision of IoT: Applications, challenges, and opportunities with China perspective, *IEEE Internet Things J.*, 1(4), 349–359 (2014).
7. Sharron, S.L. and Tuckett, N.A., *The Internet of Things: Interoperability*, Industry Standards & Related IP Licensing, Socially Aware, 2016. http://www.sociallyawareblog.com/2016/04/14/
8. Alam, S., Chowdhury, M.M.R. and Noll, J., Interoperability of security-enabled Internet of Things, *Wireless Pers. Commun.*, 61, 567–586 (2011).
9. Jing, Q., Vasilakos, A.V., Wan, J., Lu, J. and Qiu, D., Security of the Internet of Things: Perspectives and challenges, *Wireless Netw.*, 20, 2481–2501 (2014).
10. Miorandi, D., Sicari, S., Pellegrini, F.D. and Chlamtac, I., Internet of Things: Vision, applications and research challenges, *Ad Hoc Netw.*, 10, 1497–1516 (2012).
11. *ITU Internet Reports 2005: The Internet of Things*, 7th ed., 2005. http://handle.itu.int/11.1002/pub/800eae6f-en
12. Ezechina, M.A., Okwara, K.K. and Ugboaja, C.A.U., The Internet of Things (IoT): A scalable approach to connecting everything, *Int. J. Eng. Sci.*, 4(1), 9–12 (2015).
13. Accenture, *Igniting Growth in Consumer Technology, Communication, Media & Technology*, 2016. https://www.accenture.com/_acnmedia/PDF-3
14. Khan, R., Khan, S.U., Zaheer, R. and Khan, S., Future Internet: The Internet of Things architecture, possible applications and key challenges, *10th IEEE International Conference on Frontiers of Information Technology*, Islamabad, Pakistan, pp. 257–260, 2012.
15. Shaikh, F.K., Zeadally, S. and Exposito, E., Enabling technologies for green Internet of Things, *IEEE Syst. J.*, 99, 1–12 (2015).

5

Evolution of Social IoT World: Security Issues and Research Challenges

G.K.Panda

MITS, Biju Patnaik University of Technology
Odisha, India

B.K.Tripathy

VIT University
Tamil Nadu, India

M.K.Padhi

Fan's Global Social NGN LLC
Dallas, Texas

CONTENTS

5.1 Introduction

Modern virtual society is digitally forwarding toward an "always connected" paradigm, where the Internet user is shifting from persons to things, conceptualizing toward Internet of Things (IoT). The IoT integrates a large number of technologies and foresees to embody a variety of smart objects through unique addressing schemes and standard communication protocols. IoT is a hot research topic, as demonstrated by the increasing attention and the large worldwide investments devoted to it.

It is believed that the IoT will be composed of trillions of sensory elements interacting in an extremely heterogeneous way in terms of requirements, behavior, and capabilities. According to CISCO IBSG (2011), by 2020 the Internet connected devices will reach 50 billion. Unquestionably, the IoT will pervade every aspect of our world and will have a huge impact on our everyday life: indeed, as stated by the US National Intelligence Council (NIC 2008) "by 2025 Internet nodes may reside in everyday things—food packages, furniture, paper documents, and more."

Then, communications will not only involve persons but also things, thus bringing about the IoT environment in which objects will have virtual counterparts on the Internet. Such virtual entities will produce and consume services, collaborate toward common goals, and should be integrated with all other services. One of the biggest challenges that the research community is facing currently is how to organize such an ocean of devices so that the discovery of objects and services will be performed efficiently and in a scalable way. Recently, several attempts have been made to apply concepts of social networking to the IoT. There are scientific evidences that a large number of individuals linked in a social network can provide far more accurate answers to complex problems than a single individual or group of—even knowledgeable—individuals (Surowiecki 2004). Such smart objects-based principle has been widely investigated in Internet-related researches. Indeed, several schemes have been proposed that use social networks to search Internet resources, to route traffic, or to select effective policies for content distribution. To achieve the convergence among the cyber, adaptive services were addressed in the literature (Marti et al. 2004; Fast et al. 2005; Mislove et al. 2006; Mei et al. 2011).

There are various studies in the literature discussing the security issues of IoT applications. Juels (2006) and Van (2011) presented reviews on the security issues on IoT applications. The security solutions for social networks have also been presented by Tripathy and Panda (2010) and Tripathy et al. (2011a, 2011b, 2011c, 2011d). However, the future discussion on the integration of security issues in the Social Internet of Things (SIoT) system is missing. The domain of IoT applications has been strengthening in consumer products, durable goods, transportations, industrial utility components, sensors, and many more. The emergence of social network data analytics, link prediction, community grouping, recommendation systems, sentiment analysis, etc., with IoT will raise significant challenges that could stand in the way of realizing its potential benefits. In this chapter, we focus on these key issues and explore numerous views providing sufficient information therein.

The rest of the chapter is structured as follows. In Section 5.2, we present the basics of IoT and its technological evolutions. In Section 5.3, we discuss the basic concepts of social networking and its popularity. In Section 5.4, we describe the emergence of social network analytics into IoT. In Section 5.5, we discuss various security issues and research challenges pertaining to IoT, social network and SIoT. This section also exploits information in terms of questionnaires and references for further research and development applications. In Section 5.6, we discuss on pro-business and pro-people SIoT services. Finally, in Section 5.7, some concluding remarks are presented.

5.2 Technological Evolution of IoT

The first ever internet connected coke machine (at Carnegie Mellon University) could able to report its inventory and quantified the temperature of newly loaded drinks, which led to the concept of a network of smart devices (1982). The field started gathering momentum in 1999 by Kevin Ashton.

According to Cisco (Cisco IBSG 2011), the Internet of Things (IoT) was *born* sometime between 2008 and 2009. Today, IoT is well under way and would cross 50 billion connected devices (sensors to the internet) by 2020. Furthermore, the IoT is evolving toward a new era of technology in collaboration with networks of smart devices over the wired or mobile internet, mitigating with object to object or object to people interactions (see Figure 5.1). The IOT is emerging as a platform to integrate a large number of heterogeneous and pervasive electronic gadgets (sensors, actuators, RFID tags, electronic or electromechanical devices, smart phones, etc.) that continuously generate information about the physical world (Strategy, Unit 2005; Gubi et al. 2013; Guo et al. 2013). Most of this information is available through application programming interfaces (APIs) and seems to be overpopulated with intensive interactions, heterogeneous communications, and numerous services. These would lead to issues of scalability, network navigability, real-time search, minimization of local grouping, dynamic grouping, etc.

As the IoT continues to develop, its further potential is estimated by a combination with related technology approaches and concepts such as Big Data, Cloud Computing, Future Internet, M2M, Mobile Internet Robotics, Sensor Networks, Semantic Search and Data Integration, Social Network, etc. The idea is of course not new as such but becomes now evident as those related concepts have started to reveal synergies by combining them. Overcoming those hurdles would result in a better usage of the IoT potential by a stronger cross-domain interactivity, increased real-world awareness, and utilization of an infinite problem-solving space. In Sections 5.4, 5.5, and 5.6, we prompt for further allied approaches and discuss solutions to thematic questions.

As the IoT is in its early stage, enormous research is still going on to best suit applications in economic and social governance. As stated by the US National Intelligence Council (NIC 2008), "by 2025 Internet nodes may reside in everyday things—food packages, furniture, paper documents, and more." Some of the significant IoT executed ecosystems include smart cities (and regions), smart car and mobility, smart home and assisted living, smart industries, public safety, energy, environmental protection, agriculture, and tourism. At the same time, policy makers are concerned about breaking away from old

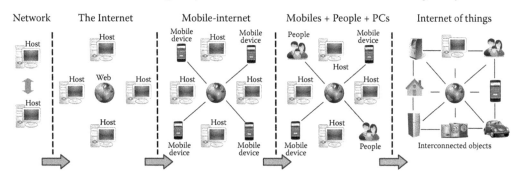

FIGURE 5.1
Evolution of the Internet of Things.

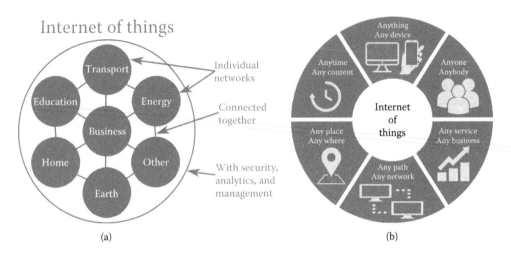

FIGURE 5.2
(a) Wide use of IoT and (b) IoT can be viewed as network of networks.

ways of thinking about data as something to be tightly controlled. It is also expected to see new international policies to open up the data moving further than "big data."

The architecture of IoT is viewed as data at motion and data at rest (White paper, CISCO, June 2014). Data initiate in the first layer from the physical devices and controllers (sensors, actuators, RFID tags, digital identifiers, etc.), which further pass through the communication and processing units (e.g., switches, gateways, Internet, and wire/wireless-3G/4G/5G communications), edge computing and data accumulation (SAN/NAS/cloud storage), data aggregation and access. The second layer further processes data through "Applications of Reporting-Analytics-Control" and "Collaboration and Process." As IoT evolves, independent individual networks will be connected with added security, analytics, and management capabilities (see Figure 5.2a) to be viewed as network of networks (see Figure 5.2b).

Researchers (Scott 2014; Ferguson 2016) are demonstrating to show that existing smartphone sensors can be used to infer a user's mood, stress levels, personality type, bipolar disorder, demographics (e.g., gender, marital status, job status, and age), smoking habits, overall well-being, progression of Parkinson's disease, sleep patterns, and happiness. Promoting privacy and data protection principles at hardware level will be paramount societal acceptance of IoT services. In this section, we limit our discussions to the evolution of technology in IoT; other challenges including security and threats are discussed in Section 5.5.

5.3 The Global Proliferation of Social Network

The Internet and its predecessors were a focal point for social interactivity. Computer networking was initially envisioned for military-centric command and control scheme, which further expanded to a great forum for discussing mutual topics of interest and perhaps even meeting or renewing acquaintances with other humans.

In the 1970s, CompuServe, a new avenue for social interaction, exploded onto the consciousness by allowing members to share files, access news and events, and interact through discussion forums. AOL (America Online) is considered as the Internet before the

Internet. During the 1980s, Bulletin Board System (BBS) used telephone lines via a modem for online meeting, which allowed users to communicate with a central system where the user could download files or games and post messages to others. BBSs were often used by hobbyists who carefully nurtured the social aspects and interest-specific nature of their projects. During the mid-1990s, Yahoo had set up shop and Amazon had begun selling books which perhaps provided encouragement for every household to have a PC. In 1995, the modern social networking concept was born through the site Classmates.com, catering the idea of virtual reunion of community. Early users could not create profiles, but they could locate themselves in their virtual community. It was a massive hit during those days. In 1997, the concept of six degrees was realized through SixDegrees.com. The theory somehow associated with actor Kevin Bacon that no person is separated by more than six degrees from another. SixDegrees.com facilitated its users to create profiles, invite friends, organize groups, and surf other user profiles. Its founders worked the six degrees angle hard by encouraging members to bring more people into the fold. Demographic-driven markets were evolved virtually through the sites like AsianAvenue.com in 1997, BlackPlanet.com in 1999, and MiGente.com in 2000. In 2002, social networking hit really its stride with the launch of Friendster. Friendster used degree of separation (similar to the SixDegrees.com) having a new concept of "Circle of Friends" dreaming to discover a different community between people who truly have common bonds. Within a year after its launch, Friendster boasted more than 3 million registered users and ample investment interests.

In 2003, LinkedIn took a decidedly more serious, sober approach to the social networking phenomenon, making a resource for business people who want to connect with other professionals. Today, LinkedIn boasts more than 450 million members through its referred term "connections." MySpace (2003) was once a favorite among young adult, demographic with music, videos, and funky, feature-filled environment. In 2004, Facebook launched as a Harvard-only exercise by university students who initially peddled their product to other university students. It remained a campus-oriented site for two full years and opened the access to public in 2006. Its innovative features, "Like" button, numerous built-in Apps made the platform user-friendly. The open API made it possible for third-party developers to create applications that work within Facebook itself. The site currently boasts more than 1.3 billion active users. Realizing the power of social networking, Google decided to launch Google+ in 2007. To date, it has more than 2.5 billion active monthly users. Over the course of the past 2 years, "Fourth screen" technology like smartphones, tablets, etc., has changed social networking and the way we communicate with one another entirely. Photo, video, and message sharing applications such as Snapchat, Instagram, and WhatsApp exist almost entirely on mobile. Users also use their smartphones to check in to various locations around the globe including matchmaking services (Foursquare). Mobile-based platforms blend social networking in an entirely different manner than their web-based counterparts.

Social network analysis uses two kinds of tools from mathematics to represent information about patterns of ties among social actors: graphs and matrices. SNA-based graphs or sociograms consist of points (or nodes) to represent actors and lines (or edges) to represent ties or relations (see Figure 5.3a, b). However, when there are many actors and/or many kinds of relations, the sociograms become so visually complicated that it is very difficult to identify the patterns as shown in Figure 5.3c. It is also possible to represent information about social networks in the form of matrices. Representing the information in this way also allows the application of mathematical and computer tools to summarize and find patterns.

According to Tripathy and Panda (2010), a social network can be modeled as a simple graph $G = (V, E, L, \zeta)$, where V is the set of vertices of the graph, E is the edge set, L is the label set, and ζ is the labeling function from the vertex set V to the label set L, that is, $\zeta: V \rightarrow L$.

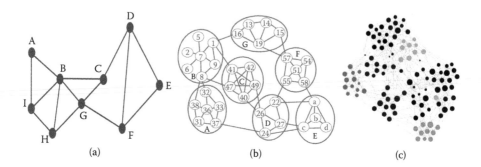

FIGURE 5.3
(a) The social network, (b) De-centralized social network of communities, and (c) Social network of section of health promotion.

Social network analysis facilitates the methodologies to uncover patterns in its connections or relations among related entities. It has been widely applied to organizational networks to classify the influence or popularity of individuals and to detect collusion and fraud.

5.4 On Emergence of Social IoT

There is scientific evidence that a large number of individuals connected in social network can provide far more accurate answers to complex problems than a single individual. Even this makes more sense in comparison to a small group of knowledgeable individuals (Surowiecki 2004). Social Network (SN) has already been solving these issues through friends' selection, community finding, dynamic interactivity, profiling, online grouping, recommendations, and sentiment analysis since three decades. Thus, it makes sense on integration of SN into the IoT solutions, with unique addressing schemes, standard communication protocols, interaction with other objects (local/ global neighbors) to reach common goals (Sarma et al. 1999; Atzori et al. 2010). This would bring tremendous opportunities of research and development in the evolution of new paradigm of SIoT.

The emergence of the SN and IoT, which were up to now thought to be in parallel by both scientific and industrial communities, is gaining momentum for integration very quickly (Atzori et al. 2012). This is due to the increase in the awareness that IoT provides the connection to the physical world by sensing and actuating while SNs contribute toward many of the daily aspects of the human world. It may lead to desirable implications in a future world populated by smart objects permeating the everyday life of human beings catering economic and social issues (see Figure 5.4). Such prolific integration needs an improvement on the connectivity of all the relationships between users and things (smart objects), and to enhance the availability of computational power via sets of things surrounding us. IoT objects should be socialized to establish *relationships* with them and to predict the link based on *trust* in the community. It is analogous to the fact that increasing the availability of processing power would be accompanied by decreasing the visibility (Weiser 1991; Strategy 2005).

FIGURE 5.4
Intelligent SIoT System.

From various empirical studies on analysis of service and application typologies, built upon the envisaged SIoT, we find some relationships of socialization (Tripathy et al. 2016) that are presented in Table 5.1 within the reference system architecture.

5.4.1 SIoT System Architecture

The three-layer architecture of SIoT system shown in Figure 5.5 is based on the simple architectural model of IoT (Zheng et al. 2011), having distinct components such as SIoT Server, the Object, and the Gateway.

In the SIoT Server, the network layer comprises various types of communication channels associated with the SIoT. The base sub-layer of application layer includes the database for the storage and the management of the data and the relevant descriptors. These records encompass the social member profiles and their relationships, as well as the activities carried out by the objects in the real and virtual worlds. Such a view is extracted through appropriate semantic engines. Indeed, these semantic services are necessary to provide a machine interpretable framework for representing functional and nonfunctional attributes and operations of the IoT devices. The top layer of the application layer consists of SIoT applications, the social agent, and the service management agent. The social agent is devoted to the communication with the SIoT servers to update profile, to update friendships, and to request services from the social network. It also implements the methods to communicate with objects. The service management agent is responsible for the interfaces with the humans that can control the behavior of the objects while communicating within their social network.

As to the Object and Gateway systems, the combination of layers may vary mainly depending on the device characteristics. The following three scenarios can be foreseen.

TABLE 5.1

Relationships of Socialization

Relationships of Socialization	Objective	Applicable Areas
Parental Object Relationship (POR)	Established among objects belonging to the same production batch	Homogeneous objects originated in the same period by the same manufacturer
Co-Location Object Relationship (CLOR)	Established among objects (either homogeneous or heterogeneous) used always in the same place (as in the case of sensors, actuators, and augmented objects used in the same environment such as a smart home or a smart city). Observe that, in certain cases, such C-LORs are established between objects that are unlikely to cooperate with each other to achieve a common goal. Nevertheless, they are still useful to fill the network with "short" links.	Heterogeneous objects originated used in the same place
Co-Work Object Relationship (CWOR)	Established whenever objects collaborate to provide a common IoT application	Objects that come in touch to be used together and cooperate for applications such as emergency response and telemedicine.
Ownership Object Relationship (OOR)	Established among heterogeneous objects which belong to the same user	Smart phones, music players, game consoles, etc.
Social Object Relationship (SOR)	Established when objects come into contact, sporadically or continuously, because their owners come in touch with each other during their lives	Devices and sensors belonging to friends, classmates, travel companions, and colleagues.

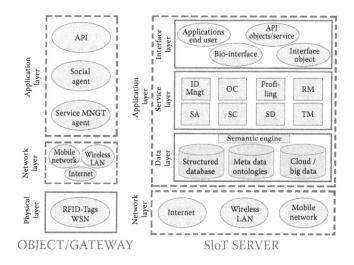

FIGURE 5.5
Architecture of SIoT.

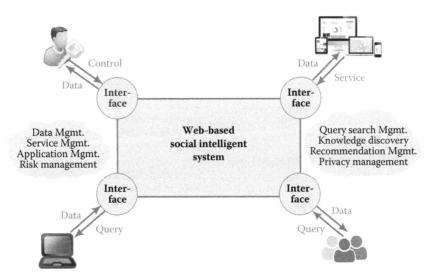

FIGURE 5.6
The future SIoT architecture.

In a simple one, a dummy Object (e.g., either an RFID tag or a presence sensing device) that is equipped with a functionality of the lowest layer is only enabled to send simple signals to another element (the Gateway). The Gateway is equipped with the whole set of functionalities of the three layers. In the second scenario, a device (e.g., a video camera) is able to sense the physical world information and send the related data over an IP network. The object would then be set with the functionality of the network layer other than that of the application one. Accordingly, there is no need for a Gateway with Application Layer functionality. An Application Layer in a server, somewhere in the Internet, with the Gateway application layer functionality would be good enough. According to the third scenario, a smart object (e.g., a smartphone) may implement the functionality of the three layers so that the Gateway is not needed, but for some communication facilities targeted to maintain the Internet connectivity of the object. This is the case of a smartphone, which has enough computational power to perform all the three-layer operations and may need a Gateway for ubiquitous network connectivity.

As stated in many research works, Social IoT appears to be the next step in the evolution of ubiquitous computing. Figure 5.6 shows the Future SIoT architecture which is based on key elements like actors, an intelligent system, an interface, and the Internet. The smart things and users represent actors. Actors' interaction is being managed by the intelligent system. Interactions among actors are facilitated through the interface. The Internet provides for open access among all the involved entities.

5.4.2 SIoT Applications

Many companies in different industrial sectors have already initiated the new trend of cutting-edge technologies to develop proprietary products and services that would drive the next wave of innovation in SIoT. These social devices ranging from low-cost sensors to powerful embedded systems can gather data and communicate these data over the Internet to social networks of people and devices who can respond to a problem, deliver a service, or sell a solution. Table 5.2 summarizes a number of off-the-shelf commercial social devices.

TABLE 5.2

List of Some Commercial Social Devices

Industry	Manufacturer	Product	Social Features	Communication	Open API
Apparel Accessories	Nike	Nike + FuelBand (https://goo.gl/NoPquU)	Fitness-tracking wristband social network. FuelBand users share their fitness data	Wireless	Yes
Appliances	Good Night Lamp	Good Night Lamp (http://goo.gl/Ad38Ko)	Share your presence and wired availability in an ambient way	Wired	No
Beverages	PepsiCo	Social Vending System (http://goo.gl/ZH7DX9)	Social networking of vending machine Gift/share a beverage to a friend with video messages	Wired	No
Health care	Corventis	Nuvant MCT (http://goo.gl/obJmJL	Noninvasive, ambulatory arrhythmia monitoring Physicians, patients, and families coordinate	Wireless	No
Smart Home	Shanghai Kingstronic Shenzhen Opiz Xiamen Hualian	http://www.globalsources.com/manufacturers/Smart-Home.html	Real-time feedback on the aspect of need in home appliances, energy conservation	Wireless	Yes
Smart City	IBM CISCO Schneider Electric	http://www.govtech.com/local/Top-10-Smart-City-Suppliers.html	Traffic navigation, innovation in green energy, waste-pollution control	Wireless	No
Smart Shopping	Johnson Plastics	https://www.engraversjournal.com/article.php/2970/index.html	Bio sensor intelligence, smart store, smart advertising feature	Wireless	No

These are very appealing products that bring the new experience of SIoT to the customers. Although, products have the provision of open APIs, most of the products available on the market are still not capable of interacting with other third-party devices and applications; this is a challenge for the future interoperability of SIoT products. We list out some of the identified common characteristics of the IoT products like Pachube, Nimbits, Paraimpu, ThingSpeak, and IoBridge.

- The objects use a Hyper Text Transfer Protocol (HTTP) protocol to send and receive data. This choice allows high interoperability among the different platforms.
- An intermediary server is used. The objects do not communicate directly with each other.
- Every object has a "data point" associated with it on the server side to keep track of the data sent.
- The methods POST and GET are used to send and request data.

- A tag is assigned to every data point.
- Data point discovery is performed using tags through an internal search engine.
- The system identifies every object with its API key.

5.5 Security Issues and Research Challenges

The overall security and resilience of the SIoT is a function of how security risks are assessed and managed at the object level, during data communication, and at the level of analysis. Security of a device is a function of the risk that a device will be compromised, the damage such compromise will cause, and the time and resources required to achieve a certain level of protection. The security issues are not new in the context of information technology, but the scale of unique challenges that can arise in SIoT implementations, as discussed in this section, makes them significant.

Privacy is a very broad and diverse notion for which the literature offers many definitions and perspectives. From a historic view, the notion of privacy shifted between media, territorial, communication, and bodily privacy. With the increasing use and efficiency of electronic data processing, information privacy has become the predominant issue. Information privacy was defined by Westin (1968) as "the right to select what personal information about me is known to what people." Westin's definition, although referred to nonelectronic environments, is still valid and general to enable the discussions about privacy in the IoT.

5.5.1 Security Challenges and Privacy Threats in IoT

Despite the immense potential of IoT in the various spheres, the whole communication infrastructure of the IoT is flawed from the security standpoint and is susceptible to loss of privacy for the end users. Some of the most prominent security issues plaguing the entire developing IoT system arise out of the security issues for information relay from one device to another. Privacy in the IoT is the threefold guarantee to the subject for:

- Awareness of privacy risks imposed by smart things and services surrounding the data subject
- Individual control over the collection and processing of personal information by the surrounding smart things
- Awareness and control of subsequent use and dissemination of personal information by those entities to any entity outside the subject's personal control domain

It is restated here that the application of IoT is fully dependent on its backbone of communication network, including Internet or IP networks, wireless device-to-device communication using Bluetooth, Z-wave, and Zigbee protocols. Out of several communication technologies, we take two most popular usages of communications, communications based on radio frequency identification (RFID) and wireless sensor network (WSN), and discuss the probable security issues relevant to them.

5.5.1.1 *RFID Technology: Security Issues*

RFID technology is mainly used in information tags interacting with each other automatically. RFID uses radio frequency waves for interacting and exchanging information between one another with no requirement for alignment in the same line of sight or physical contact. An RFID is made up of two components: transponders (RFID tags) and transceivers (RFID readers).

RFID Tag (transponder) consists of a microchip, memory embedded with an antenna. The memory unit has a unique identifier known as electronic product code (EPC). The function of the EPC in each tag is to provide a universal numerical data by which a particular tag is recognized universally. RFID tags are classified into active and passive types. Active tag houses a battery internally, which facilitates the interaction of its unique EPC with its surrounding EPCs remotely from a limited distance. Passive tags function without battery and the information relay through EPC occurs only by its activation by a transceiver from a predefined range of the tag. *Tag Readers (transceivers)* are proprietary in nature and operate in conjunction with RFID tags (active/passive). The EPC is the identifying signature of a particular tag under the scan of the reader. The RFID reader functions as the identification detector of each tag by its interaction with the EPC of the tag during its authentication. In their study, Mike and Medeiros (2007) and Qinghan et al. (2009) discussed security issues of RFID and categorized such attacks as follows: attack on authenticity, attack on integrity, attack on confidentiality, and attack on availability.

5.5.1.1.1 Attack on Authenticity (Unauthorized Tag Disabling)

Such attacks render an RFID tag to malfunction and misbehave during the scanning of a tag reader. Its EPC replies misinformation against the unique numerical identity assigned to it. This type of attack is generally exhibited remotely, allowing the attacker to manipulate the tag behavior from a distance.

5.5.1.1.2 Attack on Integrity (Unauthorized Tag Cloning)

The capturing of EPC identity information and manipulation of the tags by rogue readers falls under this category. Once the identity information of a tag is compromised, replication (cloning) of the tag is made possible which can be further used to bypass or counterfeit security measures as well as to introduce new vulnerabilities during automatic verification processes (Qinghan et al. 2009).

5.5.1.1.3 Attack on Confidentiality (Unauthorized Tag Tracking)

A tag can be traced through rogue readers, which may result in giving up of sensitive information (like a person's address). From a consumer's point of view, buying a product having an RFID tag guarantees them no confidentiality if the tag is being tracked unauthorized. This leaks the privacy.

5.5.1.1.4 Attack on Availability (Replay Attacks)

In this case, the attacker uses a tag's response to a rogue reader's challenge to impersonate the tag (Mike and Medeiros 2007). In replay attacks, the communicating signal between the reader and the tag is intercepted, recorded, and replayed upon the receipt of any query from the reader at a later time, thus faking the availability of the tag.

5.5.1.2 WSN Technology: Security Issues

Wireless sensor networks (WSN) consist of independent nodes where communication (wireless) takes place over limited frequency and bandwidth. The communicating node has sensor, memory, radio transceiver, microcontroller, and battery. Due to the limited communication range of each sensor node, multi-hop relay of information takes place in between the source and the base station. An empirical study by Guicheng and Liu (2011) showed that multi-hop transmission of data demands different nodes to take diverse traffic loads. The required data are collected by the wireless sensors through collaboration among the various nodes, which is then sent to the sink node for directed routing toward the base station. Some of the most prominent security issues plaguing the entire system arise out of the security issues for information relay from one device to another. According to Singla and Sachdeva (2013), the oppressive operations that can be performed in a WSN are due to the denial of service (DoS). This prevention of accessibility of information to legitimate users by unknown intruders can take place at different layers of a network.

5.5.1.2.1 DoS Attack: Physical Layer (WSN)

The physical layer of the WSN is attacked mainly through *Jamming* and *Node Tempering*. Jamming stems in the communication channel between the nodes (things) and thus prevents communication between nodes. Node tempering refers to the physical tempering of any node (thing) to extract sensitive information.

5.5.1.2.2 DoS Attack: Link Layer (WSN)

DoS attack in this layer is due to *Collision, Unfairness* and *Battery Exhaustion*. The Collision-type DoS attack is initiated when two nodes simultaneously transmit data packets on the same frequency channel. It results in small changes in the packet identifying a mismatch at the receiving end. It further leads to discarding of the infected data packet for re-transmission. Unfairness is a repeated collision-based attack. The DoS attack for battery exhaustion is caused by a large number of requests (requests to send) for unnecessary transmission over the channel. This leads to high traffic congestion in the channel and the efficiency of node accessibility reduces drastically.

5.5.1.2.3 DoS Attack: Network Layer (WSN)

The DoS attacker tries to disrupt while routing which is the main function of this layer due to Spoofing, Homing, Sybil, Wormhole, Hello Flood, Acknowledgment flooding, and Selective forwarding. *Spoofing* occurs due to replaying and misdirection of traffic. *Homing* is the type of attack, where the traffic is searched for cluster heads and key managers which have the capability to shut down the entire network. In a *Sybil* attack, the attacker replicates a single node and presents it with multiple identities to other nodes. *Wormhole* attack causes relocation of bits of data from its original position in the network. *Hello flood* attack causes high traffic in channels by congesting the channel with an unusually high number of useless messages. In the *Acknowledgment Flooding*, a malicious node spoofs the acknowledgments and provides false information to the neighboring destination nodes. In *Selective forwarding*, a compromised node only sends to few nodes instead of all the nodes.

5.5.1.2.4 DoS Attack: Transport Layer (WSSN)

Flooding and de-synchronization are the two types of attacks observed in the transport layer of WSSN. *Flooding* refers to the deliberate congestion of communication channels by the relay of unnecessary messages in high traffic. In *de-synchronization* attack, fake messages are created at one or both endpoints requesting retransmissions for correction of non-existent error. This results in loss of energy in one or both the endpoints in carrying out the spoofed instructions.

5.5.1.2.5 DoS Attack: Application Layer (WSSN)

Through this layer, a path-based DoS attack is initiated by stimulating the sensor nodes to create a huge traffic in the route toward the base stations. Karlof and Wagner (2003) and Chen et al. (2009) discussed the severity of attacks in WSN and classified into five types of threats.

5.5.2 Security Issues in Social Networks

Privacy of knowledge may be leaked if a social network is released improperly to public. We have experienced the fact that when an individual, organization, or a social group innovates successfully, the knowledge on which that progress is based becomes visible, at least partially, in the immediate neighborhood. As time goes on, such progress is understood and copied. Hence, there is a need for a systematic method to anonymize the social network data before it is released.

Now we would like to discuss about two basic concepts, label hierarchy and neighborhood, which are basic inputs for anonymization methods in social network.

Consider an example for *label hierarchy* as shown in Figure 5.7. The items in the label set L form a hierarchy. If the occupations are used as labels of vertices in a social network, L contains not only the specific occupations such as dentist, general physician, optometrist, high school teacher, and primary school teacher but also general categories like medical doctor, teacher, and professional. We assume that there exists a meta symbol $* \in$ L which is the most general category generalizing all labels. Similarly, in a social network G, the neighborhood of u $\in$ V (G) is the induced subgraph of the neighbors of u, denoted by Neighbor G(u) = G(Nu), where Nu = {v|(u, v) $\in$E(G)}. The components of the neighborhood graph of a vertex are the neighborhood components. The d-neighborhood graph of a vertex u includes all the vertices that are within the distance "d" from the vertex u.

5.5.2.1 Neighborhood Attack

The example in Figure 5.8 illustrates how the neighborhood attacks can take place. Let us consider the graphical representation (see Figure 5.8a) of the social network in which a group of persons are connected to each other through the relation of friendship. An edge connecting two nodes represents that they are friends.

In Figure 5.8a, if an attacker has knowledge about the 1-neighborhood of "Fred" (see Figure 5.8b), the node "Fred" can be easily identified from the basic anonymized graph of Figure 5.8c, as no other node has similar 1-neighborhoods to that of "Fred." Thus, the privacy of the social network is leaked. Once this node is identified, other private

FIGURE 5.7
Label hierarchy.

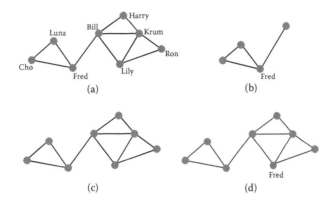

FIGURE 5.8
Illustration of neighborhood attacks: (a) original social network, (b) 1-neighborhood graph of Fred, (c) network with anonymous nodes, and (d) Fred identified in network.

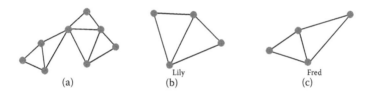

FIGURE 5.9
Illustration of 'Anonymization' as a technique to counter neighborhood attacks. (a) Anonymized network (b) 1-neighborhood of 'Lily' (c) Fred's neighborhood.

information such as connectedness, relative position, and relation with other nodes (identified in the same way) is exposed. This necessitates further anonymization processes so that the knowledge about 1-neighborhood cannot be used to identify a node uniquely. In Figure 5.8a, if an edge is added between "Luna" and "Bill," the 1-neighborhood of "Fred" and "Lily" is similar as shown in Figure 5.8b and c and it is not possible to identify 'Fred' with a confidence greater than ½ (see Figure 5.9a).

Tripathy and Panda (2010) discussed the usage of adjacency matrix and graph isomorphism to anonymize the identity of nodes or actors of the network. In their work, anonymization is done by taking the vertices from the same group. If the match is not found, the cost factor is used to decide the pair of vertices to be considered. Their algorithm adheres to the *k*-anonymity security model. More importantly, the time complexities of their anonymized algorithms are comparatively less.

5.5.2.2 Social Network Anonymization

Identifiers are the key of the entities such as the data subject's name, address, and sometimes the unique identification numbers (e.g., Social Security number or National Health Service number). These identifiers make an individual entity unique in a data set and as such highly vulnerable to re-identification. Anonymization is a technique that is used to shield, remove, or aggregate the basic identifiers in the data sets. Anonymizing social networks data is much more challenging than anonymizing relational data due to many problems, some of which are listed below.

- It is much more challenging to model the background knowledge of adversaries and attacks about social network data than that about relational data. In a social network many pieces of information can be used to identify individuals, such as labels of vertices and edges, neighborhood graphs, induced subgraphs, and their combinations. So, all these make modeling social networks difficult.
- It is also challenging to measure the information loss in anonymizing relational data. Unlike for relational data for which the sum of the information loss for tuples solves the purpose, it is hard to compare two social networks by comparing their vertices and edges individually.
- Anonymizing a social network is much more difficult since changing labels of vertices and edges may affect the neighborhoods of other vertices, and removing or adding vertices and edges may affect other vertices as well as the properties of the network.

To protect the privacy satisfactorily, the models make guarantee that any individual cannot be identified correctly in the anonymized social network with a probability higher than 1/k, where k is user-specified parameter in the k-anonymity model (Tripathy and Panda 2010). An adversary with the knowledge of neighborhood of a vertex cannot identify any individual from this anonymous graph with a confidence greater than 1/k.

Furthermore, Tripathy et al. (2011a, 2011b, 2011c, 2011d) improved the clustering stage of the One Pass K-Means (OKA) algorithm, clustering Minimum Mean Roughness (MMeR) algorithm, and introduced *l*-diversity algorithms and variances using soft computing techniques like rough set theory. Such algorithms satisfy the privacy of individual nodes to the extent of anonymity in public social networks. These *k*-anonymity and *l*-diversity algorithms have the highest possibility of convergence of integration in the SIoT security threats.

5.5.3 Research Opportunities in SIoT

Social IoT is going to experience massive deployment of sensors and consumer objects which would affect the potential quantity of interconnected links between these devices (Ortiz et al. 2014; Alam et al. 2015). Many of these devices will be able to establish links and communicate with other devices on their own in an unpredictable and dynamic manner. Therefore, existing tools, methods, and strategies associated with IoT security may need new consideration. Many SIoT deployments will consist of collections of identical or near-identical devices. This homogeneity magnifies the potential impact of any single security vulnerability by the sheer number of devices that all have the same characteristics. Some IoT devices are likely to be deployed in places where physical security is difficult or impossible to achieve. Attackers may have direct physical access to IoT devices or wirelessly hack such devices. Anti-tamper features and other design innovations will ensure security in SIoT. The social network-based collaborative approach will be an effective solution to industry, government, and public authorities to secure the Internet and cyberspace, including the SIoT.

We list out some questionnaires related to the security and probable threats and attacks in IoT/SIoT in Table 5.3. The list may be helpful to the students, academicians, and researchers who have an interest in this topic for further research investigations.

TABLE 5.3

Research Questionnaires on Security Issues in IoT/Social IoT

- What is the role of technical and operational standards for the development and deployment of secure, well-behaving IoT devices?
- How do we effectively identify and measure characteristics of IoT device security?
- How do we measure the effectiveness of IoT security initiatives and countermeasures? How do we ensure security best practices are implemented?
- Would it be possible for regulation to keep pace and be effective in light of evolving IoT technology and evolving security threats?
- How should regulation be balanced against the needs of permission-less innovation, Internet freedom, and freedom of expression?
- What is the optimal role of data encryption with respect to IoT devices?
- Is the use of strong encryption, authentication and access control technologies in IoT devices an adequate solution to prevent eavesdropping and hijacking attacks of the data streams these devices produce?
- Which encryption and authentication technologies could be adapted for the IoT, and how could they be implemented within an IoT device's constraints on cost, size, and processing speed?
- Are the end-to-end processes adequately secure and simple enough for typical consumers to use?
- With an extended service life expected for many IoT devices, should devices be designed for maintainability and upgradeability in the field to adapt to evolving security threats?
- New software and parameter settings could be installed in a fielded IoT device by a centralized security management system if each device had an integrated device management agent. But management systems add cost and complexity; could other approaches to upgrading device software be more compatible with widespread use of IoT devices?
- Are there any classes of IoT devices that are low risk and therefore don't warrant these kinds of features?
- Are the user interfaces IoT devices expose (usually intentionally minimal) being properly scrutinized with consideration for device management (by anyone, including the user)?
- What is the right approach to take with obsolete IoT devices as the Internet evolves and security threats change?
- Should IoT devices be required to have a built-in end-of-life expiration feature (bio-decomposable) that disables them? Such a requirement could force older, non-interoperable devices out of service and replace them with more secure and interoperable devices in the future. Certainly, this would be very challenging in the open marketplace.
- What are the implications of automatic decommissioning of IoT devices?
- How should we protect data collected by IoT that appears not to be personal at the point of collection or has been "de-identified", but may at some point in the future become personal data (e.g. because data can be reidentified or combined with other data)?

From a research point of view, it is wise to know about the institutions, organizations, and government alliances, who work and address issues on IoT, SIoT, and its variants (in particular privacy and threat issues). We list out in Table 5.4 such additional information sources which may serve as a starting point for further investigations to the interested researchers concerned.

5.6 Pro-Business and Pro-People Social IoT Services

Social IoT services can be modeled as pro-business and pro-people types. Facebook, Google, LinkedIn, and similar services are categorized as pro-business models, where users have to trade privacy for a free service and must agree to be tracked or snooped to generate a business intelligence database. In pro-people SIoT service, which we will

TABLE 5.4

List of Organizations Working for Security Issues in IoT/SIoT

Organizations	Web Link for Further References	Expertise in IoT/SIoT Consultant/ Security Solutions
AIOTI	https://ec.europa.eu/digital-agenda/en/alliance-internet-thingsinnovation-aioti	Development of European IoT ecosystem, standardization policies
AllSeenAlliance	https://allseenalliance.org/	Applications in IoT
ETSI	http://www.etsi.org/technologies-clusters/clusters/connectingthings	Smart object-based data management, security
IEC 62443/ISA99	http://isa99.isa.org/ISA99%20Wiki/Home.aspx	Secure industrial automation, control system
IEEE (P2413)	http://iot.ieee.org/	Research and implementation
IERC	http://www.internet-ofthings-research.eu/	IoT research cluster
IETF	https://trac.tools.ietf.org/area/int/trac/wiki/IOTDirWiki	Standardization of IoT directorate
IIC	http://www.industrialinternetconsortium.org/	IoT architecture
IGF	http://www.intgovforum.org/cms/component/content/article?id=1217:dynamiccoalition-on-the-internet-of-things-component	Dynamic collaboration
IoT Consortium	http://iofthings.org/#home	IoT consumer research and market education
IPSO	http://www.ipso-alliance.org/	IoT education, research, and promotion
ISO/IECJTC-1	http://www.iso.org/iso/internet_of_things_report-jtc1.pdf	Report formation
Internet of Food SIG	http://internet-offood.org/	IoT technicality in food industry
ITU	http://www.itu.int/en/ITU-T/studygroups/2013-2016/20/Pages/default.aspx	IoT applications
MAPI FOUNDATION	https://www.mapi.net/research/publications/industrie-4-0-vsindustrial-internet	Industrial applications
Ministry of Comm. & IT, Govt. of India	http://deity.gov.in/content/internet-things	IoT industry ecosystem in India
OASIS	https://www.oasis-open.org/committees/tc_cat.php?cat=iot	Developing protocols for IoT interoperability
OneM2M	http://www.onem2m.org/	Development in M-T-M communications architecture
Online Trust Alliance	https://otalliance.org/initiatives/internet-things	Application development on security, privacy in IoT
OIC	http://openinterconnect.org/	Application development on open-source software for device-to-device IoT
OMG	http://www.omg.org/hottopics/iot-standards.htm	DDS, IFML, threat modeling for real-time embedded system developments
OWASP	https://www.owasp.org/index.php/OWASP_Internet_of_Things_Top_Ten_Project	Project sponsors for manufacturers, developers and consumers of IoT
SGIP	http://sgip.org/focus-resilience	Consults IoT solutions for energy industry
Thread Group	http://threadgroup.org/About.aspx	Integrator for IoT-based home appliances and security systems

discuss next, the users don't have to trade their privacy for free service. They can opt in or opt out from available hundreds of services in one platform with a sign on system.

The bulky Privacy and Terms of Usage pages hardly anyone ever reads. In this pro-business model, users can't easily opt out to be snooped or tracked and must trade some privacy for one or more free services. Some free Apps like WhatsApp even target customer's phone contact list to shove them with targeted advertisements. All popular browsers are designed to track users by default. Business intelligence is generated from volunteered opt-in information and the legal Privacy and Terms of Usage agreements can fit in one paragraph or maximum one page that can be read once.

A typical web user, over his or her life time, could use many online services for numerous activities like buying products, learning skills, paying bills, etc. If the user registers such service sites through e-mail, Facebook, or LinkedIn APIs (if supported by the website), then the user would most likely be blasted with e-mails from those websites or will be followed to user's timeline page. Either way, the user can't escape tracking in any typical pro-business SIoT websites.

In addition to privacy, there are security and fake news concerns. Figure 5.10 shows up fake alerts during a search of a popular search engine that disables further browser activity. A typical user won't know that these are rogue websites but the way search engines are designed, they can't detect and block the stuff. Of late, during recent US presidential elections, credibility of Facebook also suffered because of "fake news." Now, even if there is negative, but "true" news, supporters argue that as fake news.

Fan's Global Social NGN LLC has been researching on a pro-people SIoT services since couple of years to develop a single sign on (SSO) platform to host multiple domains—around 100 different services—where the names of websites contain the service names. All these websites share the same web platform and hence the platform is named as "Web Platform as Service" (WPaS). The common functionalities such as social button clicks, event logging, alerts and messaging are shared among all domains. If a user registers in any one of the services (say, Service2Buy.com), the credentials would be valid for all the services like Malls2Go.com, Videos4Rent.com, or Dating2Wedding.com that are hosted under the WPaS platform.

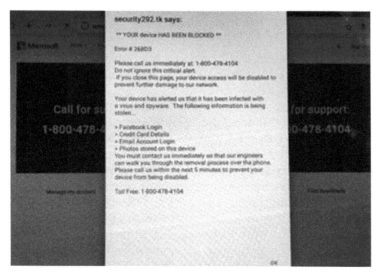

FIGURE 5.10
Fake alerts during browsing.

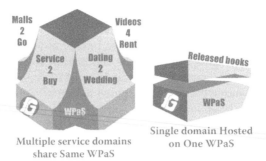

FIGURE 5.11
Next generation Social IoT services.

With WPaS (see Figure 5.11), many existing popular web services, who lack their own social features and are not attractive to advertisers or subscribers, can simply bring their services to a WPaS platform and get all the platform features automatically and focus on core service and compete among other peers in WPaS. For example, two dating websites, Dating2Wedding.com and VideoMatrimony.com, can compete for subscribers by providing better features while not bothering about getting subscribers, authentication, and social features.

Because of the WPaS-based control of services, the user database is no more visible to service providers. When the user opts in to a business, product, or services, the respective advertisers get alert. The RDMBS-integrated Search based on opted information from users would be much better than that of the current search that can't stop virus-infected URLS. The upcoming pro-user-based services of 100 domains would help users to get an alternative secure and private web space which would be free from fake news, virus, rogue security alerts, and above all much better search with easy-to-remember service names.

5.7 Conclusion and Future Perspectives

The concept of integrating sensors, mobile devices, computers, and networks to monitor and control devices has been around for decades. The recent development of key technologies and market trends is evolving into a new reality for the concept of IoT, which promises to usher in a revolutionary, fully interconnected "smart" world, with social relationships between objects and their environment. Independent objects and people are becoming more intertwined socially to evolve the concept of SIoT. As the potential ramifications to the evolution of SIoT are significant, a number of challenges also knock in the way of this vision, particularly in the areas of security, privacy, interoperability, and standards like legal, regulatory, and demographic rights issues. Security breaches, link threats, and identity attacks may be serious hindrances to emerging global economies and demographic issues. Hence, through this chapter, we tried to address such challenges while reducing its risks in the domain of IoT. The Internet and social network theory provide numerous integrated solutions to IoT because of the growing aspects of how people and institutions are binding with and best utilize the Internet and network connectivity to their personal, social, and economic lives. This chapter justifies to these key issues and explores numerous views for further investigations by interested researchers providing sufficient information therein.

References

Alam, K.A., Saini, M., and Saddik, A.E. 2015. Toward social internet of vehicles: Concept, architecture, and applications. *IEEE Access*, 3, 343–357.

Atzori, L., Iera, A., and Morabito, G. 2010. The Internet of Things: A survey. *Computer Networks*, 54(15), 2787–2805.

Atzori, L, Lera, A, Morabito, G, and Nitti, M., 2012. The Social Internet of Things (SIoT)—When social networks meet the internet of things: Concept, architecture and network characterization, *Elsevier*, 56(16), 3594–3608.

Chen, X., Yen, K.M.K., and Pissinou, N. 2009. Sensor network security: A survey. *IEEE Communications Surveys and Tutorials*, 11, 52–73. DOI: 10.1109/SURV.2009.090205.

Cisco IBSG. 2011. *The Internet of Things How the Next Evolution of the Internet Is Changing Everything*, White Paper, pp. 1–11. www.cisco.com/c/dam/en_us/about/ac79/docs/innov/IoT_IBSG_0411FINAL.pdf.

Clauberg, R. 2004. *RFID and Sensor Networks: From Sensor/Actuator to Business Application*. RFID Workshop. University of St. Gallen, Switzerland, September 27.

Fast, A., Jensen, D., and Levine, B.N. 2005. Creating social networks to improve peer-to-peer networking. In *The Proceedings of the ACM KDD'05*, August 21–24, Chicago, IL pp. 568–573.

Ferguson, A.G. 2016. The Internet of Things and the fourth amendment effects, *California Law Review*, Vol. 104(4), 805–818.

Gubbi, J., Buyya, R., Marusic, S., and Palaniswami, M. 2013. Internet of Things (IoT): A vision, architectural elements, and future directions. *Future Generation Computer Systems*, 29(7), 1645–1660.

Guicheng, S., and Liu. B. 2011. The visions, technologies, applications and security issues of Internet of Things. In: *International Conference on E-Business and E-Government (ICEE)*. IEEE, May 6–8, Shanghai, China, pp. 1–4.

Guo, B., Zhang, D., Wang, Z., Yu, Z., and Zhou, X. 2013. Opportunistic IoT: Exploring the harmonious interaction between human and the internet of things. *Journal of Network and Computer Applications*, 36(6), 1531–1539.

Juels A. 2006. RFID security and privacy: A research survey. *Selected Areas in Communications, IEEE Journal*, 24(2), 381–394. DOI:10.1109/JSAC.2005.861395.

Karlof, C., and Wagner, D. 2003. Secure routing in sensor networks: Attacks and countermeasures, Special Issue on Sensor Network (SNPA), *Elsevier's AdHoc Networks Journal*, 1(2–3), 293–315.

Marti, S., Ganesan, P., Garcia-Molina, H. 2004. SPROUT: P2P routing with social networks, In *The Proceedings of the EDBT 2004 Workshops*, LNCS 3268, Springer, March 14, pp. 425–435.

Mei, A., Morabito, G., Santi, P., and Stefa, J. 2011. Social-aware stateless forwarding in pocket switched networks. In *The Proceedings of the IEEE Infocom–Miniconference*, April 10–15, Shanghai, China, pp. 251–255.

Mike, B., and Medeiros, B.D. 2007. RFID security: Attacks, countermeasures and challenges. *The 5th RFID Academic Convocation, The RFID Journal Conference*, pp. 1–10. https://www.cs.fsu.edu/~burmeste/133.pdf, Accessed July 22, 2017.

Mislove, A., Gummadi, K.P., and Druschel, P. 2006. Exploiting social networks for Internet search. In *The Proceedings of 5th Workshop on Hot Topics in Networks (HotNets-V)*, Irvine, CA, pp. 79–84.

NIC., 2008. *Disruptive Civil Technologies Six Technologies with Potential Impacts on US Interests Out to 2025*. IDTechEx. http://www.dni.gov/nic/NIC home.html, Accessed July 22, 2017.

Ortiz, AM., Hussein, D., Park, S., Han, SN. 2014. The cluster between internet of things and social networks: Review and research challenges. *IEEE Internet of Things Journal*, 1(3), 206–215.

Prodanoff, Z.G. 2009. Optimal frame size analysis for framed slotted ALOHA based RFID networks. *Computer Communications*, 33(5), 648–653. DOI:10.1016/j.comcom.2009.11.007.

Qinghan, X., Gibbons, T., and Lebrun, H. 2009. *RFID Technology, Security Vulnerabilities, and Countermeasures*. Supply Chain the Way to Flat Organization, Intech, Vienna, Austria, pp. 357–382.

Sarma, S., Brock, D., and Ashton, K., 1999. *The Networked Physical World: Proposals for the Next Generation of Computing Commerce, and Automatic Identification.* AutoID Center White Paper, MIT Auto-ID Center, MIT, Cambridge, MA, pp. 1–16.

Scott R.P., 2014. Regulating the Internet of Things: First steps towards managing Discrimination, Privacy, Security, and Consent. *Texas Law Review*, 93(85), 86–176.

Singla, A., and Sachdeva, R. 2013. Review on security issues and attacks in wireless sensor networks. *International Journal of Advanced Research in Computer Science and Software Engineering*, 3(4), 387–391.

Strategy, I.T.U., and Unit, P. 2005. *ITU Internet Reports 2005: The Internet of Things.* International Telecommunication Union (ITU), Geneva.

Surowiecki, J. 2004. *The Wisdom of Crowds.* Penguin Random House, NY, pp. 1–336.

Tripathy, B.K., and Panda, G.K. 2010. A new approach to manage security against neighbourhood attacks in social networks. *The International Conference on Advances in Social Networks Analysis and Mining*, ASONAM 2010, University of Southern Denmark, Denmark, 2010, IEEE Computer Society, pp. 264–269.

Tripathy, B.K., Kumaran, K., and Panda, G.K. 2011d. An improved *l*–diversity anonymisation algorithm, *5th International Conference on Information Processing, Springer Verlag, Communications in Computer and Information Sciences (CCIS)*, Bangalore, India, August 5–7, vol. 157, pp. 81–86.

Tripathy, B.K., Panda, G.K., and Kumaran, K. 2011a. A fast *l*-diversity anonymisation algorithm. *Third International Conference on Computer Modeling and Simulation-ICCMS 2011*, Mumbai, India, January 7–9, vol. 2, pp. 648–652.

Tripathy, B.K., Panda, G.K., and Kumaran, K. 2011c. A rough set approach to develop an efficient *l*-diversity algorithm based on clustering. *ICADABAI-2010*, IIM Ahmedabad, 43(8).

Tripathy, B.K., Panda, G.K., and Kumaran, K. 2011b. A rough set based efficient *l*-diversity algorithm. *International Journal of Advances in Applied Science Research, Pelagia Research Library*, 2(3), 302–313.

Tripathy, B.K., Dutta, D., and Tazivazvino, C. 2016. On the research and development of social Internet of Things. *Internet of Things (IoT) in 5G Mobile Technologies, Modeling and Optimization in Science and Technologies* 8, 153–173.

Van, D.T. 2011. *50 Ways to Break RFID Privacy. Privacy and Identity Management for Life*, IFIP Advances in Information and Communication Technology 352. Springer, Boston, MA pp. 192–205.

Weiser, M. 1991. The computer for the 21st century. *Scientific American*, 265(3), 94–104.

Westin A.F. 1968. Privacy and freedom, *Washington and Lee Law Review*, 25(1), Article 20, pp. 1–487.

Zheng, L. 2011. *Technologies, Applications, and Governance in the Internet of Things, Internet of Things—Global Technological and Societal Trends*. River Publisher Ed, Gistrup, Denmark.

6

User Authentication: Keystroke Dynamics with Soft Biometric Features

Soumen Roy and Devadatta Sinha
University of Calcutta
Calcutta, India

Utpal Roy
Visva-Bharati
Santiniketan, India

CONTENTS

6.1 Introduction

Knowledge-based user authentication technique is a common and easy access control mechanism. But people are uninspired while choosing a healthy password or PIN. It increases the probability of guessing attacks. In this situation, to minimize these attacks, keystroke dynamics is a good choice; here users are not only identified by the password but their typing style is also accounted for. Keystroke dynamics is the method of analyzing typing pattern on a computer keyboard or touch screen and classifying the users based on their regular typing rhythm. It is a behavioral biometric characteristic which we have learned in our life and relates to the issues in human identification/authentication. This is the method where people can be identified by their typing style similar to hand writing or voice print. Being noninvasive and cost-effective, this method is a good field of research. But the performance of keystroke dynamics is less than other popular morphological biometric characteristics like face print, iris, and finger print recognition due to high rate of intraclass variation or high Failure to Enroll Rate (FER). So, this technique demands higher level of security. In this chapter, we are interested in investigating the integration of the soft biometric features, gender and age group, with the existing keystroke dynamics user authentication systems proposed by [1–3].

We have investigated the probability of predicting gender and age group based on typing pattern. This is possible if all the patterns of certain peer are similar and dissimilar from one peer to another. As per the research direction, it is possible to predict the gender with 88.55% to 95.04% accuracy based on typing pattern on the keyboard. Similarly, we also obtained 84.75% accuracy with regard to gender based on typing pattern on touch screen. If we use only gender information as an extra feature, then we can achieve 3.5% to 7.72% of accuracy. Similarly, the age group (18–30/30+ years) can also be predicted based on typing pattern. We obtained 86.87% to 94.68% accuracy with regard to prediction of age group (18–30/30+ years) based on typing pattern on keyboard, whereas 84.75% of accuracy was obtained by analyzing the typing pattern on touch screen with regard to age group (7–29/30–65 years). We also analyzed the age group (≤18/18+ years) and obtained 89.2% to 92% accuracy based on typing pattern on touch screen.

These two biometric traits have low discriminating power but can be used as additional soft biometric features to reduce the error rate in keystroke dynamics user authentication systems. This technique can also be used in e-commerce sites to reach out to the right client to avoid adverse products more efficiently based on the gender and age group. There are many application areas where automatic gender and age group identification methods can be applied like any surveillance system, online automatic user account profiling, and protection of minors from online threats.

The main objective of this study is to develop a model that can identify the gender and age group of users through the way of typing on a computer keyboard and touching a computer screen for a predefined text, and it increases the accuracy by recognizing this soft biometric information as additional features in keystroke dynamics user authentication systems.

The major goal and contributions of this chapter are as follows:

- Develop an efficient model to recognize the gender and age group automatically from typing pattern.
- Validate the model on keystroke dynamics dataset collected through a touch screen device.

- Compare our results with other leading approaches.
- Integrate both soft biometric features with keystroke dynamics user authentication systems and show the impact and effectiveness of our approach.

In essence, this study is one of the approaches to recognize the gender and age group of Internet users. The secret behind this technique is physical structure, hand weight, finger tips' size, and neuro-physiological and neuro-psychological factors reflecting on the computer keyboard which discriminate the gender and age group of the users.

We have used published and authentic CMU keystroke dynamics dataset [4] along with the datasets collected through android hand held devices [5]. The classification results to determine the gender and age group by using FRNN-VQRS showed that more than 94% accuracy can be achieved through a computer keyboard while a touch screen device offered more than 84% accuracy. The details of the datasets are summarized in Table 6.1.

We have used Weka GUI 3.7.4 to evaluate and compare the leading machine learning algorithms on public keystroke dynamics datasets. Obtained results are reported with default parameter values in Weka.

This chapter is organized as follows. Related works have been described in Section 6.2. Section 6.3 describes the basic idea about keystroke dynamics. Section 6.4 compares the performance of keystroke dynamics with other behavioral biometric systems. Section 6.5 represents the details of the datasets which have been used in our experiments. Our proposed methodology has been clearly explained in Section 6.6. All experimental results are reported in Section 6.7. Results and Discussion have been explained in Section 6.8. The last section compares our system with that of others and highlights its achievements.

6.2 Related Works

Keystroke dynamics technique started in 1980. Many journal articles, conference articles, and master theses were published. Figure 6.1 clearly indicates the increasing trends in keystroke dynamics research. Many datasets have been created considering different types of texts with different lengths from different subjects; many methods have been applied and many innovative ideas have come out from the previous studies. Some studies showed that keystroke dynamics holds better performance when using common words used daily than strong password-type texts. Modi and Elliott [6] showed that nonfamiliar words do not give

TABLE 6.1

Evaluation of Behavioral Biometric Techniques

Parameters	Keystroke Dynamics	Signature	Voice	Gait
Universality	L	L	M	M
Uniqueness	L	L	L	L
Permanence	L	L	L	L
Collectability	M	H	M	H
Performance	L	L	L	L
Acceptability	M	H	H	H
Circumvention	M	L	L	M

Note: L, Low; M, Medium; and H, High.

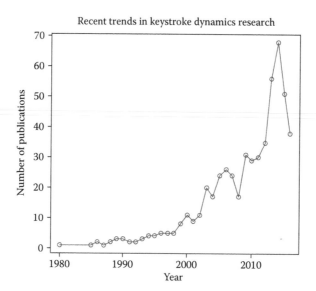

FIGURE 6.1
Published articles on keystroke dynamics by year.

interesting results. Giot et al. [7] showed that the gender of an individual user can be identified based on who types a predefined text. Epp et al. [8] showed that it is possible to get the emotional state of an individual through keystroke dynamics. Khanna and Sasikumar [9] showed that 70% of users decrease their typing speed while they are in a negative state, and 84% of users increase their typing speed when they are in a positive emotional state. Joyce and Gupta [3] observed that shorter and easy to type login texts were easier to impersonate. Killourhy et al. [4] conducted experiments to investigate the effect of clock resolution on keystroke dynamics. They observed that the equal error rate (EER) increased by approximately 4.2% when using a 15 ms resolution clock instead of a 1 ms resolution clock. Ru and Eloff [10] observed that password and user ID with normal "English-like" text seemed less discernible from each other than string combining special characters such as &, %, @, ! etc. Roy et al. [11,12] showed that it can be used as a password recovery mechanism and also can be applied in cryptosystem. In another paper, Roy et al. [12,13] applied 22 different classification algorithms on keystroke dynamics and showed that distance-based algorithms, namely Canberra, Lorentzian, scaled Manhattan, and outlier count, are the suitable classifiers on keystroke dynamics in identification/authentication.

Data acquisition technique is the primary and most essential stage in keystroke dynamics; here subjects are required to type only character-based text, purely numeric-based text, or alpha-numeric-based text. Character-based text can be further sub divided into short text, long text, and paragraph. Alpha-numeric text can be further sub divided into strong text or password-type text and logically strong text. Figure 6.2 indicates the percentage distribution of the type of texts used in literature.

Most of the time, simple, common, fixed-size words used daily were used along with multiple predefined words (long text) in literature. As per the experiment, if we consider familiar words for all subjects in our experiment, we can get a consistent typing style across different sessions and each repetition from all the subjects. Performance of keystroke dynamics in user authentication/identification depends on type of text: familiar words are suitable than password-type texts and password-type texts achieve more

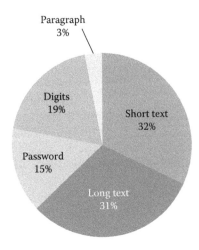

FIGURE 6.2
Percentage distribution of type of texts used in literature.

accurate results than texts containing only digits [14]. Bleha et al. [1] showed that longer words minimize the misclassification rate.

Many type of keystroke features have been used in literature. A percentage distribution of keystroke features has been indicated in Figure 6.3, where combination of key hold time of a single key and latency times between two subsequent entered characters are the mostly used features. Some researchers say that di-graph time is the most elementary timing feature in keystroke dynamics, but only 12% of di-graph timing features have been used. Key pressure is also applied as an effective feature in keystroke dynamics where a special keyboard or a pressure-sensitive keyboard is required. As technology evaluation grows, many advance and sensitive hardwares have been used in smart phones or portable devices. Through the advance sensing devices, we can get some additional effective features such as finger tips size, velocity of finger movements in different directions [15].

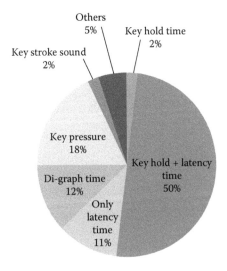

FIGURE 6.3
Percentage distribution of keystroke features used in literature.

The methods were used on keystroke dynamics datasets, where statistical measurements are common like mean, median, and standard deviation. Many distance-based algorithms were also used as pattern recognition techniques such as Canberra, Chebyshev, Czekanowski, Gower, Intersection, Kulczynski, Lorentzian, Minkowski, Motyka, Ruzicka, Soergel, Sorensen, Wavehedges, Manhattan Distance, Euclidean Distance, Mahanobolis Distance, Z Score, and KMean. Some machine learning methods also have been applied like support vector machine (SVM), naïve bayes, multi layer perceptron, fuzzy set, K-nearest neighbor, OneR, hidden Markov model (HMM), Gaussian Markov model (GMM), and random forest. Other methods like direction similarity measure (DSM), degree of disorder, and array disorder were also used. ACO, PSO, Best First, and GA were used as optimization techniques.

Many classification methods have been applied over the last 30 years. Where statistical methods were common, over the last 10 years, strong machine learning and distance-based methods are common approaches. Figure 6.4 indicates the percentage distribution of the type of methods used in literature.

To test the biometric system, few parameters (EER, FAR, FRR, etc.) are used to evaluate the performance of the system. Figure 6.5 indicates the percentage distribution of average EER previously recorded. The European standard for access control specifies that FAR must be less than 1% and FRR must not be more than 0.001% [16]. But in literature, only 1.36% of studies [17,18] reached those acceptable results. In [17], large samples were collected from each subject in training session which is impractical in real life, whereas Ali and Salami [18] used only key pressure as keystroke features and took data from only seven subjects, with the scalability of the study being very low. Hence, further research has to be done on keystroke dynamics for identification/verification of users since the application area of this technique is very large.

To summarize, most of the works have been studied on datasets collected through a computer keyboard than a touch screen device. Many researchers have created the dataset and applied classification algorithms and obtained the results. Some of the researchers worked on optimization techniques and endeavored to enhance the keystroke dynamics user authentication performance. Giot et al. [7] extracted the gender feature only from the typing pattern on computer keyboard, but did not work on the typing pattern on touch screen. They also used this soft biometric feature with the timing features and obtained gain accuracy of up to 20%.

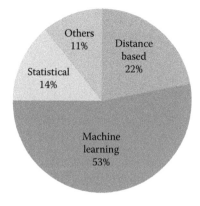

FIGURE 6.4
Percentage distribution of different classification methods used in literature.

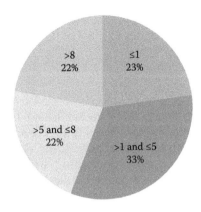

FIGURE 6.5
Percentage distribution of performance (EER) recorded in literature.

6.3 Keystroke Dynamics

6.3.1 Basic Idea

Keystroke dynamics is the method of analyzing the way users type on a computer keyboard and classifying the users based on their regular typing rhythm. Here, users are not only identified by their corresponding used ID and password, but their typing style is also accounted for. In this study, people can be recognized by their typing style much like hand writing or voice print. Recognizing typing style promises a parameter like behavioral biometric characteristics in biometric science to solve the issues in identification and authentication technique. It facilitates nonintrusive, cost-effective, and continuous monitoring. It is a behavioral biometric characteristic which may vary during the course of a day or between 2 days; it depends on the mental state of the person, educational level, position of keyboard, and many more. This technique, as such, is prone to issues related to accuracy. Thus, in order to realize this technique in practice, a higher level of security and performance together with convenient and low-cost version is demanded with an acceptable level of accuracy. So, controlling parameters that optimize accuracy to a great extent is necessary.

6.3.2 Science

Placement of fingers on keyboard, hand weight, hand geometry, and neuro-physiological factors reflect on keyboard indicating a unique typing style for each individual just like each person's written signature.

6.3.3 Features

The features of the keystroke dynamics are Key Duration (KD) Key Latency, Up Up (UU) Key Latency, Down Down (DD) Key Latency, Up Down (UD) Key Latency, Down Up (DU) Key Latency, Total Time (T_Time) Key Latency, Tri-graph Time (T), and Four-graph Time (F). Some new features can also be incorporated such as key pressure (pressure-sensitive keyboard is required), finger tips' size (touch screen keypad is needed), finger placement on keyboard (camera is needed), keystroke sound (microphone is needed) which were used in [6], along with error correcting mechanism, sequence of left–right control keys. Dwell time and UD Key Latency times were used most of the time; Janakiraman et al. [19] used only di-graph and tri-graph time.

6.3.4 Keystroke Dynamics as User Authentication

There are many ways in which a user can be authenticated. However, all of these ways can be categorized into one of three classes: "Something we know," for example, password; "Something we have," for example, token; "Something we are born with," for example, physical biometric characteristics; and "Something we have learned in our life," for example, behavioral biometric. Keystroke dynamics behavior is behavioral biometric characteristics what we have learned in our life.

6.3.5 Advantages

In this technique, no extra security apparatus is needed; here the keyboard is enough to recognize the typing behavior. It is cost-effective and easy to implement with the existing system. This characteristics cannot be lost or stolen and nothing we have like password to remember. We cannot mimic the typing style of others.

6.3.6 Disadvantages

External factors such as injury, fatigue, type of keyboard, and position of keyboard affect the system. The typing pattern of a person may vary during the course of a day or between 2 days and is dependent on his or her mental state. More datasets are required to fine-tune the system. It takes a lot of time.

6.3.7 Application Areas

This technique can be effectively applied in student or employee attendance system, distance-based examination, password recovery mechanism, emotion recognition, private data encryption, continuous user verification, criminal investigation, backdoor accounts identification, free-text user authentication, gender identification, age group recognition, etc.

6.3.8 Security Issues

Keystroke dynamics is a behavioral characteristic which is unique and can be effectively implemented with the existing system with minimal alternation. It can be used to safeguard our password from different types of attacks.

6.3.9 Factors Affecting Performance

Some of the factors that affect the working of keystroke dynamics is as follows: text length, sequences of character types, word choice, number of training sample, statistical method to create template, mental state of the user, tiredness or level of comfort, keyboard type, keyboard position and height of the keyboard, hand injury, weakness of hand mussel, shoulder pain, education level, computer knowledge, and category of users.

6.4 Evaluation of Behavioral Biometric Techniques

Jain [20] presents an extensive comparison of various behavioral biometric techniques in Table 6.1. He proposes a table that presents evaluation of those parameters for behavioral biometric technique in the following scale: H = high, M = median, L = low.

The following parameters are generally used to evaluate the biometric technique.

6.4.1 Universality

It defines how commonly biometric is found individually. The number of users of computer-/handheld-device-enabled Internet connection is rapidly increasing and the keyboard has become a mandatory device. But entered texts may not have the same sequence for all the users. Universality of keystroke dynamics is the ratio between the number of users having same sequence of characters and total number of users [19].

6.4.2 Uniqueness

It is a measurement of how well biometric separates individuals from others. It is globally accepted that typing rhythm is unique for all. Most of the time, experiments were done in a controlled environment with impressive results. But due to a large number of affecting factors, it is not accepted in practice.

6.4.3 Permanence

It defines that how much complexity to measure in acquisition. Timing parameters are very easy to calculate in keystroke dynamics. But to measure key pressure and acceleration, extra sensors are required. Nowadays, advance sensing device and accelerometer are embedded in each smart phone to measure pressure and acceleration.

6.4.4 Collectability

It represents how well it resists the aging problem. Typing behavior varies during the course of a day or between 2 days. So updating the mechanism is necessary.

6.4.5 Performance

Performance of all the behavioral biometrics is low. Many papers have been published with impressive results in a controlled environment. But in practice, it is very hard to achieve impressive results.

6.4.6 Acceptability

It describes the degree of approval of a technology. Many commercial products have been introduced; trademarks of the products are BioPassword, AuthenWare, TypeSence, Phylock, etc.

6.4.7 Circumvention

How easy it is fraud specific biometric characteristics. Keystroke dynamics is not easy to mimic even if you observe the typing style of others many times.

6.5 Benchmark Soft Biometric Datasets on Keystroke Dynamics

In this section, we describe the datasets used in our experiment. Many variants of authenticated datasets on keystroke dynamics are available on the Internet, which can be downloaded or accessed on request. In this chapter, we have used four datasets for different predefined texts as well as different environments for the prediction of gender and age group. Table 6.2 represents the details of the publicly available authentic and recognized datasets. All the datasets are given by a name for the purpose of identification throughout this chapter.

We summarized the different keystroke dynamics datasets in Table 6.2 which are used in our study. Most of the researchers have performed different experiments to develop a model to recognize the user with these datasets. Some of them obtained impressive results to identify the user, but they are not acceptable in practice. Datasets A, B, and C were collected through a computer keyboard. Dataset A was collected from 38 male and 27 female users where 38 were aged 18–30 years and 27 were aged 30+ years; Dataset B was collected from 25 male and 13 female users where 23 were aged 18–30 years and 15 users were aged 30+ years;. Dataset C was collected from 21 male and 21 female users where 24 were aged 18–30 years and 18 were aged 30+ years; and Dataset D was collected from 26 male and 25 female users using a touch screen device where 11 were aged 7–18 years, 30 were aged 19–29 years, and 10 were aged 30–65 years.

6.6 Proposed Methodology

Biometric systems are not 100% accurate as per Jain et al. [21] due to various problems in data acquisition methods or interclass variations. As per previous studies, accuracy can be improved by using the soft biometric information as additional features with the typing pattern in keystroke dynamics. Giot et al. [7] used gender as additional information to predict gender with 91% accuracy. In order to realize this technique in practice, we have used FRNN-VQRS to predict the gender as well as the age group of the users based on the typing pattern of different predefined texts which elicited more than 94% accuracy on CMU keystroke dynamics dataset. The performance metric, area under curve (AUC), is always high and proved FRNN-VQRS to be an efficient approach. In this study, we applied gender and age group as additional features and obtained 94.37% accuracy in keystroke dynamics authentication which is a gain of 6.29% using the same algorithm as a recognition method for predefined text ".tie5Roanl." The proposed methods are described in the following subsections:

6.6.1 Data Acquisition and Feature Extraction

It is the most fundamental and essential part in any biometric system. Here, key press time (P) and key release time (R) in millisecond unit are recorded while typing user ID and

TABLE 6.2

Public Datasets Available on the Internet

Datasets	Study	Texts	Subject Size	Session	Repetition	Sample	Features	Downloaded Links
Dataset A	Killourhy et al. [4]	".tie5Roanl"	65	8	50	26,000	KD, DD,UD	http://www.cs.cmu.edu/~keystroke
Dataset B	Killourhy et al. [4]	"4121937162"	42	4	50	8,400	KD, DD,UD	http://www.cs.cmu.edu/~keystroke
Dataset C	Killourhy et al. [4]	"hester"	38	4	50	7,600	KD, DD,UD	http://www.cs.cmu.edu/~keystroke
Dataset D	El-Abed et al. [5]	"rhu.university"	51	3	15–20	951	DD, DU, UD, UU	http://www.coolestech.com/download/14441/

password as raw data in keystroke dynamics technique. Then some common features are calculated by the following equations:

Key hold duration time (KD)=R_i–P_i (6.1)

Interval time between two subsequent keys released (RR)=R_{i+1}–R_i (6.2)

Interval time between two subsequent keys pressed (PP)=P_{i+1}–P_i (6.3)

Interval time between one key released and next key pressed (RP)=P_{i+1}–R_i (6.4)

Interval time between one key pressed and next key released (PR)=R_{i+1}–P_i (6.5)

Interval time between first key pressed and last key released (*t*-time)=R_n–P_1 (6.6)

Interval time between one key pressed and third key released
    (Tri-graph-time)=R_{i+2}–P_i (6.7)

Interval time between one key pressed and fourth key released
    (Four-graph-time)=R_{i+3}–P_i (6.8)

In our experiment, we have used only KD, DD, and UD for Dataset A, B, and C, respectively, as typing features, but we have used all the above features while working with Dataset D which we have extracted using Equations 6.1 to 6.8.

Key pressure, finger tips size, finger movements, choice of control keys, type of frequent errors, and choice of error correction mechanisms also can be measured for better performance in identification/authentication [5]. As per the study, these features could discriminate the gender and age group as well.

6.6.2 Normalization and Feature Subset Selection

Normalization is the first preprocessing step where we standardized the data within the range {–1, 1} for faster processing. Feature selection method is used to find out the optimal or close to optimal subsets of features when some irrelevant features are captured. It optimizes the accuracy rate along with computational speed. However, in our study, we have not used any feature selection methods.

6.6.3 Gender and Age Group Recognition

We have evaluated 11 leading machine learning methods and calculated the accuracy listed in Tables 6.3 and 6.4. We have divided total instances into 10 folds for cross-validation; here, in each stage, 1 fold will be treated as a test set and others will be treated as training sets. All the evaluation processes have been done with the supplied default parameter values by Weka. Then, we have added this additional information to each sample as additional features by assigning 0 for male and 1 for female; 0 for the age group 18–30 years and 1 for the age group 30+ years to learn about the system manually.

SVMs a popular supervised machine learning method has been introduced by Vapnik et al. [29] in 1995. Nowadays, SVMs have been widely studied in recognition and classification techniques to balanced datasets and have shown tremendous success in handwriting

TABLE 6.3

Accuracy to Predict the Gender

Classification Algorithms	Accuracy (%)			
	Dataset A	Dataset B	Dataset C	Dataset D
FRNN-VQRS [22, 23]	**94.81**	**88.55**	**95.04**	**84.75**
FRNN [22]	**94.81**	**88.55**	**95.04**	**84.75**
Fuzzy Rough NN [22]	93.16	85.93	93.45	84.75
Random Forest [24]	92.75	87.54	93.11	79.1
Bagging [25]	91.34	85.59	91.42	76.97
Fuzzy NN [22]	88.92	81.85	92.64	76.45
IBK (Euclidean) [26]	88.71	80.83	91.72	81.07
J48 [27]	86.33	80.34	88.28	71.82
MLP [28]	82.15	75.71	85.89	68.24
SVM [29]	71.47	69.38	79.16	65.72
Naive Bayes [30]	64.37	63.9	72.11	56.7

TABLE 6.4

Accuracy to Predict the Age Group

Classification Algorithms	Accuracy (%)			
	Dataset A	Dataset B	Dataset C	Dataset D
FRNN-VQRS [22, 23]	**94.31**	**86.87**	**94.68**	**84.75**
FRNN [22]	**94.31**	**86.87**	**94.68**	**84.75**
Fuzzy Rough NN [22]	92.81	83.10	93.24	79.70
Random Forest [24]	92.13	86.40	92.47	79.81
Bagging [25]	90.65	84.63	90.09	75.60
Fuzzy NN [22]	88.03	78.62	92.04	72.45
IBK (Euclidean) [26]	88.00	76.79	91.22	77.92
J48 [27]	86.35	77.89	86.53	64.88
MLP [28]	79.74	70.39	84.55	73.08
SVM [29]	65.41	58.71	70.59	66.67
Naive Bayes [30]	59.13	56.18	66.99	59.41

recognition to text classification. Due to the remarkable success rate, SVMs are also used in keystroke dynamics not only to identify the user but also to recognize the soft biometric information. A support vector–based machine distinguishes imposter pattern by creating margin which separates other patterns from that of the imposter, which provides a learning technique for pattern recognition and regression estimation. It is commonly used and effective for large practical problems. To predict the gender, Giot et al. [7] used libSVM—a library of SVM. But in our study, we have used FRNN.

As a recognition method, FRNN classification algorithm with vaguely quantified rough sets is more suitable. This method is an alternative to Sarkar's fuzzy rough ownership function (FRNN-O) approach [31]. FRNN uses the nearest neighbors to construct lower and upper approximations of decision classes, and classifies test instances based on their membership to these approximations [22]. FRNN-VQRS is a new approach to FRNN. The hybridization of rough sets and fuzzy sets has focused on creating an end product that extends both contributing computing paradigms in a conventional way.

6.6.4 Classification and Decision

Machine learning methods are widely used in pattern recognition domain. The purpose of classification is to find the closest or the near-closest class to the claimed class. Statistical methods such as mean, median, and standard deviation; distance-based algorithms such as Euclidean, Manhattan, scaled Manhattan, Mohanobolis, z score, Canberra, and Chebycev; and some machine learning algorithms such as SVM, multi layer perceptron, OneR, J48, naïve bayes, nearest neighbor, fuzzy, neural network, and random forest can be used. But in our experiment, FRNN-VQRS has proved that it is an efficient approach in this domain.

Here, the claimant's feature data are compared to the reference template using classification algorithm, and a final decision will be made based upon the classification accuracy. To increase the user authentication accuracy, we have integrated gender and age group as soft biometric features with timing features.

6.7 Experimental Results

In this section, we present the results obtained from our evaluation process. Eleven machine learning algorithms were applied on each dataset and accuracy with 10-fold cross-validation were listed to predict the gender identity in the Table 6.3 and to predict the age group identity in the Table 6.4. As per obtained results, FRNN and FRNN-VQRS have proved that they are suitable learning methods to predict the gender as well as age group in both desktop and android environments. Accuracies were recorded by Weka 3.7.4 simulator with default parameter values.

From the literature survey, it has been observed that gender information as additional feature improves the performance of keystroke dynamics user recognition. Figure 6.6 indicates that the age group information can also be used to improve performance. Further, if

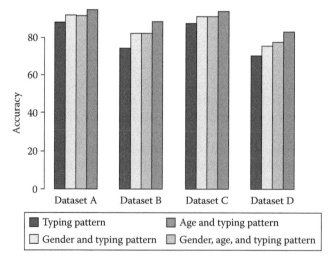

FIGURE 6.6
User authentication accuracy combining gender, age group, and typing pattern.

TABLE 6.5

Gain Accuracies Using Soft Biometric Information

	Gain Accuracy (%)			
Features	Dataset A	Dataset B	Dataset C	Dataset D
Gender + timing features	3.5	7.72	3.55	5.05
Age group+ timing features	3.38	7.52	3.56	7.15
Age group + gender+ timing features	**6.29**	**14.38**	**6.38**	**12.52**

we used both gender and age group as additional features instead of only gender, then the accuracy of the system will be improved. The gain accuracies are described in detail in Table 6.5.

6.8 Discussion

In order to solve the problem of gender and age group prediction, we have employed Fuzzy Rough K-NN and FRNN-VQRS, mainly because of higher and consistent accuracy status. This could help to learn about the system and also could be used to improve the classification accuracy in keystroke dynamics user authentication systems. The searched input is the key parameter to check the gender as well as age group. As per the results of our experiments, simple, commonly used words or password-type words are suitable to predict the gender and age group than only numeric text. It is also observed that desktop environment is more accurate than android platform since simple text is concerned. This accuracy rate will be impressive if enrollment phase (type of keyboard, timing resolution of the system, screen size of android device, etc.) is extremely accurate. This method will be more reliable and consistent if we include some additional features like mouse dynamics, key pressure proportional to force, and hand weight which may be a good factor in desktop environment. In android platform, key pressure, acceleration, and finger tips' size may be included where advance sensing device, accelerometer are embedded in each smart phone; so this technique can achieve promising results and can be used to predict the gender and age group of Internet users for smooth, fake-free, and loyal social networking sites and can be used as additional features to improve the identity of the user through the typing pattern. We have not compared our approach with previous studies, because Giot et al. [7] used a different dataset where soft biometric information is not supplied. They used only gender as additional information whereas we have taken both gender and age group as additional information. Generally speaking, gender prediction is a bit difficult of users in the 18-year age group due to intra-class variations. We have to take care of this.

In Figures 6.7 to 6.14, we can see that prediction of gender or age group by FRNN-VRQS is possible based on the typing style on a computer keyboard or touch screen, and it does not depend on the type of text. But numeric text pattern is not much suitable than others.

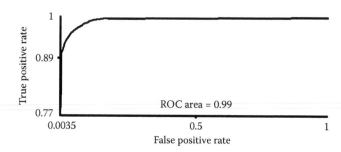

FIGURE 6.7
ROC analysis to predict the gender from Dataset A.

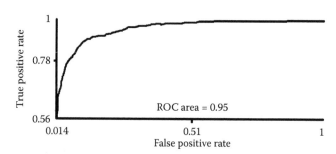

FIGURE 6.8
ROC analysis to predict the gender from Dataset B.

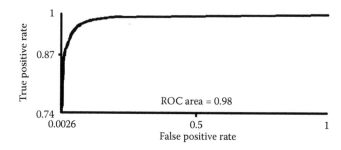

FIGURE 6.9
ROC analysis to predict the gender from Dataset C.

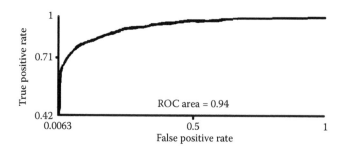

FIGURE 6.10
ROC analysis to predict the gender from Dataset D.

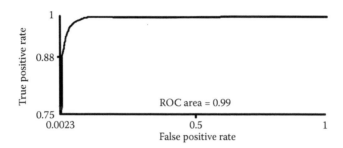

FIGURE 6.11
ROC analysis to predict the age group from Dataset A.

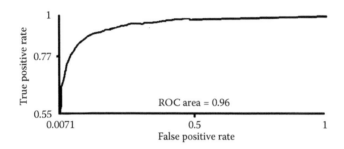

FIGURE 6.12
ROC analysis to predict the age group from Dataset B.

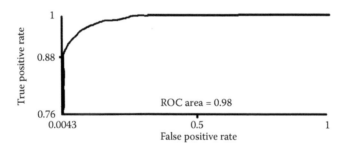

FIGURE 6.13
ROC analysis to predict the age group from Dataset C.

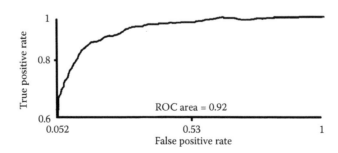

FIGURE 6.14
ROC analysis to predict the age group from Dataset D.

6.9 Conclusions

The chapter employs machine learning methods to develop a model that predicts gender and age group based on keystroke dynamics features which significantly improve accuracy. Gender and age group alone are not sufficient features to identify the individual user, but they can be used as additional features. We have used three public authentic datasets on keystroke dynamics through keyboard, and one dataset through touch screen to verify whether or not this technique is applicable in both environments. Our proposed approach, FRNN-VQRS, a new approach to FRNN, achieved a gender and age group prediction accuracy of more than 94% in desktop environment, and 84.74% accuracy in android environment. We have used paired *t*-test where FRNN-VQRS is most significant than previously used libSVM by Giot et al. [7]. This is a very positive outcome in keystroke dynamics system for a single predefined text which can be used as soft biometric additional features in identification/authentication technique which improves the gain accuracy by 3.5% to 14.38%.

As per the obtained results listed in Tables 6.3 and 6.4, gender as well as age group of the users can be extracted from the typing pattern. It is also observed that both gender and age group information can be extracted from the dataset collected through the touch screen device. This is the first time we have used FRNN as per our knowledge on keystroke dynamics datasets instead of the very popular machine learning method, libSVM. In this chapter, we have also fused these two soft biometric scores with the timing features to enhance the performance of keystroke dynamics user authentication systems. It is also observed that gender and age group information as extra features increase the user authentication performance instead of using only gender information. So, this technique can be used to predict the gender and age group of the Internet users as it is evident from our experiment, as keystroke dynamics is a common measurable distance-based activity to monitor the use of the Internet through keyboard/touch screen. It could be used to deal with the problem of fake accounts and would facilitate creation of a more loyal and authentic social networking sites. This keystroke Dynamics user recognition with inclusion of personal traits as additional features is the modest and efficient approach.

References

1. Bleha S., Slivinsky C., and Hussien B. Computer-access security systems using keystroke dynamics. *IEEE Transactions on Pattern Analysis and Machine Intelligence*, 12(12):1217–1222, 1990.
2. Gaines R., Lisowski W., Press S., and Shapiro N. *Authentication by keystroke timing: Some preliminary results.* Technical Report Rand Rep. R-2560-NSF, RAND Corporation, Santa Monica, CA 90406, 1980.
3. Joyce R., and Gupta G. Identity authentication based on keystroke latencies. *Communications of the ACM*, 33(2):168–176, 1990.
4. Killourhy K., Maxion R., *A scientific understanding of keystroke dynamics.* Carnegie Mellon University, Pittsburgh, PA, 2012.
5. El-Abed M., Dafer M., El Khayat R. RHU Keystroke: A mobile-based benchmark for keystroke dynamics systems. *48th IEEE International Carnahan Conference on Security Technology (ICCST)*, Rome, Italy, 2014.
6. Modi S., and Elliott S.J. Keystroke dynamics verification using a spontaneously generated password. In *Proceedings of the 40th Annual IEEE International Carnahan Conference on Security Technology (ICCST '06)*, USA, October 2006, pp. 116–121.

7. Giot R., and Rosenberger C. A new soft biometric approach for keystroke dynamics based on gender recognition. *International Journal of Information Technology and Management*, Special Issue on Advances and Trends in Biometrics, 11(1/2):1–16, 2012.

8. Epp M.L., and Mandryk R.L. Identifying emotional states using keystroke dynamics. In *Annual Conference on Human Factors in Computing Systems (CHI 2011)*, May 7–12, Vancouver, BC, Canada. ACM, New York, pp. 715–724, 2011.

9. Khanna P., and Sasikumar M. Recognising emotions from keyboard stroke pattern. *International Journal of Computer Applications*, 11(9):0975–8887, 2010.

10. de Ru W.G., and Eloff J.H.P. Enhanced password authentication through fuzzy logic. *IEEE Expert*, 12(6):38–45, 1997.

11. Roy S., Roy U., and Sinha D. Rhythmic password-based cryptosystem. In *2nd International Conference on Computing and System*, University of Burdwan, West Bengal, India, pp. 303–307, 2013.

12. Roy S., Roy U., and Sinha D. Performance perspective of different classifiers on different keystroke datasets. *International Journal of New Technologies in Science and Engineering (IJNTSE)*, 2(4):64–73, 2015.

13. Roy S., Roy U., and Sinha D. Distance based models of keystroke dynamics user authentication. *International Journal of Advanced Engineering Research and Science (IJAERS)*, 2(9):89–94, 2015.

14. Roy S., Roy U., and Sinha D.D. Free-text user authentication technique through keystroke dynamics. In *IEEE International Conference on High Performance Computing & Application (ICHPCA-2014)*, IEEE, Bhubaneswar, India, pp. 1–6, 2014.

15. Antal M., Szabó L.Z., and László I. Keystroke dynamics on android platform. *Procedia Technology*, 19:820–826, 2015.

16. CENELEC. *European Standard EN 50133–1: Alarm systems. Access control systems for use in security applications. Part 1: System requirements, 2002.* Standard Number EN50 1 3 3–1: 1 9961A l: 2002, Technical Body CLClTC 79, European Committee for Electrotechnical Standardization (CENELEC), UK.

17. Obaidat S., and Sadoun B. Verification of computer users using keystroke dynamics. *IEEE Transactions on Systems, Man, and Cybernetics B*, 27(2):261–269, 1997.

18. Ali, W.W., and Salami M.J.E. Keystroke pressure based typing biometrics authentication system by combining ANN and ANFIS-based classifiers. In *Proceedings of the 5th International Colloquium on Signal Processing and Its Applications (CSPA '09)*, Kuala lumpur, Malaysia, pp. 198–203, March 2009.

19. Janakiraman R., and Sim T. Keystroke dynamics in a general setting. In *Advances in Biometrics, Proceedings*, vol. 4642, Springer, Berlin, Germany, pp. 584–593, 2007.

20. Jain, A.K. Biometric recognition: How do I know who you are?, In *Image Analysis and Processing—CIAP* 2005, F. Roli and S. Vitulano (Eds.), Springer-Verlag, Berlin, pp. 19–26, 2005.

21. Jain A.K., Flynn P., and Ross A.A. *Handbook of Biometrics*, Springer US, 2008.

22. Jensen R., and Cornelis C. Fuzzy-rough nearest neighbour classification. In *Transactions on Rough Sets XIII*. Springer, Berlin, pp. 56–72, 2011.

23. Cornelis, C., De Cock M., and Radzikowska A.M. Vaguely quantified rough sets. In *Rough Sets, Fuzzy Sets, Data Mining and Granular Computing*. Lecture Note in Computer Science, Vol. 44, P2 Springer, Berlin, pp. 87–94, 2007.

24. Breiman, L. Random forests. *Machine Learning*, 45(1):5–32, 2001.

25. Breiman, L. Bagging predictors. *Machine Learning*, 24(2):123–140, 1996.

26. Peterson L.E. K-nearest neighbor. In *Scholarpedia*, M. Eugene and Izhikevich (Eds.), *Open Access Encyclopedia* 4(2):1883, 2009.

27. Quinlan R. *C4.5: Programs for Machine Learning*. Morgan Kaufmann, San Mateo, CA, 1993.

28. Gardner M.W., and Dorling S.R. Artificial neural networks (the multilayer perceptron)—A review of applications in the atmospheric sciences. *Atmospheric Environment*, 32(14):2627–2636, 1998.

29. Vapnik V. *Statistical learning theory*. Wiley, New York, 1998.

30. Koch K.R. *Bayes' Theorem*. Springer, Berlin, 1990.

31. Sarkar M. Fuzzy-rough nearest neighbor algorithms in classification. *Fuzzy Sets and Systems*, 158:2134–2152, 2007.

7

Internet of Nano-Things Forensics: Performing Digital Forensics in Nanoscale Systems

Ezz El-Din Hemdan and Manjaiah D.H.

Mangalore University

Mangalore, India

CONTENTS

7.1 Introduction

Presently, it is the era of nanotechnology which has introduced the concept and idea of nanodevices or nanomachines on a scale ranging from one to several nanometers in dimension. Nanotechnology has led to new nanomaterials with new properties and characteristics that will usher in novel developments in nanodevices like nanosensor and nanorouter. These nanodevices are integrated to perform tasks such as sensing, acquisition, or transferring data through nanonetwork, which will cover unmatched locations to perform additional in-network processing. The emergency development technology that develops and creates nanomachines defines a new networking paradigm that consists of an interconnection of nanoscale devices with existing communication networks and ultimately the Internet of nano-things (IoNT). The IoNT adds a new dimension to the Internet of things (IoT) technology by embedding nanosensors inside devices to enable them to communicate through the nanonetwork via the Internet for global connection among a lot of devices around the world. "Technavios analysts forecast the global internet of Nano-Things (IoNT) market to grow at a CAGR of 24.25% during the period 2016–2020" (Technavio 2016). The concept of IoNT was introduced by Ian Akyildiz and Josep Jornet who defined an architecture for electromagnetic nanodevice communication that includes protocols, channel modeling, and information encoding (Akyildiz and Jornet 2010; Balasubramaniam and Kangasharju 2013). On the IoNT, nanonetworks are connecting nanodevices which can sense, collect, process, and store information. These nanodevices communicate through nano-communication process which means transfer of data and information between different nanodevices inside the nanonetwork. This nano-communication process consists of two types, namely molecular and electromagnetic communication, as follows (Akyildiz et al. 2008):

1. *Molecular Communication (MC):* This type of communication is defined as the exchange of information through the transmission and reception of molecules. These molecules will interact with nanodevices in a biological environment such as human body.

2. *Electromagnetic Communication (EM):* This type of communication is defined as the exchange of information through the transmission and reception of electromagnetic radiation from nanodevices in nanonetworks. These radiations will be emitted in specific bandwidth for allowing nanodevices to interact and communicate with each other.

The IoNT enabling technologies could pose new, severe security threats if managed with pernicious intent against IoNT infrastructure; therefore, security on the IoNT plays a vital role in providing safe and reliable communication environment between nanodevices in nanonetworks, which consist of nanodevices that can connect together to exchange information. Attackers can exploit vulnerabilities and weaknesses in these nanonetworks. An attacker can also exploit the weakness of crucial health and safety equipment or the communication channel and trigger malicious instructions to jeopardize a patient's life. For example, in the area of Internet of bio-nano-things (IoBNT), malicious people could hack bio-things which are used to access human body and create health problems by introducing new types of viruses that can cause new diseases. Current security mechanisms and techniques cannot secure nanodevices in the nanonetwork from malicious attacks and crimes because nanodevices work in terahertz band. To protect the IoNT infrastructure, there is a serious need to propose and develop new security solutions to prevent crimes related to the IoNT. The existing security solutions cannot be used directly for securing the IoNT infrastructure. Some of the suggested solutions for securing the IoNT environment are checking the integrity of data by using checksum algorithms, using encryption algorithms to encrypt data before transferring between nanodevices, using data hiding algorithms for hiding critical data, and using multi layer authentication to guarantee that only the user can access nanonetworks.

To investigate such attacks in the IoNT paradigm, there is a need to execute digital forensics procedures to find any digital evidence about criminal or illegal activities. Unfortunately, performing digital forensics investigation in the IoNT brings a new challenge for examiners and digital investigators as the existing digital forensics tools and procedures do not cope with the IoNT environment to collect and extract digital evidence from nanodevices inside nanonetworks. The huge number of nanodevices will generate a massive amount of possible evidence, which will bring new challenges for all aspects of data acquisition for the forensic purpose, so that the investigators will find it a complex challenge for collecting evidence from the highly distributed IoNT infrastructure.

The rest of this chapter is organized as follows: Section 7.2 introduces a brief background about digital forensics, IoT forensics, and IoNT while the new area of IoNT forensics (IoNTF) is presented in Section 7.3. Section 7.4 provides the proposed IoNTF investigation model and finally the chapter conclusion and future directions are presented in Section 7.5.

7.2 Background

This section introduces a brief overview about digital forensics, IoT forensics, and IoNT.

7.2.1 Digital Forensics

This section presents the concept of digital forensics and an introduction into current digital forensic investigation process to build the reader's understanding of the discipline.

7.2.1.1 Definition

The first Digital Forensic Research Workshop (DFRWS) defined digital forensics as:

> The use of scientifically derived and proven methods toward the preservation, collection, validation, identification, analysis, interpretation, documentation, and presentation of digital evidence derived from digital sources for the purpose of facilitating or furthering the reconstruction of events found to be criminal, or helping to anticipate unauthorized actions shown to be disruptive to planned operations. (Palmer 2001).

7.2.1.2 Digital Forensics Investigation Life Cycle

From the digital forensics definition, the digital forensic investigation process involves several stages and steps to handle and manage the digital evidence that can be extracted from the crime scene as follows (Figure 7.1):

- *Identification:* This process involves the identification of an incident and the evidence thereof, which will be required to prove the incident.
- *Collection:* In this process, an examiner and a digital investigator collect digital evidence from the crime scene.

FIGURE 7.1
Digital forensic investigation process.

- *Extraction:* In this phase, a digital investigator extracts digital evidence from different types of media, for example, hard disk, cell phone, e-mail, and much more.
- *Analysis:* In this phase, a digital investigator interprets and correlates the available data to arrive at a conclusion, which can prove or disprove an incident.
- *Examination:* In this phase, an investigator extracts and inspects the data and their characteristics.
- *Report:* In this process, a digital investigator makes an organized report stating his or her findings of the incident which have to be appropriate enough to present to the jury.

7.2.2 Internet of Things Forensics

Currently, the IoT has become an attractive research topic where interconnected devices known as "things" or "objects" with embedded processing abilities are employed to extend the usage of Internet capabilities to several application domains such as medical, industry, and military. The IoT is considered as a new environment that provides rich sources of data such as sensors that generate data. These sources can be used in conjunction with one another in the same IoT environment for certain purpose. The variety of these sources provides different challenges to the various forensics communities, especially the investigators who will be required to interact with this new technology to investigate crimes related to the IoT environment. In the IoT environment, a lot of devices or machines operate, such as wireless sensors, radio frequency identification (RFID), the Internet connection, intelligent or smart grids, cloud computing, and vehicle networks that can integrate to each other in an intelligent manner. However, interconnecting of various "machines" also refers to the possibility of interconnecting various different threats and attacks. For example, a malware can easily propagate through the IoT at an unprecedented rate. In the following four design aspects of the IoT system, there may be various threats and attacks as follows (Giuliano et al. 2015):

1. *Data Perception and Collection:* In this aspect, typical attacks include data leakage, sovereignty, and authentication.
2. *Data Storage:* The following attacks may occur: denial-of-service attacks (attacks on availability), integrity attacks, impersonation, and modification of sensitive data.
3. *Data Processing:* In this side, there may exist computational attacks that aim to generate wrong data processing results.
4. *Data Transmission:* During the transmission process, severe types of attacks may occur like session hijacks, routing attacks, flooding, and channel attacks. Therefore, efficient and effective defense procedures and strategies are of extreme importance to ensure the security of the IoT infrastructure.

In this remaining part, we will present the previous work in the area of IoT forensics that may help researchers and scientists in the digital forensics field to introduce and propose new procedures and techniques in digital forensics in the field of IoNT because very little work was done in the field of IoNT Fx till writing this work. Some work has been done to explain the concept of "IoT forensics." Also, new procedures and methodologies have been proposed and introduced to perform the digital investigation process in the IoT environment. Oriwoh et al. (2013) proposed two methods for digital investigation in the IoT environment which are 1-2-3 Zones Digital Forensics and Next-Best-Thing Triage:

1. *1-2-3 Zones Digital Forensics:* This approach divides the IoT infrastructure into three areas or zones to help in performing digital investigation process. These zones are zone 1, zone 2, and zone 3, as shown in Figure 7.2:

    a. *Zone 1:* This zone is called the internal zone that includes all IoT smart devices like a smart refrigerator and TV that can contain valuable data about a crime committed in the IoT infrastructure.

    b. *Zone 2:* This zone includes all intermediate components that reside between the internal and external networks to support the communication process. These devices may be protection devices such as intrusion detection and prevention systems (IDS/IPS) and firewalls. The digital investigators can find evidential data that help them to extract facts about a crime committed in relation to IoT.

    c. *Zone 3:* This zone includes hardware and software components that reside in the external part of the IoT Infrastructure such as cloud services and other service providers that use IoT devices and users. These components with hardware devices and software in zone 1 and zone 2 will help digital practitioners to perform their investigation mission in a timely fashion manner.

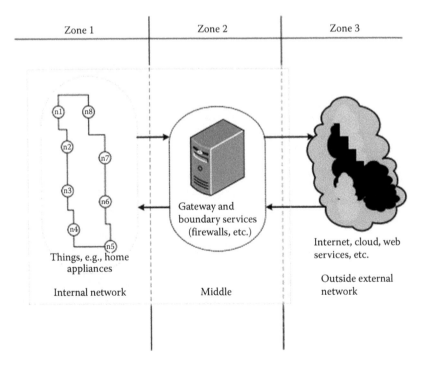

FIGURE 7.2
1-2-3 Zones of digital forensics. (From Oriwoh, E., et al. Internet of things forensics: Challenges and approaches. *Collaborative Computing: Networking, Applications and Worksharing (Collaboratecom), 2013 9th International Conference on IEEE*, 2013.)

This approach reduces the challenges that will be encountered in IoT environments and ensures that investigators can focus on clearly identified areas and objects in preparation for investigations.

2. *Next-Best-Thing Triage:* The next-best-thing triage (NBT) can be used in conjunction with the 1-2-3 zones approach. This approach discusses to find an alternative source in the crime scene if it is unavailable after a crime occurred in the IoT environment. The NBT approach can be used to determine what devices were connected to the objects of forensic interest (OOFI) and find anything which was left behind after the devices were removed from the network. Direct access to the OOFI may not always be possible. Therefore, in such situations, the option of identifying and considering the next best source of relevant evidence may have to be taken. The design of a method of systematically deciding what this next best thing might be in different scenarios and situations can be the subject of further research.

Zawoad and Hasan (2015) introduced a forensics-aware IoT (FAIoT) model for supporting digital forensics investigations in the IoT environment in a reliable manner. The FAIoT model provides secure evidence preservation module and secure province module as well as access to evidence using application programming interface (API) that will reduce the challenge in performing the investigation. To facilitate the digital investigators, a centralized trusted evidence repository in the FAIoT is used to ease the process of evidence collection and analysis. The IoT devices need to register this secure evidence repository service. The FAIoT architecture shown in Figure 7.3 is as follows:

- *Secure Evidence Preservation Module*: This module will be used to monitor all the registered IoT devices and store evidence securely in the evidence repository. Also, segregating of the data according to the IoT devices and its owner will be done in this module. Hadoop distributed file system (HDFS) will be used to handle a large volume of data.

- *Secure Provenance Module:* This module ensures the proper chain of custody of the evidence by preserving the access history of the evidence.

- *Access to Evidence through API:* In this model, a secure read-only APIs to law enforcement agencies is proposed. Only digital investigators and the court member will have access to these APIs. Through these APIs, they can collect the preserved evidence and the provenance information.

Perumal et al. (2015) proposed an integrated model which is designed based on triage model and 1-2-3 zone model for volatile based data preservation. This model started with the following authorization, planning and obtaining a warrant as the fundamental steps in the digital forensic investigation process as shown in Figure 7.4. Then, it starts to investigate the IoT infrastructure and finally after seizing the IoT device from the selected area or zone, the investigator completes the digital forensic procedure which includes a chain of custody, lab analysis, result and proof, and archive and storage.

7.2.3 Internet of Nano-Things

The concept and idea of the IoNT are proposed and introduced by Akyildiz and Jornet (2010). The IoNT consists of connected nanodevices through the existing

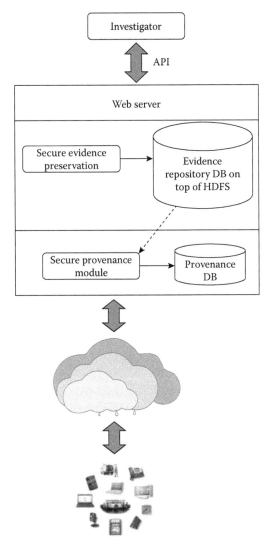

FIGURE 7.3
Forensics-aware IoT (FAIoT) model. (From Zawoad, S. and R. Hasan, FAIoT: Towards building a forensics aware eco system for the Internet of things, *Services Computing (SCC), 2015 IEEE International Conference on IEEE,* 2015.)

telecommunication and network systems. The envisioned IoNT is shown in Figure 7.5. The IoNT added a new dimension for the IoT by embedding nanosensors inside things/ devices to enable them to communicate together through the nanonetwork and the Internet for global connection among devices around the world. The IoNT describes how the Internet will get bigger as nanosensors are connected to physical things such as physical assets or consumer devices for collecting, processing, and sharing of data with the end users. IoNT has various important applications such as smart agriculture, health care, military, logistics, aerospace, industrial control systems, manufacturing, and smart cities. There are new domains promised from the IoNT technology such as IoBNT and internet of multimedia nano-things (IoMNT) which will make new developments in

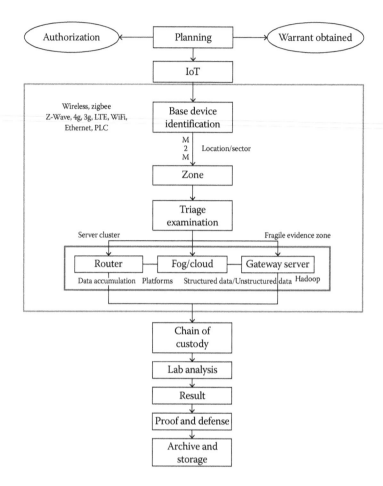

FIGURE 7.4
IoT-based digital forensic model. (From Perumal, S., et al., Internet of things (IoT) digital forensic investigation model: Top-down forensic approach methodology, *Digital Information Processing and Communications (ICDIPC), 2015 Fifth International Conference on* IEEE, 2015.)

health care and multimedia fields. The terms of IoBNT and IoMNT can be briefly introduced as follows:

1. *Internet of Bio-Nano-Things:* The IoBNT refers to a paradigm-shifting concept for communication and network engineering, where new complex challenges are faced to develop efficient and safe methods for information exchange, interaction, and interconnecting within the biochemical area, while enabling an interface to the electrical domain of the Internet (Akyildiz et al. 2015). The IoBNT stems from synthetic biology and nanotechnology tools that allow for biological engineering for embedded processing and computing machines. Depending on the basics of biological cells' architecture and its functionality, from the biochemical viewpoint, bio-nano-things promise to enable applications and services such as actuation networks, environmental control of toxic agents, pollution, and intrabody sensing.

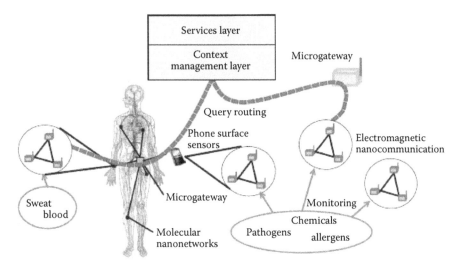

FIGURE 7.5
Envisioned IoNT. (From Balasubramaniam, S., and J. Kangasharju, *Computer*, 2, 62–68, 2013.)

2. *Internet of Multimedia Nano-Things:* Currently, nanotechnology is able to help in the development of new devices in nanoscale which are able to generate, process, and transmit multimedia content. The interconnection of pervasively deployed multimedia nanodevices with existing communication networks and ultimately the Internet to introduce a new field called the internet of multimedia nano-things (IoMNT). The IoMNT is a powerful field that can be used in many applications in the areas of security, defense, environment, and industry, among others (Jornet and Akyildiz 2012).

7.2.3.1 Network Architecture in IoNT

In the IoNT, the network is the interconnection of nanodevices to each other to facilitate communication between them through the existing communication networks and systems. Akyildiz and Jornet (2010) introduced a network architecture for the IoNT for two applications which are intrabody nanonetworks for remote health care and future interconnected office as shown in Figure 7.6:

1. *Intrabody Networks:* In intrabody networks, nanodevices like nanoactuators and nanosensors deployed inside the human body are remotely controlled over the Internet by an external operator as a health-care provider. The nanoscale is the natural domain of molecules, proteins, DNA, organelles, and the major components of cells. Among others, existing biological nanodevices can provide an interface between biological phenomena and electronic nanodevices, which can be exploited through this new networking paradigm.

2. *Interconnected Office:* In the interconnected office, each element found in an office and even its internal components are provided a nano-transceiver which allow them to be constantly connected to the Internet. As a result, a user can keep track of the location and status of all its belongings in an easy manner.

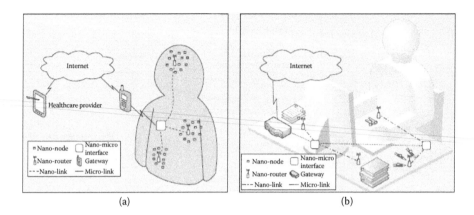

FIGURE 7.6
Network architecture for IoNT: (a) Intrabody nanonetwork for health care and (b) interconnected office. (From Akyildiz, I. F., and J. M. Jornet, *IEEE Wireless Communications*, 17(6), 58–63, 2010.)

Besides this, Akyildiz and Jornet (2010) introduced the following components in the network architecture of the IoNT:

- *Nano-Node:* This is considered as the smallest and simplest nanodevice. It can perform a computation and processing task, has limited memory, and can only transmit over very short distances due to its reduced energy and limited communication capabilities.

- *Nanorouter:* These nanodevices have comparatively larger computational resources than nano-nodes and are suitable for collecting information coming from limited nanodevices. In addition to this, it can also control the behavior of nano-nodes by exchanging very simple control commands such as on/off and sleep.

- *Nano-Micro Interface Devices:* This is able to collect information coming from nano-routers and convey it to the microscale, and vice versa.

- *Gateway:* This device gives the ability to control the entire system remotely over the Internet. For example, in an intrabody network scenario, an advanced cell phone can forward the information that is received from a nano-micro interface in our wrist to our health-care provider. In the interconnected office, a modem-router can provide this functionality.

7.2.3.2 IoNT Security

Nano-things can be vulnerable to physical and wireless attacks. For example, nano-things will be unattended most of the time, so they can face physical attacks. In addition to this, because of the involuntary physical damage, it is also likely to occur because of their almost indiscernible dimension. Different types of wireless attacks that include both traditional and new nano-things are possible and relatively simple, despite the fact that the nanodevices communicate at ultra-high transmission rates and over very short transmission distances. Existing security solutions cannot be directly used in the IoNT because they do not capture the peculiarities of the terahertz band physical layer and the heterogeneity in the capabilities of diverse nano-things. For this, future research trends should be along the following three main directions (Jornet and Akyildiz 2012):

- *Develop Novel Authentication Methods:* In various envisioned applications, it is critical to certify the identity of the transmitting or receiving nanodevices. Due to this, a very large number of nanodevices in the IoNT environment like IoMNT, traditional authentication solutions that are based on complex authentication infrastructures and servers, are not suitable for nanodevices. For this, novel authentication methods that exploit the network hierarchical structure of the IoNT are required. For example, nano-things might need to only authenticate themselves to the closer nanorouter or nano-to-micro interface; this can be done by means of a unique EM signature, which is a well-established property of terahertz radiation.

- *Develop New Data Integrity Mechanisms:* In communication networks, it is important to guarantee that an adversary cannot change the information during the transmission process. Data can be modified either when stored or when being transmitted. Due to the expectedly very limited memory of miniature nanothings, the first type of attack is unlikely. However, new techniques to protect the information in nanomemories will be developed by exploiting the quantum properties of single-atom memories to implement practical solutions from the realm of quantum encryption. In its turn, despite the information being transmitted at very high bit-rates, guaranteeing the data integrity while the information is being transmitted requires the development of novel safe communication techniques for IoNT environment.

- *Develop Novel User Privacy and Security Mechanisms:* Nano-things can be used to detect, measure, and transmit very sensitive and confidential information, which in any case should be available to non-intended addressees. Moreover, due to their miniature size, nano-things will be usually imperceptible and omnipresent. So, new mechanisms to guarantee the privacy in the IoNT are required. Among others, methods to guarantee that a user can determine and limit the type of information that nano-things can collect and transmit are needed. Moreover, physical-layer security methods need to be explored to prevent problems like eavesdropping.

Novel security and privacy mechanisms will also be needed to protect sensitive data gathered by nanosensors, which can include detailed chemical and biological samples from individuals. For example, molecular nanonetworks could gather data about people infected with a harmful virus to shed light on the nature and severity of the disease. Safeguards must be in place to ensure that such data does not fall into the wrong hands. There are many challenges like information collected from nanosensors might include individuals' molecular and genetic data. On the other side, there are also some solutions to solve it as a solution by implementing safeguards to ensure that sensitive IoNT data do not fall into the wrong hands (Balasubramaniam and Kangasharju 2013).

7.2.3.3 Data Management and Analysis in IoNT

In traditional sensor networks, the data acquisition and collection process commonly occurs via a static tree where each node in the tree senses the data and then passes it along the tree to the sink node at the root. This way of sensing could lead to enormous data traffic during transmission, particularly if sensing process is periodic; due to this, each microgateway connects with many nanosensors in the nanonetworks. A suitable solution for this problem is to think about another way that cannot be static but depends on the dynamic process in data collection in the tree; therefore, it is needed for interactions from nano-node to another nano-node among microgateways. In both molecular and EM nanonetworks, the microgateway has to integrate data from different nanosensors before sending it down the tree. However, the timing difference in data propagation between nanomachines could lead to long delays for reaching messages to the sink. For example, in molecular nanonetworks, information transmission could take a large time, especially when queries expect feedback. And also in electromagnetic nanonetworks, energy harvesting is a major constraint, as the harvesting process can take up to a minute before transmission can occur. An ideal, time-delayed data fusion process must be implemented at the microgateway to process all information before further transmission along the data collection tree (Balasubramaniam and Kangasharju 2013). Figures 7.7 through 7.9 can help to understand the data collection, management, and analysis in nanoscale networks that have new properties. This can provide a better understanding of the IoNT environments that help in developing novel methods and procedures to secure the IoNT environment.

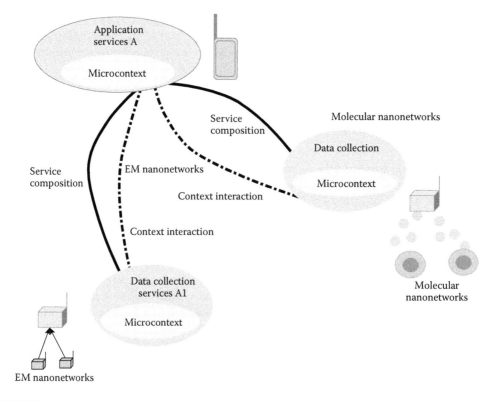

FIGURE 7.7
To deal with nanonetworks' large quantity and a variety of data, IoNT services can be subdivided into application and data collection layers, each with clustered service composition and discovery models. (From Balasubramaniam, S., and J. Kangasharju, *Computer*, 2, 62–68, 2013.)

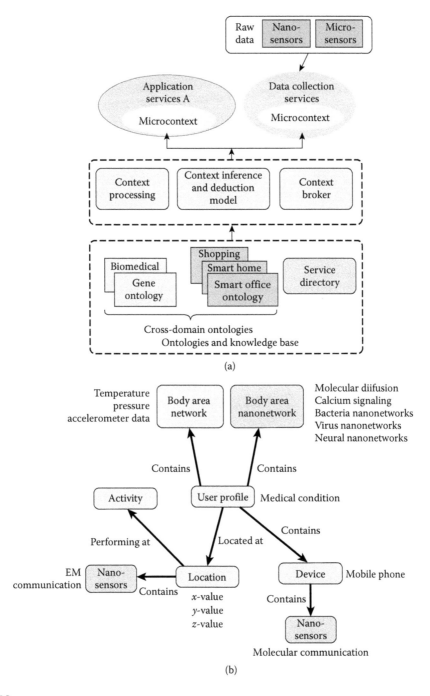

FIGURE 7.8
The wide range of data collected by nanonetworks will require cross-domain reasoning that spans multiple specialized domains. (a) Example context model that uses smart-space and gene ontologies to process data from molecular and EM nanonetworks. (b) An example definition of integrated IoNT ontology. (From Balasubramaniam, S., and J. Kangasharju, *Computer,* 2, 62–68, 2013.)

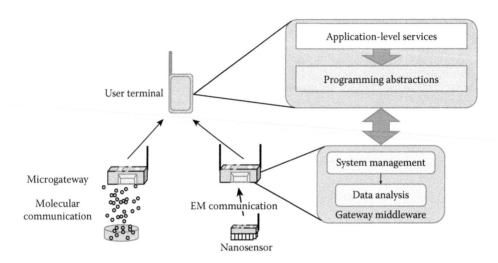

FIGURE 7.9

IoNT middleware system architecture. Microgateways contain system management and data analysis modules. On the user end, programming abstractions link to the microgateway middleware, and application services use data from the nanonetworks. (From Balasubramaniam, S., and J. Kangasharju, *Computer* 2, 62–68, 2013.)

7.3 Internet of Nano-Things Forensics

The digital forensic investigation process of internet of nano-things is called "IoNT Forensics." The IoNT represents a big challenge for digital investigators and examiners to collect and extract digital evidence as admissible evidence about committed crime in the IoNT environment. To the best of the author's knowledge, in this area, very little work is done to cover the IoNTF so that researchers and scientists who are working in the digital forensics area have to work to develop and design new methods, procedures, techniques, and tools to cope with new types of cybercrimes, attacks, and malicious and illegal activities. The IoNTF can identify itself as a special branch of IoT forensics, where the identification, collection, acquisition, analysis, examination, and presentation phases of digital investigation process will apply to the IoNT infrastructures to establish the facts about a crime that occurred. The digital forensic can be performed in the IoNT environment as well as its new sub domains such as IoMNT and IoBNT, as shown in Figure 7.10.

7.3.1 IoNT Forensics Definition

The IoNTF is the process of performing a digital forensics investigation in the IoNT. The IoNTF is a cross-discipline that combines the IoNT and digital forensics. We identify IoNTF as a combination of three digital forensics levels: nanodevice forensics level, internet forensics level, and IoNT services/application forensics level, which are illustrated in Figure 7.11. Therefore, The IoNTF can follow the main phases of digital forensics according to each level in the architecture of IoNT.

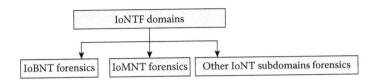

FIGURE 7.10
IoNTF domains.

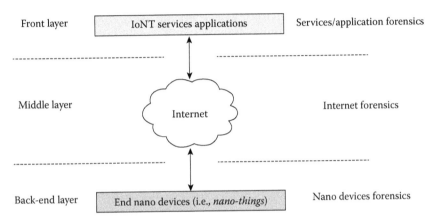

FIGURE 7.11
Conceptual IoNTF model.

7.3.2 Hypothetical Malicious Scenario

Alex is a doctor who monitors a patient's case remotely through his mobile phone. This patient stays in a hospital where there are various smart nanodevices connected for monitoring his/her case. Alex uses his mobile to control smart devices in his home, such as television, refrigerators, air condition, and several other smart devices remotely, as well as for monitoring his patient. Alice is a malicious attacker who developed a program to steal and control mobile phones remotely to perform illegal activities. Alice hacked Alex's phone and started to control his phone and browse applications as well as control smart devices in home and hospital. John is a digital investigator who received a task to investigate this crime case and find facts about these illegal activities. When John started the digital investigation process, he faced various complex challenges to investigate this case as millions of smart devices and nanodevices are connected together through the Internet. John begins to think how to deal with the complex and huge connected devices, especially nanodevices that have new features and properties that are new to the forensics community. He starts by dividing his investigation to three levels; the first level to investigate devices and nanodevices, the second level to investigate the nanonetwork and the Internet connections through log files. Thirdly and finally investigate the application and services which are available in Alex's mobile phone. This division will help John in his investigation task. The question here is how will John deal with these new nanodevices that have new unique features? Also, the existing digital forensics methods, techniques, and procedures cannot cope with these types of devices. This led us to think about introducing this topic to the forensics community by providing some challenges and opportunities.

7.3.3 Challenges

The IoNT contains a large-scale source of potential evidence. However, due to the heterogeneous nature of the IoNT devices, the ways in which data are distributed, aggregated, and processed present challenges to digital forensics investigations. New methods, techniques, and tools are required to overcome these challenges and leverage the architecture and processes employed in IoNT to gain access to this rich source of potential evidence. Current tools and techniques that are used for digital investigation are not suitable to investigate crimes in the IoNT environment due to new communication and standard protocols that will be used to adapt to this environment. There are many challenges and implications in the IoNT for performing the digital forensics in a timely fashion and effective manner. The digital investigation process consists of six main stages and various steps. These six steps are identification, collection, extraction, analysis, examination, and reporting, as shown in Figure 7.1, where each step has the following new challenges in the IoNT domain:

1. *Identification:* In the IoNT environment, there are billions of nanodevices which will generate enormous amount of data. This huge amount of data needs more time and special processing tools to identify evidential data that will help the examiners and digital investigators to find any traces about an incident. This amount of generated data requires big data tools such as Hadoop for data processing and handling.

2. *Collection:* In this process, an investigator identifies and collects digital evidence from the crime scene. The digital investigators collect evidence to extract valuable information about crimes committed in the IoNT environment and finds facts that will help build evidence against the attackers. The collection process is very important for any digital investigator because it will complete the subsequent phases of the digital investigation process, with errors, if any, propagating to the remaining investigation stages; hence, this phase is considered the most crucial phase. In the IoNT, the collection process becomes more difficult and harder due to the massive amount of nanodevices that are connected together. The IoNT environment can contain large numbers of nanodevices that are connecting and communicating using new nano-electromagnetic and MC systems. A digital investigator may find it an intricate challenge to collect data from this environment which has new properties. Within IoNT, expect to see a large amount of generated data from interconnected nanomachines that cannot be managed using traditional mining methods and procedures.

3. *Extraction:* In this phase, investigators extract digital evidence from different evidential sources as well as preserve the integrity of the evidence. In the IoNT, evidential data extracted from collected digital evidence will take more time, especially volatile data from massive and complex connected nanodevices.

4. *Analysis:* This phase is a vital phase in the digital investigation for interpreting and analyzing of extracted evidential data to make a conclusion about an incident that occurred in the IoNT environment. The massive amount of data in the IoNT needs new processing and analysis tools and techniques to deal with these data in a timely fashion and forensically sound manner.

5. *Examination:* In the examination phase, an investigator inspects the data and their characteristics. In this phase, we expect to develop new tools to examine digital evidence that extract from nano-things such as bio-things and multimedia nano-things.

6. *Report:* The final phase of the digital forensic process is to make an organized report about findings to present in a court of law. There is a challenge in IoNTF to present these types of digital evidence that the jury may have less knowledge about.

7.3.4 Opportunities

There are many sources which will help digital practitioners to identify valuable information about IoNT-related crimes. These locations and sources include all hardware and software in IoNT environment. Studying IoT forensics will help reduce the complexity and challenges in performing IoNTF in a timely fashion and forensically sound manner. There are some solutions proposed in Balasubramaniam and Kangasharju (2013) to make data collection possible in an IoNT environment; these solutions may help the investigators and examiners understand how to collect evidential data related to an incident or crime in an efficient and effective manner. The challenges and proposed solutions are presented in Table 7.1.

7.4 IoNTF Investigation Model

This section introduces an IoNTF investigation model for assisting and supporting reliable digital forensics investigations in the IoNT environment. From the above definition of IoNTF, we identify that the digital investigation process in the IoNT could be done in three digital forensics levels as follows:

1. Nanodevice forensics level
2. Internet forensics level
3. IoNT services/application forensics level

TABLE 7.1

Data Collection in IoNT

Category	Challenge	Proposed Solution
System architecture	A high ratio of nanosensors to microgateways could lead to swift energy depletion if microgateways must process information from every nanosensor.	Distribute the sink architecture and develop a two-layered hierarchy consisting of microgateways and nanonetworks.
Routing technology	Molecular nanonetworks: Information-carrying molecules could move very slowly between nodes as well as become lost.	Opportunistic routing through multihop relays of nanodevices; base the topology on random or unstructured graphs.
	EM nanonetworks: Limited memory, computational power, and energy will constrain data transmission between nodes.	Single-hop transmission to microgateways through a star topology; incorporate query-based routing, with queries routed between microgateways.
	With only one microgateway per nanonetwork, bulk data transmission could be difficult.	Incorporate unconventional routing technologies such as mobile delay-tolerant networks to carry bulk data.

Source: Balasubramaniam, S., and J. Kangasharju, *Computer*, 2, 62–68, 2013.

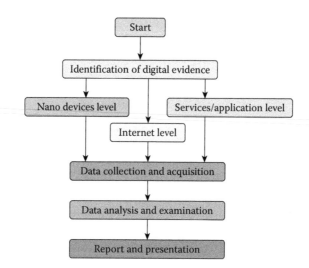

FIGURE 7.12
IoNTV investigation model.

The proposed IoNTF investigation model that can be adopted to investigate crimes related to IoNT is shown in Figure 7.12. This model consists of four main phases, namely identification of digital evidence, data collection and acquisition, data analysis and examination, and finally reporting and presentation of a summary of the entire investigation process.

7.5 Conclusion and Future Directions

Recently, the advanced development of IoNT to provide new services and applications in several areas such as manufacturing, health care, industry, military, multimedia, and agriculture prompted attackers to launch new types of crimes and attacks against IoNT infrastructure. The new types of severe attacks are increasing against nanodevices and nanonetworks. Therefore, there is a serious need for developing new digital forensic procedures and techniques to prevent and trace these types of attacks in a timely and forensically sound manner. This chapter systematically presented the challenges and opportunities for the IoNTF field as well as the IoNTF investigation model for assisting and supporting reliable digital forensics in the IoNT environment. Some of the future directions in this innovation area can be summarized as:

- Develop and implement the proposed IoNTF model with real case studies.
- Study and solve IoNTF challenges that can open the opportunity for identifying numerous new insights that were not possible before.
- Development of new practical forensic tools and procedures to facilitate forensic examinations of the IoNTF.
- Apply data analytic methods for various types of forensic data collection from IoNT environments.

Bibliography

Akyildiz, I. F., F. Brunetti, and C. Blázquez. Nanonetworks: A new communication paradigm. *Computer Networks* 52(12) (2008): 2260–2279.

Akyildiz, I. F., and J. M. Jornet. The internet of nano-things. *IEEE Wireless Communications* 17(6) (2010): 58–63.

Akyildiz, I. F., et al. The internet of bio-nano-things. *IEEE Communications Magazine* 53(3) (2015): 32–40.

Atzori, L., A. Iera, and G. Morabito. The Internet of Things: A survey. *Computer Networks* 54(15) (2010): 2787–2805.

Balasubramaniam, S., and J. Kangasharju. Realizing the Internet of nano-things: Challenges, solutions, and applications. *Computer* 2 (2013): 62–68.

Giuliano, R., et al. *Security and privacy in Internet of Things (IoTs): Models, algorithms, and implementations*. CRC Press, Boca Raton, FL, 2015.

Jornet, J. M., and I. F. Akyildiz. The Internet of multimedia nano-things. *Nano Communication Networks* 3(4) (2012): 242–251.

Oriwoh, E., et al. Internet of Things forensics: Challenges and approaches. *9th International Conference on Collaborative Computing: Networking, Applications and Worksharing (Collaboratecom)*, IEEE, 2013, pp. 608–615.

Palmer, G. A road map for digital forensic research. *First Digital Forensic Research Workshop*, Utica, New York, 2001.

Perumal, S., N. M. Norwawi, and V. Raman. Internet of Things (IoT) digital forensic investigation model: Top-down forensic approach methodology. *Fifth International Conference on Digital Information Processing and Communications (ICDIPC)*, IEEE, 2015, pp. 19–23.

Technavio. *Global Internet of Nano-Things Market 2016–2020*, http://www.researchmoz.us/global-internet-of-nano-things-market-2016-2020-report.html [Accessed 30 October 2016].

Zawoad, S., and R. Hasan. FAIoT: Towards building a forensics aware eco system for the Internet of Things. *IEEE International Conference on Services Computing (SCC)*, IEEE, 2015, pp. 279–284.

8

Aspects of Ambient Assisted Living and Its Applications

J. Anuradha and S. Vandhana

VIT University

Vellore, India

CONTENTS

8.1 Introduction

The Internet of Things (IoT) is a subject of much attention these days. IoT is all about various devices or anything with embedded sensors that can be interconnected and communicated. These devices acquire intelligence and can respond well on the stipulated scenario. Active research is being carried out in every dimension of this field to handle the demand and challenges of the vendor. It includes architecture, platform interoperability, communication, security, sensor data analytics, and cognitive computing. The advancement in technology is to serve the society. IoT has a wide range of applications from health care to industry where the solutions are brought to people to help deal with chaos in their daily lives. This chapter focuses on how IoT devices are used for assisted living.

IoT is a technology trend which produces trillions of data through connecting multiple devices and sensors with the cloud and business intelligence tools. It is thus causing a large number and distinct types of products to emit colossal amount of data at a phenomenal rate. According to Manyika et al. (2013) in Global Institute report, IoT is one of the 12 technologies that can bring huge economic transformations and have a massive impact on life, business, and the global economy in the coming years. The report further states that it can make a potential economic impact of around $3 trillion to $6 trillion by 2025. Similarly, Gartner predicts that the IoT market is going to bloom in the years to come. It is estimated to generate $300 billion in revenue by 2020, with an assessment of connecting devices ranging from 25 billion to more than 200 billion. These connecting devices need sensors, networks, back-end infrastructure, and analytics software to extract useful information from the data produced.

Advancement in wireless communication technology and improved quality services made it possible to collect data from the sensors installed at different locations which can be collected any time. Still, the IoT is in an evolutionary stage and its widespread adoption will take time to mature. However, it is the time for company decision makers to understand the potential impact of IoT and its robust future prospects. Precisely, the IoT is the network of devices and sensors, which we use in our daily life; and the data it produces, gathers, stores, and processes help to make the right decisions for a business to be more productive and successful.

8.2 Introduction to AAL

As the life expectancy of humans has increased due to rapid growth of health care, the average life span is increased to around 70 years. In this scenario, the longevity and health issues in old age pressurize caregivers to provide continuous support to the aged in their day-to-day activities. Rapid growth in technology transformation can empower them by using sensor-based IoT devices that facilitate to overcome chaos in their daily lives and allow them to feel comfortable and independent. AAL consists of technical systems, infrastructure, and services to support elderly people to have an independent lifestyle through seamless integration of automated home appliances and its communicating devices. These solutions ensure the safety and security of the person who is being continuously monitored in the background by software systems and smart sensors with good connected network.

Numerous such applications have been developed to significantly improve the quality of life. Some of them are listed below.

- Smart homes that can monitor the home environment and human behavior
- Tracking the health of an individual by monitoring blood pressure, ECG, oxygen intake, weight, and food intake
- Emergency response with GPRS to locate loved ones
- Pain-care system that can electronically locate pain and accurately report its intensity to the physician
- Enabling people to sleep well by playing soothing music, with the option of changing the sound track by merely shaking the phone
- Sensors can also record the sleeping frequency and report abnormalities, if any
- Increasing the memory of the elderly and detecting disorders like Alzheimers
- Rehabilitation for specially abled people
- One touch solution in clinical care, nursing, health care, home care, and other services that is required for daily activities

Most of the applications concentrate on activity recognition, mobility of the person, and monitoring of changes in their environment. The functions of these applications also include tracking health records and medications, locating new friends, enhancing mental sharpness, and providing soothing music for users to sleep.

8.3 Sensors and Devices for AAL

It is long since IoT started its journey, and the objective of AAL is to apply technology and communication to assist the elderly to lead an independent and well-connected life. A wide range of sensors are available for this purpose like those monitoring a person's activity, reading environment conditions, assessing a person's behavior response, and ensuring their health care. Sensors are connected to a network, and wireless communications exchange the data recorded between heterogeneous nodes. Sensors' reliability, miniaturization, power economy, network capability, and cognitive computing have moved the research of AAL and their usage to be part of our daily life.

Some of the basic categories of the sensors are wearable, dense, and sparse sensors. Wearable sensors are integrated with some devices or other materials that we wear the whole day. These sensor signals are in contact with our body to recognize the human characteristics like temperature and activities like walking, sleeping, falling, and also other movements. The difficultly here will be the viability of such devices and the adaptability of people to such devices. However, the dense sensors can be placed externally at different places in the house to capture any movement and to monitor the environmental changes at the residence. The sensors used for this purpose include RFID, accelerometers, and those that enable acoustic and motion capturing. Sparse sensors are placed at particular places at home. It is used for behavioral recognition using probabilistic or fuzzy analysis.

Figure 8.1 indicates that the sensors could respond at different intervals for monitoring activity on monthly, weekly, daily, hourly, or minute basis. There are also indicators that

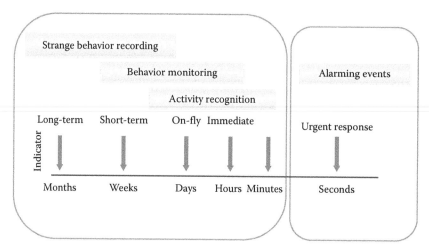

FIGURE 8.1
Sensor indicators for seniors and caretakers.

could respond immediately and are in continuous monitoring every second. Urgent response will be helpful in abnormal activities, especially in the case of emergency. Table 8.1 shows selection criteria for sensors based on the requirements of the living environment. These sensors can be chosen based on the necessity, activity for which it is installed, and its efficiency and strength.

8.3.1 Fall Detection

This is one of the primary and the most basic requirement in AAL. The elderly are exceptionally fragile and because of diminished elements of the nerve frameworks, or due to age-related maladies, they frequently are dependent, slow to respond, and confused. This increases the danger of falls resulting in wounds. Frequently, the elderly aren't aware that they have fallen and require help. They may lie on the floor for quite a long time without help, and in numerous cases this could lead to a long drawn out bout of hospitalization or treatment, and now and then even to death. A ton of exertion has been focused on making a fall identifier that would trigger an alert on the occasion of a fall. The primary prerequisites for the equipment and programming execution of such a fall locator are:

- Not to miss the occasion of a fall, that is, high affectability
- To minimize the quantity of false-positive cautions (the enlistment of a fall when there is no fall), that is, high specificity

The principal prerequisite is coherent as the point of a fall locator in the perfect case is to enlist the falls, that occur. In addition, as the fall could bring about a state of defenselessness or even debilitation, a quick response to such occasions is very important. In the second case, a high number of false-positive alerts could trade-off between the framework and the decrease of acknowledgment by the end clients. Diverse creators utilize distinctive calculations and sensors to recognize the fall. The regularly utilized sensors are:

- Gyroscopes or accelerometers (Bagnasco et al. 2011; Bourke et al. 2010; Kangas et al. 2008; Klenk et al. 2011; Li et al. 2009; Luštrek et al. 2009)

TABLE 8.1

Sensor Selection Criteria for Home

Sensor Type	Location	Targeted Activity	Robustness	Efficiency
FSR	Under bed	Lying, sleeping	High	High
	Under couch	Sitting, lying	High	High
	Under chair	Sitting	Low	Medium
Photocell	In drawer	Kitchen activities	High	High
	Cupboard/ wardrobe doors	Bathroom activities, changing clothes	High	High
Digital distance	Back of chair	Sitting	Medium	High
	Toilet seat cover	Bathroom activities	Medium	High
	Above water tap	Bathroom/kitchen activities	Medium	Medium
Sonar distance	Walls	Activity related to presence in a room	High	Medium
Contact	Regular door	Activity related to leaving/entering room/house, showering	High	High
	Sliding door	Showering, changing clothes	Medium	Medium
	Drawer	Bathroom/kitchen activities	Low	Medium
Temperature	Above oven	Cooking	High	Medium
	Near stove	Cooking	Medium	Low
Infrared	Around TV	Watching TV	High	Medium
Humidity	Near shower sink	Showering	Medium	Low
Pressure mat	On bed	Lying, sleeping	Medium	High
	On couch	Sitting, lying	High	High
	On chair	Sitting	Medium	Medium
Vibration	In drawer	Kitchen activities	Medium	Low

Source: Tunca, C., et al., *Sensors (Basel)*, 14(6), 9692–9719, 2014. Doi: 10.3390/s140609692.

Note: Infrared (IR) sensors—to track multiple people in a closed environment.
Sensors in smart phones—accelerometer, gyroscope, magnetometer, and barometer—can capture the multiple activities like walking, climbing, running, and moving. It can locate the person in the environment. RGB-D sensor—senses a person's skeleton in a RGB video.
Motion sensor, door sensors, light switch sensor, power usage sensors are used in a smart home.

- Cameras-static or wearables (Rougier et al. 2011; Tabar et al. 2006)
- Combination of several strategies (Grassi et al. 2010)

In spite of all the examinations devoted to fall recognition, there isn't a 100% solid calculation that captures all falls without issuing false cautions.

8.3.2 Activity Classification

The goal of this classification is to develop classifiers to recognize human activities and exercises. The distinction between activity and action is that activity is a basic development or change in the stance of the client (e.g., gets up, lies, and strolls), while movement could be a mix of a few activities speaking to a complex, more elevated amount of reflection, for example, cooking, cleaning, and eating. Distinctive machine learning strategies are used for activity classification (Jalal et al. 2011).

8.3.3 Location Tracking

Area tracking is an imperative segment of a teleobserving framework, particularly for people with any shade of dysfunctional behavior or dementia, for example, Alzheimer sickness. Distinctive methodologies for handy acknowledgment exist (Schindhelm et al. 2011), some of which utilizes:

- Identification of radio frequency
- Cameras
- Multimodal approaches

8.3.4 Telemonitoring of Vital-Parameters

Teleobserving is an imperative part of the well-being administrations in AAL. Parameters, for example, heartbeat, pulse rate, blood oxygen immersion, and ECG, are observed (De Capua et al. 2010). Most often the sensors are wearable; however, this may affect the convenience of the clients. That is the reason endeavors are engaged on creating other method for getting the key parameter's value.

8.3.5 Wireless Sensors Networks

Remote sensor systems are a fundamental segment of the AAL frameworks. Every wearable sensor in the framework is bound to be battery-fueled and transmits information remotely. Other environmental sensors and gadgets are likely to transmit their information to a remote server. There are a few entrenched remote norms, for example, Bluetooth, ZigBee, and WiFi (IEEE 802.11), which are generally utilized as a part of AAL frameworks. A considerable measure of research has been directed to the field of WSN; however, issues, for example, movement improvement, arrange administration, security and unwavering quality, are yet to be completely explored and established (Lopez et al. 2011).

8.3.6 Innovative User Interface Development

A few segments of complicated AAL frameworks may be intuitive, for example, in an informal organization the client ought to have the capacity to impart and set up the framework. Given that the elderly are frequently hesitant to utilize PCs, a few novel interfaces ought to be created which are more qualified for their specialized information and capacities (Aghajan et al. 2010).

Some of such interfaces are:

- Computer brain interface
- Recognition of gesture
- Recognition of speech
- Touchscreen

8.3.7 Behavior Determination

It is centered on building a behavioral profile of the client and observing whether there are deviations from the model. Behavioral assurance and areas following it are vital parts of the client's standard life. The target of the behavioral frameworks is to distinguish atypical conduct which may be created by diminished well-being status, and advancing ailment or crisis circumstance (Chung and Liu 2008).

8.4 Wearable Sensors for AAL

The following figures depict some of the wearable devices used for assisting and monitoring the elderly. Figure 8.2 shows the sensor woven in a cloth for recording the temperature. It is a smart temperature sensing textile (Swiss Federal Institute of Technology Zürich, Electronics Laboratory—Wearable Computing). Figure 8.3 is a patch used for diabetics sensing and to give therapy based on the readings. Graphene-based electrochemical

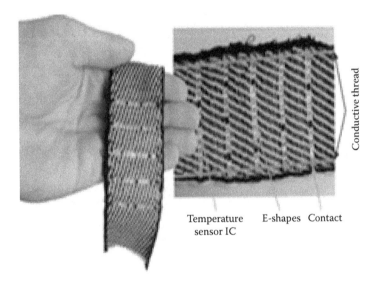

FIGURE 8.2
Smart temperature sensing textiles.

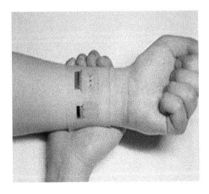

FIGURE 8.3
Diabetes monitoring and therapy.

device is used as a patch with thermo-responsive micro-needles for monitoring blood level glucose and delivers medication through micro needle. (Hui Won Yun, Seoul National University).

Smart wrist band (Figure 8.4) is used to measure pulse rate and calories burning throughout the day. It has OLED display for text and calendar notifications. It can track the duration and quality of sleep and wake you up with a silent vibrating alarm. Overall, it can monitor the fitness level to improve and personalize cardio fitness score with personalized breathing sessions based on one's heart rate. Figure 8.5 shows a remote and digital ECG monitoring system. The ECG signals are sent to remote devices from which analytics are possible through distributed and parallel computing. Figure 8.6 shows the sleep tracker that can even record even minute movements during sleep. The broad strip is placed below the bed and a comfortable wrist-based sleep monitoring device is set up for sleep detection. It works based on the LED light at the bed-side unit which is passed through fiber optic cables. When the measure of light supply is lost, it indicates the movement of the person in the bed.

Another dimension of this is independent living which is adopted for smart route planning, concerning mobility, orientation, and intelligent guidance for indoor and outdoor environments. It is useful in public buildings such as hospitals, museums, offices, shopping malls, and for pedestrians and transportation. AAL has also given a solution

FIGURE 8.4
Wrist band for heart rate tracking and calorie measuring.

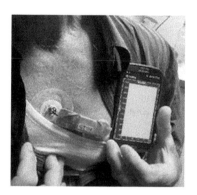

FIGURE 8.5
Digital ECG monitoring system.

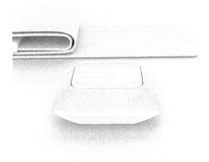

FIGURE 8.6
Sleep tracking system.

FIGURE 8.7
Assisted walking.

for the specially abled persons in performing their day-to-day chores. A smart interactive system has been developed for rehabilitation of neuro-cognitive disorder. Systems that use ontologism and analytics gains knowledge will be rational in making decisions.

8.5 AAL Architecture

This section discusses the system architecture adopted by some applications. Every system has actuators and sensors that perform actions and recognizes the environment, respectively. Through different types of sensors and camera, data are acquired and transmitted through wired or wireless networks. IoT systems manage everything from data acquisition to protocols used for communication. Data packets transmission, filtering, synchronization, and processing are carried out in every sensor. The generated data from the environment is applied for processing. To derive intelligence from the data some of the machine learning techniques or analytics is performed based on which rational decisions can be made. The feedback of the decision is studied by the caregivers or the doctors based on which further actions will be taken. Figure 8.8 shows the processes that are carried out at different levels.

AAL systems that work for emergency treatment are rigidly time constrained. These systems should be accurate with early prediction of abnormalities which are of concern for human safety. The system developed by Benghazi has time trace semantic technique methodology based on UML–RT model. With this model, one can develop a complex AAL system designed through UML-RT constructs represented graphically by software engineers (Benghazhi et al. 2012). It also provides a set of transformation rules for specification in the formal language that can be used for verification. Figure 8.9 shows the communication between the smart home care system and the caretaker or the hospital. The sensor data are recorded and communicated to the server of the home care system either through wired or wireless transmission or through local or Internet connections. The server communicates the emergency message to the hospital with the details of the patient. Thus, persons left alone can be treated in emergency through smart IoT systems that could monitor them continuously.

Communication plays a vital role in AAL for immediate response for assisting elderly people. Wireless communication technologies have developed to a greater extent and satisfy the requirements of the applications of AAL. Figure 8.10 shows various communication standards with their range varying from smaller to a larger distance and strength. It also supports the limited and full mobility and their operating strength based on the range. Protocols such as ZigBee, Bluetooth, and ultra-wide band suit for a smaller range of communication. However, Wifi and Wifi max can be used for a wide range of communication that can cover up to an entire city. Protocols such as GSM, GPRS, EDGE, and HSOPA can be used for communicating across the region, country, etc.

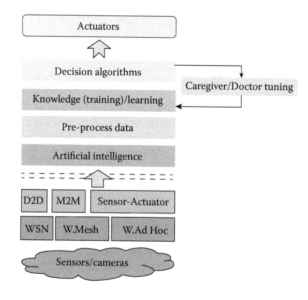

FIGURE 8.8
AAL architecture.

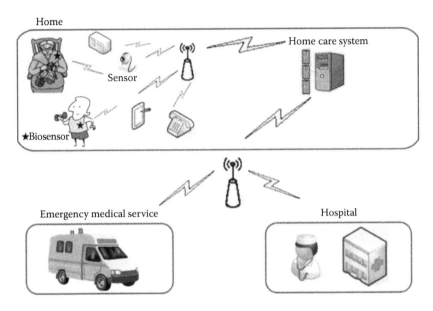

FIGURE 8.9
Emergency assistance for home care.

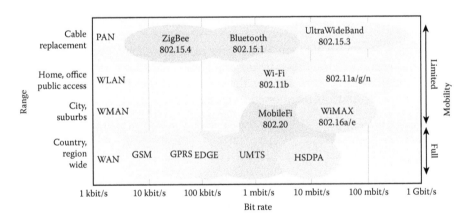

FIGURE 8.10
Wireless communication standards system.

Micro Electro and Mechanical (MEMS) devices integrate the sensors, and mechanical and electronic elements on a chip using micro-fabrication technology. MEMS play a crucial role in integration, sensor, communication, storage, and computational capability on a single chip thereby reducing the size of the wearable device with sophisticated facilities in a small device that can be worn with no disturbances. These sensors are reliable and can also work on low power consumption (WSN 2009).

The architecture in Figure 8.11 consists of data acquisition, model to understand the state of the person at any instance of time and handling timely activity to make decisions. It includes hardware and software functional systems. Hardware system includes various sensors fixed at different places for various purposes. Device drivers are installed at

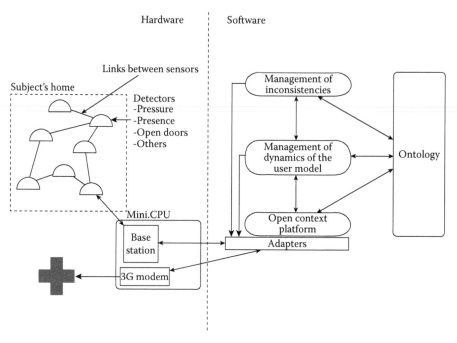

FIGURE 8.11
Architecture on ontology-based communicating emergency response system.

the base stations for the management of the sensors. These sensors are interconnected by wireless communication in the base station and connected to the remote places through 3G modem (Botia et al. 2012). It maintains the communication from small range to larger range using different protocols mentioned in Figure 8.10.

8.6 Analytics in IoT

Sensor, mobile, and wireless technologies are driving the evolution of the IoT. The true business value of the IoT lies in analytics rather than hardware novelties. The first thing to understand about analytics on IoT data is that it involves datasets generated by sensors, which are now cheap and sophisticated to support a seemingly endless variety of use cases.

The potential of sensors lies in their ability to gather data about the physical environment, which can be analyzed or combined with other forms of data to detect patterns.

8.6.1 The IoT Analytics in Health Care

The IoTA in health care encompasses heterogeneous computing, wireless communication systems, and applications and devices that help patients and providers for monitoring, tracking, and storing patients' vital statistics or medical information. Examples of such systems are smart meters, RFID, wearable health monitoring sensors, and smart video cameras. Also, smart phones, intelligent vehicles, and robotics are considered to be part of IoT. These IoT devices produce enormous amounts of data which become a challenge for providers to deal with efficiently. To harness this huge data in a technological way and make sense of it, IoTA is implemented. Data mining, data management, and data analytics techniques are used to make this deluge of data useful and also to extract medically relevant information. In fact, it has been predicted that by 2017 more than 50% of analytics techniques will make a better use of this influx of data generated from instrumented machines and applications.

In the United States, physicians and other providers are using new wireless technologies to monitor patients remotely for early detection of health problems and for timely recovery. The influx of data from Internet-connected devices is a valuable source for health systems. The analytics empowers caregivers to optimize it fully in order to provide better care, cut down costs, and to reduce other inefficiencies in health care.

8.6.2 Transforming Health Care with the IoT

In the US health care industry, the small and big health care organizations are using the IoT tools and devices that are revolutionizing medical care in unique ways. IoT devices range from headsets that measure brainwaves to clothes that include sensing devices, Google Glass, BP monitors, etc. All these have moved personal health monitoring system to the next level.

According to ABI Research, in 2016 the sale of wireless wearable medical devices will bloom and the sales will reach more than 100 million devices annually. Another report by IMS Research—the research partner of wearable technologies—states that the devices which are wearable works in close proximity with the human body and produces more realistic results. There has been clinical evidence that the physiological data received from wireless devices has been a valuable contributor for managing and preventing chronic diseases and monitoring patients post hospitalization. As a result, a growing number of medical devices are becoming wearable nowadays, including glucose monitors, ECG monitors, pulse oximeters, blood pressure monitors, and so on.

The market for wearable technologies in health care is expected to exceed $2.9 billion in 2016. The IoT enables health organizations to achieve superior technology interoperability, gather critical data from multiple sources in real time, and a better decision-making capability. This trend transforms the health care sector by increasing its efficiency, lowering costs, and providing better performance in patient care.

8.6.3 IoT—A Health-Care Game Changer

The IoT has already brought in significant changes in many areas of health care. It is rapidly changing the health care scenario by focusing on the way people, devices, and applications are connected and interact with each other. IoT is emerging as the most promising information and communication technology (ICT) solution which enables providers to improve health care outcomes and reduce health care costs by flawlessly collecting, recording, analyzing, and sharing myriads of new data streams in real time. Moreover, as the widespread adoption of IoT grows, many of the inefficiencies in health care will be reduced. For example, sensors embedded in medical devices such as diagnostic equipment, drug-dispensing systems, surgical robots, implantable devices, and personal health and fitness sensors will perform data collection and measurements, and conduct tests digitally in no time which are currently administered and recorded manually.

This is extremely important for gaining new insights and knowledge on various issues in health care. For example, to study a patient's response to a specific therapy or drugs, traditionally, health care providers study different samples taken from the patients. These samples sometimes are not up to the required standard to derive a clear conclusion. But, the IoT has made it possible for the first time to collect real-time data from unlimited number of patients for a definite period of time through connecting devices. It is anticipated that it will also improve health care services for people in remote locations as monitoring systems provide a continuous stream of data that enables health care providers to make better decisions. IoT is gaining momentum among health care providers and the IT sector in emerging as a major technology trend for improved health care.

8.6.4 The Future of IoT in Health Care

According to market researchers like Gartner and McKinsey, the IoT will "add $1.9 trillion to the global economy by 2020," or it will have a "potential impact of $2.7 to $6.2 trillion by 2025." This shows that the market for IoT is rising and gaining popularity in the United States, especially after President Obama's comments on health care industry that the "nation should create a Smart Manufacturing infrastructure and approaches that leads operators make real-time use of 'big data' flows from fully instrumented plants."

IoT has been identified as one of the fast emerging technologies in IT according to Gartner's IT Hype Cycle which shows the emergence, adoption, and maturity of specific technologies, and their impact on applications. It has been predicted that it will take 5–10 years for a widespread market adoption of IoT. The IoT revolution which has arrived in health care and medicine is making a strong impact on health care systems worldwide. New types of sensor technology, rapidly growing data analytics, and the new health care structures are formed due to the outbreak of IoT. The health IoT has a tremendous potential to create a more revolutionary archetype for health care industry if developed on a privacy/security model. In addition, it will have a major impact on health economy in the near future.

8.7 Stages of Transforming Raw Data into Meaningful Analytics

Understanding the tools and data acquisition from IoT devices, analysts need to transform data into the required format to infer knowledge from it. Any type of data, including health care data, goes through three stages before an analyst can use it to achieve sustainable, meaningful analytics which is shown in Figure 8.12.

1. Data capture
2. Data provisioning
3. Data analysis

Stage 1: Data Capture

An analyst's job is impacted by the way people, processes, and devices produce and capture data. These three entities are responsible for the appropriateness of the data—did they capture the right stuff? discreteness—did they capture it in the right format? and ease of data extraction—was the data captured in an accessible way?

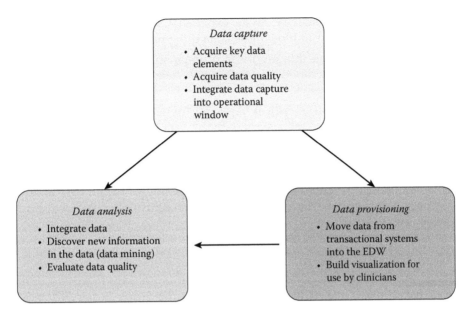

FIGURE 8.12
IoT analytics.

Stage 2: Data Provisioning

Analysts need data from multiple source systems throughout the organization to produce meaningful insights. For example, an analyst assisting a team of clinicians on a quality improvement issue needs a variety of data from multiple source systems:

- EMR data (for clinical observations and lab results)
- Billing data (for identifying cohorts using diagnosis and procedure codes, charges, and revenue)
- Cost data (for determining the improvements' impact on margins)
- Patient satisfaction data

Aggregating data manually—pulling all of the data into one location in a common format and ensuring datasets are linked to each other (through common linkable identifiers such as patient and provider identifiers)—is extremely time-consuming. It also makes data more susceptible to errors. There are more effective ways to gather data.

Stage 3: Data Analysis

The data analysis starts after the appropriate data have been captured, pulled into a single place, and tied together. The analysis process consists of several parts:

- *Data quality evaluation:* Analysts need to understand the data by taking time to evaluate the same. They also need to note their method of evaluation with the reference when they share their findings with the audience.
- *Data discovery:* Before attempting to answer a specific question, analysts should take time to explore the data and look for meaningful oddities and trends. It is a critical

component of good data analysis. In my experience, more than 50% of acted-upon analyses resulted from stumbling upon something in the discovery process.

- *Interpretation:* Most people think of the interpretation step when they think about analyzing data, but it's actually the smallest sub-step in a long process in terms of the time analysts spend on it.

- *Presentation:* Presentation is critical. After all the work for getting data to this point, the analyst needs to tell a story with the data in a consumable, simple way that caters to the audience. Presentation principles need to be considered.

These three stages of insightful data analysis will drive improvements. Each step alone, however, is not enough to create sustainable and meaningful health care analytics. It's equally important to empower data analysts to focus on analyzing data, not just capturing and provisioning data.

8.8 Common Health-Care Analytics Solutions and Their Pitfalls

Many health systems opt to implement one of the following three types of analytics solutions. Although these solutions may initially show promise, they inevitably fall short of expectations.

8.8.1 Point Solutions

When developing an analytics platform, some health systems deploy one or more best-of-breed or point solutions. These applications focus on a single goal and a single slicing of the data. For example, the solution might focus solely on reducing surgical site infections or central-line-associated blood infections. One problem with this approach is something called sub-optimization. While the organization may be able to optimize the specific area of focus, these point solutions can have negative impacts both upstream and downstream. They also don't offer much in the way of insight outside of the specific area of focus.

Another problem is what is called the "technology spaghetti bowl." When a hospital or group practice has only a few point solutions in its dish, a small IT shop can provide adequate support. But with additional point solutions (consider them noodles, if you will), an IT department finds it all but impossible to unravel all the disparate noodles in the spaghetti bowl. Imagine there are 10 different point solutions, and it's necessary to update coding standards from multiple source systems in each of these point solutions. You'll end up with sauce on your face by the time you're done.

Eventually, one or two of the senior IT employees may own this spaghetti bowl mess with a huge dependency placed on these individuals, creating an unstable house built out of playing cards. While this situation works well initially, if either of the individuals leaves the organization, then the house of cards will crumble, leaving a mess for someone else to clean up.

In addition, point solutions typically result in multiple contracts, multi-cost dependencies, and multiple interfaces. These, in turn, lead to complexity, confusion, and organizational chaos that impede improvement and continued success.

8.8.2 Electronic Health Record System

Implementing an electronic health record (EHR) is clearly a necessary step toward data-driven care delivery, and an EHR system alone is insufficient to enable an enterprise-wide, consistent view of data from multiple sources. The conversion of clinical data from paper to an electronic format is a necessary step—it allows for the use of data to improve care. However, without a way of organizing all sources—clinical, financial, patient satisfaction, and administrative data—into a single source of truth, a health care organization is unable to harness the analytic power of the data. Some EHR vendors are beginning to come out with data warehouse offerings that run on top of the EHR's transactional database. However, these data warehouses still have the limitation that they don't aggregate data from a variety of external sources—because the vendor can't or won't. Some EHR vendors are becoming willing to integrate some external data, but they are years behind analytics vendors in terms of their performance.

8.8.3 Independent Data Marts in Different Databases

Independent data marts that live in different databases throughout a health system provide limited analytics capabilities because they can only deliver little sources of truth from the different soiled systems. Take, for example, the ADT (admission, discharge, transfer) information that lives in the EMR. When there's a need to analyze the ADT information and the role it plays on costs, analysts move the data over to the costing system with the independent data mart model. Requests like this one can happen over and over for many different types of scenarios, which ends up becoming time-consuming. It also slows down the entire system as analysts repetitiously bombard the system with requests for each new use case.

8.9 Ambient Intelligence for AAL

The research on Aml was initiated by European Commission in 2001 (I.A. Group 2001). Aml is characterized by sensitivity, responsivity, adaptability, transparency, ubiquity and intelligence. It's interaction with other technologies is shown in Figure 8.13.

The project PERSONA named after PERceptive Space prOmoting iNdependent Aging was developed with a scalable open standard platform to building wired range of applications in AAL. It has an efficient infrastructure with self-organizing middleware technology having extensibility of components or device ensembles. The components of this system communicate with each other based on the distributed coordination strategies using PERSONA middleware. It has different types of communication buses: input, output, context, and service buses that enable interoperability with different components using middleware. The middleware of PERSONA was implemented using the OSGi platform with communication through UPnp, Bluetooth, and R-OSGi. In this model, various reasoning algorithms are adopted for modeling, activity prediction and recognition, decision-making, and spatial temporal reasoning (Cook et al. 2009).

Modeling—it deals with the design that separates the computing and user interaction. This can customize the Aml software toward user requirement. These models have capability of recognizing the changes and adopting itself based on the changes in the patterns.

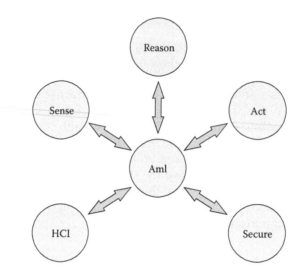

FIGURE 8.13
Interaction of AmI system with other technologies.

Modeling can be classified based on the data used to build the model, type of model built, and algorithms used for nature of model building (Cook et al. 2009).

Activity prediction and recognition—different type of sensors can recognize different activities. It is based on the user requirement and the type of sensors used in the application for recognizing various activities. Figure 8.14 recognizes falling of a person. Table 8.1 shows the different sensors and the activity that it can recognize.

Activity model—different machine learning algorithms are applied to understand the activity and to adapt to the changes based on the behavioral patterns. The pattern recognition can be in batch processing or steam data or can be for conditional patterns. Probabilistic models, neural network, and fuzzy models are some of the techniques applied for activity recognition.

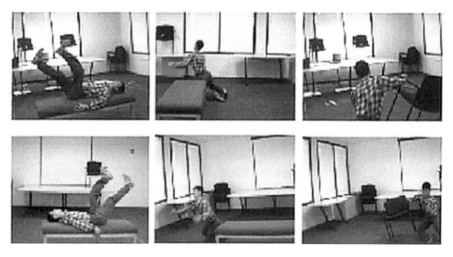

FIGURE 8.14
Activity recognition—detecting falling of a person.

Decision-making—Artificial Intelligence (AI) planning can be applied for the daily activity data and also on the complete task recorded earlier for deriving any conclusion. This type of analysis on previously recorded data will facilitate in recognizing and decision-making in emergency situations. The following is an example for deriving knowledge from the rule generated.

Subject(?x) ∧Location (?x, ?y) ∧MovementSensor(?s) ∧detectMovement(?, true) ∧inside(?s,?r) ∧HabitableSpaceInBuilding(?r) ∧haveConnection(?r,?c) ∧connectinWith(?c,?c1) ∧equal(?y,?c1) ⇒state(?x, Active) ∧location(?x,?r)

Spatial temporal reasoning—the spatial characteristics and temporal behavior of sensor data give additional information about an activity. For example, after, before, meets, over-lapped, started by, finishes, and during are some of the temporal relationships. Some of these will be required for medication in health care while suggesting pills right after food, before food, etc. Capturing these characteristics require special techniques and algorithms. The algorithm, Jakkula's TempAI recognizes nine intervals with such temporal relationship used for prediction (Jakkula et al. 2007). Figure 8.15 states the temporal relationships such as before, contains, overlaps, meets, and visualization of two activities X and Y with respect to time.

8.10 Software Engineering Approach for AAL Systems

Probabilistic Symbolic Model Checker (PRISM) is a tool used to build AAL system based on dependability analysis (Kwiatkowska et al. 2004). This tool is highly successful for the reason that it uses various probabilistic models like discrete time Markov chains (DTMCs), continuous time markov chains (CTMCs), and Markov decision process (MDP). It can also derive model based on state-based language reactive model formalisms. This tool also uses temporal logic like Probabilistic Computational Tree Logic (PCTL) (Rodrigues et al. 2012).

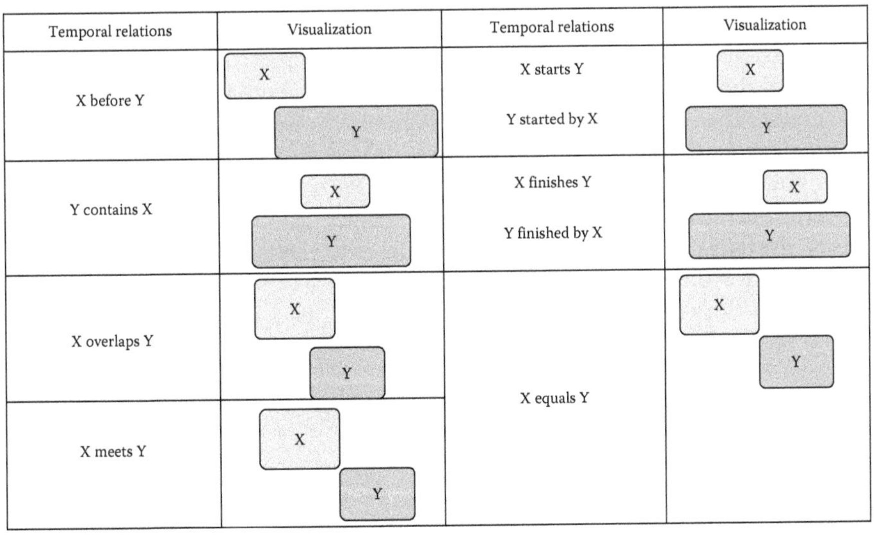

FIGURE 8.15
Temporal interval of two activities X and Y.

Analyzing the AAL system development with dependability as a function of reliability involves the following eight steps to assure qualitative and quantitative dependability analysis.

    a. Specify UML behavior models
    b. Annotate the models.
    c. Convert to PRISM language.
    d. Compile and check PRISM model.
    e. Simulate the model.
    f. Define the properties for dependability.
    g. Run the properties.
    h. Analyze the result.

Steps d to step h use PRISM tool to compose the dependability analysis. Every activity is modeled after the UML activity diagram (AD). It consists of decision and action nodes, each representing the execution scenarios as a sequence diagram. Once the ADs and SDs are completed, the next step is to build the annotation with reliability components and transition probabilities. The assumptions of the reliability of the components are based on (i) the reliability of the services performed where the execution time of the invocation is not considered as a factor, (ii) state of transition which depends on the source state, and the available transition and its transfer of control between the components is estimated through Markov Chain property, and (iii) failures of the transition.

An example of annotation in PRISM with component reliability is as follows and shown in Figure 8.16.

$$[notify]s = 0 \rightarrow R_c : (s' = 1) + (1 - R_c) : (s' = 2);$$

To ensure whether the system is working properly, software design is very essential which is set before it gets developed. Besides the functional requirements defined through activity and sequence diagram, it also requires nonfunctional requirements to assure the quality of the system developed. AAL systems require activities that are time constrained and critical. To ensure the nonfunctional requirements of the system, Benghazi et al. (2012) has modeled a design based on UML-RT that uses timed traces semantic and methodology to check the timeliness and safety of AAL activity. Petri Nets is used as tool for modeling semantic and interactive graphical language to develop and verify nonfunctional requirements of the system (Hein et al. 2009; Palanque et al. 2007). Benghazi has used formal language CSP+T for defining the timed traces semantics of the operations. CSP+T is the subset of CSP developed by Hoare in 1985. Let P and Q be a process that has defined a set of events Σ with $a \in \Sigma$ and $B \subseteq \Sigma$ in a time interval [t,t+T].

$$P ::= STOP \,|\, SKIP \,|\, 0.^* \rightarrow P \,|\, a \infty v \rightarrow P \,|\, I(T,t).\, a \rightarrow P$$

$$|\, I(T,t) \rightarrow P \,|\, P \Box Q \,|\, P \sqcap Q \,|\, P;Q \,|\, P\|Q \,|\, P \,|||\, Q$$
$$_{B}$$

The system is modeled in such a way that the process of target system is divided into subsystems, each designed through an UML-RT context diagram and interconnected

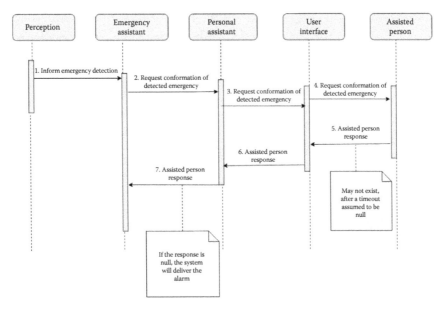

FIGURE 8.16
Sequence diagram for emergency detection AAL system.

by a predefined protocol and interaction between actors and the system. The following procedure is applied recursively for the subsystem.

1. Its architecture is defined by UML-RT context diagram
2. Its internal structure is described by composite structure diagrams
3. For each component in the subsystem
   a. if the component is composite, then jump to step i
   b. else, if the component is basic, then capture its timed behavior by timed state diagram (Benghazi et al. 2012).

Figure 8.17 shows an example model setup using MEDISTAM_RT. It consists of two capsules C1 and C2 for which activity is defined. The system also considers the temporal consistency of the basic capsules and their protocols for coordination between them through message passing. Thus, the behavior of the ports (edges of communication point) determines the pattern of interaction in time-constrained message passing to the environment.

8.11 Cognitive AAL

Cognitive architecture uses a distributed intelligent system for controlling and decision-making of an ambient intelligence system (Peter Ivanov et al. 2015). Neurodegenerative diseases (Antonio Coronato and Giuseppe De Pietro 2012) are progressive and very hard to diagnose in its early stages. Progressive impairment in the activities of daily living, as well

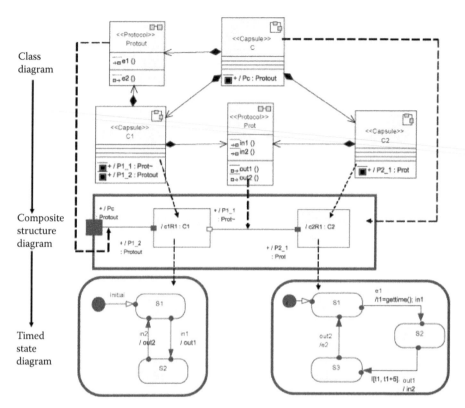

FIGURE 8.17
Modeling process of MEDISTAM—RT.

as cognitive deterioration, leads to an increase in patient dependency. Neuropsychiatric symptoms are common features and include psychosis (delusions and hallucinations), depressive mood, anxiety, irritability/lability, apathy, euphoria, dis-inhibition, agitation/aggression, aberrant motor activities, sleep disturbance, and eating disorder (Segal-Gidan et al. 2011). In particular, cognitive impaired patients need constant monitoring to ensure their well-being and that they do not harm themselves. Additionally, they have a tendency to wander and misplace objects. Situation awareness requires the environment be able to analyze the behavior of patients with cognitive impairments, which is extremely challenging. Indeed, traditional behavioral and situation analysis techniques assume "rational" behaviors, in the sense that behaviors are led by clear goals, and rely on matching the observed activities with predefined behavioral patterns (Ye et al. 2012). The cognitive sciences began as an intellectual movement in the 1950s, called the cognitive revolution, arguably initiated by Noam Chomsky. Cognitive science is mainly concerned with artificial intelligence, attention, knowledge and processing of language, learning and development, memory, perception and action, and consciousness.

Cognitive AAL is a cognitive development and function involving conscious intellectual activity to support the elderly with their daily activities in order to help them be healthy and safe while living independently. However, most current systems are ineffective in actual situation, difficult to use, and have a low acceptance rate. There is a need for an intelligent assisted living solution and also to resolve practical issues such as user acceptance and usability in order to truly assist the elderly. Small, inexpensive, and low-powered consumption sensors are now available which can be used in assisted living applications to provide sensitive and responsive services based on users' current environments and situations.

Many activity recognition systems are developed for assisted living. The three major requirements for an assisted living system which need to be met in order to fulfill its purpose and potential to assist vulnerable people are: (1) High acceptance: system needs to be ambient and unobtrusive. (2) Adaptation: able to adapt to changing situations or abilities of the individual and environment to satisfy individual needs. (3) High usability: services must be provided in an accessible way (Kleinberger et al. (2007). The three characteristics can be viewed from two perspectives, namely practical and technical aspects. The aim of the practical aspect is to satisfy the needs in term of practicality, that is, high usability and acceptance. Several studies on perceptions toward assisted living technology (Chernbumroong et al. 2010; Demiris et al. 2004, 2008) have identified concerns arising from the use of technology for elderly care such as user acceptance, system cost, usability, and privacy issues. For example, a system which requires users to wear special equipment may be perceived as too complicated to use, resulting in low acceptance. In a mobility-aided system (Yu et al. 2003), for example, the user interface is a critical requirement as it has direct physical interaction with the users. An interview-based investigation by Demiris et al. (2004, 2008) also showed that the elderly were concerned about privacy violation, visibility, and accuracy of the assisted living systems. Even if systems could deliver the best services for assisting people, unless they are easily accessible and usable and address the real needs and concerns of the users, they will not be accepted. The cost of assisted living systems is another important issue. For a practical solution in assisted living, the systems need to be cost-effective to make it affordable for the general population, especially the elderly who may be on benefits or pensions. From a technical aspect, the aim is to enable assisted living systems to become intelligent in order to adaptively assist the elderly in changing a dynamic environment. Systems with such aims are also referred to as ambient intelligent systems or smart homes (Aarts 2004). Ambient intelligence technology used in assisted living solutions can provide some hands-on support based on current user status to maintain elderly people's independence, reducing the health care cost while increasing quality of life (Kleinberger et al. 2007). To offer intelligent services, the key is to understand user environments such as surrounding temperature and users activities in order to provide adaptive assistance to users. In this study, we are interested in understanding the user's current activities which are often referred to as activity recognition.

AAL is a new and extending zone of research concentrating on the arrangement of in-home care, that is, unimportant for patients requiring support with day-by-day exercises. AAL makes utilization of the most recent innovation and sensors accessible to give solid data. These data are focused at enhancing the level of care which the patient gets and empowering them to live inside their own home with least danger.

Dementia is likewise created by other subjective ailments, for example, Parkinson and Creutzfeldt–Jakob sicknesses. Dementia causes dynamic corruption of neurological procedures related specifically to memory. Subsequently, patients can turn out to be effortlessly befuddled and their conduct can get to be distinctly eccentric. Because of the impacts new surroundings may have on a patient, it is desirable to keep the patient inside well-known surroundings for whatever length of time that is conceivable.

There are a few issues to consider when observing patients inside their own homes. The first and foremost reason is that observing ought to be encompassing and inconspicuous. To encourage surrounding observing a few gadgets and sensors have been outlined.

Metcalf et al. (2009) introduced a near survey of 30 diverse conductive yarns sewed utilizing two unique machine techniques. The utilization of such yarns is centered on measuring strain and development in patients.

Agreeable wearable pieces of clothing give a subtle and compelling method for observing patients. Be that as it may, while checking patient's protection and the affectability of the information got from the sensors ought to be considered. Another approach to decide development is the utilization of accelerometers which measure changes in development. Slyper and Hodgins (2008) portrays the procedure of activity catch utilizing accelerometers to copy developments performed by the subject utilizing movement. Accelerometers are greatly little, cheap, and generally inconspicuous. Five accelerometers are sewn into a shirt and associated utilizing conductive string. Doukas and Maglogiannis (2008) makes this one stride additionally laying out a framework which utilizes accelerometers and mouthpieces to recognize a fall in an aged people.

Virone et al. (2006) drew a model framework which utilizes a remote sensor system to screen a patient's well-being continuously with a view to giving an ALL framework. The framework additionally keeps a total history of permitting approved clients, for example, specialists or medical attendants, to perform remote discussion in light of the information got. The framework utilizes both wearable and living space sensors to gather a wide range of data. Remote sensors are utilized to decrease the requirement for retrofitting. A caution framework portrayed has been (Giansanti 2008) looked into to screen epileptic emergency. This framework utilizes GPS and the GSM to send a caution message specifying the patient's area when an epileptic emergency happens.

A framework is proposed in Oregon health (2010) for distinguishing and checking the movement of dementia in patients. The framework plans to screen sensors situated in entryways and around the home. The thought is to endeavor to distinguish changes in memory and utilize these sensors. Comparable undertakings aim at creating a computerized in-home constant demonstrative apparatus to screen the intellectual degeneration of patients experiencing dementia.

8.12 Applications of AAL

Assisted living has become important these days for the reason that every individual is engaged in his or her own work and has little time to spend with others. People spend most of the time in the digital world and are connected by it. This situation leads to the finding of a solution to attend and assist the old aged and the specially abled individuals to live an independent and comfortable life. AAL facilitates this and has a lot of solutions ranging from a small requirement like BP monitoring to critical health care. Below are some of the applications on assisted living.

- Temperature, blood pressure, diabetes, oxygen, calorie recording, and monitoring system
- Digitalized ECG monitoring system with remote access
- Sleep monitoring system with sensors set on the bed
- Mobility assistance system that supports in activities like walking
- Medication reminders and step count

- Personal emergency response devices that are connected to the caretaker and doctors
- Monitoring the daily routine activity patterns from which we can find the change in diet, sleeping, medications, etc.
- Fitness tracker system
- Smart security devices for homes
- Smart home with automated response for door bell, temperature control, smart stove, light sensing, smart lock, etc.
- Smart devices that can predict the fall of the elderly, which are quiet common, and assist them

Advantages involved in installing these devices are:

- Comfort and independent life style
- Cost efficiency
- Reduce the physical demand, assistance, and home maintenance
- Reduces the anxiety about the elderly during office hours
- Secure and safe home for independent living
- Medical support in absence of caretaker

8.13 Conclusion

AAL is a boon of technology for the elderly that can support their life. It makes their life hassle-free, easy, simple, and sophisticated giving them much needed confidence and happiness. Wearable devices designed for all-time wearing without discomfort will acquire the data and sent to remote servers for further processing. Analytics performed on the acquired data will support decision-making in the critical situation or in absence of the caretaker. Designing such a system needs extensive care as it deals with real life situations. Software engineering concepts are applied for effective modeling of AAL systems to match with the requirements of the application in user and system perception. Communicative technologies and system capabilities play an important role in developing a flawless application. The main challenges are using reliable sensors with cost-effectiveness, uninterrupted power supply for the device with less consumption, and selection of good communication medium with appropriate protocol. However, supremacy of the AAL relies on the above points still the design of device bounces an impression of comfort to the end user's.

References

Aarts, E. (2004), Ambient intelligence: A multimedia perspective. *IEEE Multimedia* 11(1): 12–19. DOI: 10.1109/MMUL.2004.1261101.

Aghajan, H., et al. (2010). *Human-Centric Interfaces for Ambient Intelligence*. Academic Press, Cambridge, MA.

Bagnasco, A., et al. (2011, October). Design, implementation and experimental evaluation of a wireless fall detector. In *Proceedings of the 4th International Symposium on Applied Sciences in Biomedical and Communication Technologies,* p. 65. ACM, San Diego, CA.

Benghazi, K., et al. (2012), Enabling correct design and formal analysis of Ambient Assisted Living systems. *The Journal of Systems and Software* 85: 498–510.

Botia, A. J., et al. (2012). Ambient assisted living system in home monitoring of healthy independent elders. *Expert Systems with Applications* 39: 8136–8148.

Bourke, A. K., et al. (2010). Evaluation of waist-mounted tri-axial accelerometer based fall-detection algorithms during scripted and continuous unscripted activities. *Journal of Biomechanics* 43(15): 3051–3057.

Chung, P. C., and Liu, C. D. (2008). A daily behavior enabled hidden Markov model for human behavior understanding. *Pattern Recognition* 41(5): 1572–1580.

Ciuti, G., Ricotti, L., Menciassi, A., and Dario, P., (2015), *MEMS Sensor Technologies for Human Centred Applications in Health Care, Physical Activities, Safty and Environmental Sensing: A Review on Research Activities in Italy, Sensors,* 15, pp. 6441–6468. MDPI, Switzerland.

Cook, J. D., et al. (2009). Ambient intelligence: Technologies, applications and opportunities. *Pervasive and Mobile Computing* 5: 277–298.

De Capua, C., et al. (2010). A smart ECG measurement system based on web-service-oriented architecture for telemedicine applications. *IEEE Transactions on Instrumentation and Measurement* 59(10): 2530–2538.

Demiris, G., Rantz, M. J., Aud, M. A., Marek, K. D.,Tyrer, H. W., Skubic, M., and Hussam, A. A. (2004). Older adults attitude towards and perceptions of smart home technologies: A pilot study, *Medical Informatics and the Internet in Medicine* 29(2): 87–94.

Demiris, G., et al. (2008). Senior residents' perceived need of and preferences for "smart home" sensor technologies. *International Journal of Technology Assessment in Health Care* 24(1): 120–124. DOI: 10.1017/S0266462307080154.

Demiris, G., and Hensel, B. K. (2008). Technologies for an Aging Society: A systematic Review of Smart Home Applications. *IMIA Year Book of Medical Informatics* 3: 33–40.

Doukas, C., and Maglogiannis, I. (2008, January). Advanced patient or elder fall detection based on movement and sound data. In *2008 Second International Conference on Pervasive Computing Technologies for Healthcare,* pp. 103–107. IEEE, Piscataway, NJ.

Giansanti, D., et al. (2008). Toward the design of a wearable system for the remote monitoring of epileptic crisis. *Telemedicine and e-Health* 14(10): 1130–1135.

Grassi, M., et al. (2010). A multisensor system for high reliability people fall detection in home environment. In P. Malcovati, A. Baschirotto, A. D`Amico and C. Natale (Eds.), *Sensors and Microsystems.* and Vol. 54, Lecture Notes in Electrical Engineering, pp. 391–394. Springer, The Netherlands.

Hein, A., et al. (2009). A service oriented platform for health services and ambient assisted Living. In *Proceedings of the International Conference on Advanced Information Networking and Applications Workshops,* IEEE Computer Society, Washington, DC, pp. 531–537.

http://saviance.com/whitepapers/internet-health-industry

https://www.healthcatalyst.com/healthcare-analytics-best-practices

I.A. Group. (2001). *Scenarios for Ambient Intelligence in 2010.* European Commission, CORDIS, IST, Belgium.

Ivanov, P., Nagoer, Z., Pshenokova, I., and Tokmakova, D. (2015). Forming the multi-modal situation content in Ambient Intelligent Systems on the Basis of Self Organizing Cognitive architectures. In *5th World Congress on Information and Communication Technologies (WICT),* Marrakesh - Morocco, December 14–16, pp. 7–12.

Jakkula, V., et al. (2007). Knowledge discovery in entity based smart environment resident data using temporal relations based data mining. In *Proceeding of the ICDM Workshop on Spatial and Spatio— Temporal Data Mining.* IEEE, Piscataway, NJ.

Jalal, A., et al. (2011, June). Daily human activity recognition using depth silhouettes and mathcal {R} transformation for smart home. In *International Conference on Smart Homes and Health Telematics,* pp. 25–32. Springer, Berlin, Heidelberg.

Kangas, M., et al. (2008). Comparison of low-complexity fall detection algorithms for body attached accelerometers. *Gait & Posture* 28(2): 285–291.

Kleinberger, T., et al. (2007). Ambient Intelligence in Assisted Living: Enable elderly people to handle future interfaces, In C. Stephanidis (ed.), *Universal Access in HCI, Part II, HCII, LNCS 4555*, pp. 103–112. Springer, New York.

Klenk, J., et al. (2011). Comparison of acceleration signals of simulated and real-world backward falls. *Medical Engineering & Physics* 33(3): 368–373.

Kwiatkowska, M., et al. (2004). PRISM 2.0: A Tool for Probabilistic model checking. In *Proceedings of the First International Conference on Quantitative Evaluation of Systems (QEST'04)*, pp. 322–323. IEEE Computer Society, Washington, DC.

Li, Q., et al. (2009, June). Accurate, fast fall detection using gyroscopes and accelerometer-derived posture information. In *2009 Sixth International Workshop on Wearable and Implantable Body Sensor Networks*, pp. 138–143. IEEE, Piscataway, NJ.

Lopez, F., et al. (2011, June). Cognitive wireless sensor network device for AAL scenarios. *International Workshop on Ambient Assisted Living*, pp. 116–121. Springer, Berlin.

Luštrek, M., et al. (2009). Fall detection and activity recognition with machine learning. *Informatica* 33(2): 205–212

Manyika, J., et al. (2013). *Disruptive Technologies: Advances that will Transform Life, Business, and the Global Economy*. Global Institute report. McKinsey, San Fransisco, CA.

Metcalf, C., et al. (2009). *Fabric-based strain sensors for measuring movement in wearable tele monitoring applications*. IET Conference on Assisted Living, p. 13. London, UK.

Oregon Health. (2010). *In-Home Sensors Sport Dementia Signs in Elderly*, Oregon Health and Science University. http://www.ohsu.edu/ohsuedu/newspub/releases.

Palanque, P., et al. (2007). Improving interactive systems usability using formal description techniques: Application to Health care. In *Proceedings of 3rd Human—Computer Interaction and Usability Engineering of the Austrian Computer Society Conference HCI and Usability for Medicine and Health Care*, pp. 21–40. Springer-Verlag, Heidelberg.

Rodrigues, G. N., et al. (2012), Dependability analysis in the ambient assisted living domain: An exploratory case study. *The Journal of System and Software* 85: 112–131.

Rougier, C., et al. (2011, June). Fall detection from depth map video sequences. *International Conference on Smart Homes and Health Telematics*, pp. 121–128. Springer, Berlin.

Schindhelm, C. K., et al. (2011). Overview of indoor positioning technologies for context aware AAL applications. In *Ambient Assisted Living*, pp. 273–291. Springer, Berlin.

Segal-Gidan, F., et al. (2011). Alzheimer's disease management guideline: Update 2008. *Alzheimer's and Dementia* 7(3): e51–e59. DOI:10.1016/j.jalz.2010.07.005.

Slyper, R., and Hodgins, J. K. (2008, July). Action capture with accelerometers. In *Proceedings of the 2008 ACM SIGGRAPH/Eurographics Symposium on Computer Animation*, pp. 193–199. Eurographics Association, San Diego, CA.

Tabar, A. M., et al. (2006, October). Smart home care network using sensor fusion and distributed vision-based reasoning. In *Proceedings of the 4th ACM International Workshop on Video Surveillance and Sensor Networks*, pp. 145–154. ACM, San Diego, CA.

Tunca, C., et al. (2014), Multimodal wireless sensor network-based ambient assisted living in real homes with multiple residents. *Sensors (Basel)* 14(6): 9692–9719. DOI: 10.3390/s140609692.

Virone, G., et al. (2006, April). An assisted living oriented information system based on a residential wireless sensor network. In *1st Transdisciplinary Conference on Distributed Diagnosis and Home Healthcare. D2H2*, pp. 95–100. IEEE, Piscataway, NJ.

Ye, J., et al. (2012). Situation identification techniques in pervasive computing: A review. *Pervasive and Mobile Computing* 8(1): 36-66. DOI: 10.1016/j.pmcj.2011.01.004.

Yu, H., et al. (2003). An adaptive shared control system for an intelligent mobility aid for the elderly. *Autonomous Robots* 15: 53–66.

9

Naming Services in the Internet of Things

T.R. Sooraj, R.K. Mohanty, and B.K. Tripathy

VIT University

Vellore, India

CONTENTS

9.1 Introduction

The term the Internet of things (IoT) is becoming an increasingly growing topic of conversation both in workplace and outside of it. IoT is a system of interrelated computing devices, mechanical and digital machines, objects, animals, or people that are provided with unique identifiers and the ability to transfer data over a network without requiring human-to-human-to-computer interaction. Here we need to know how the devices or things in IoT are communicating and how the name service will translate the meaning-ful names to a machine-understandable form. Name services means to translate human-understandable names into network identifiers that can be used for communications in the computer networks. Mainly we are using Internet Protocol (IP) address for the communi-cation of hosts over the network. But it will be very difficult for humans to memorize the IP address, which is a 32-bit number and type it in the URL field to access the web pages. This difficulty raised the need of using memorable words (for example, www.google.com) to access the web pages. Li et al. introduced a naming, addressing, and profile (NAPS) server for the naming and addressing in IoT.

One of the main issues in the existing methods is the lack of de facto standard in the NAPS server, as a middleware [1] interoperable with heterogeneous platforms [2]. So if there exists a homogenous naming and addressing convention, then the application develop-ers can easily retrieve data from sensors and control the actuators of different networks. Otherwise the application developers need to spend much time to learn about different pro-tocols and standards. So a higher layer of device naming–addressing mapping should be provided to integrate with legacy systems and different platforms. For the device naming, the convention should contain key elements of meta-data, such as device type and domain information, while for addressing its format allows the granularity of efficient accessibil-ity and addressability to the physical world. Profile services are also needed to aid the application query and system configurations, like device status and presence. Furthermore, sensing tasks are always achieved by a group of devices with similar sensing capabilities, and thus NAPS should provide device group management functionalities, such as to cre-ate, update, read, and delete groups. As we can create groups, only a device group name is needed to share the information. In this way, application development logic is greatly simplified where only a device group name is needed and NAPS handles the internal map-ping. As a middleware, it should extend its usability by providing abundant external inter-faces. IPv4, IPv6, and Domain Name System (DNS) are usually considered as the candidate standard for naming and addressing; however, due to the lack of communication and pro-cessing capabilities of many small and cheap devices (like Radio Frequency Identification (RFID) tags), it is quite challenging to connect every "thing" with an IP. Furthermore, with the increasing amount of end devices, even IPv6's address space may not be enough. On the other hand, industry standards have put much effort in each application domain. EPCglobal [3] uses a 96-bit binary sequence to identify each RFID tag, and the object naming service (ONS) for URL translation. OPC- Unified Architecture [4] defines client-server-based mod-els for industrial production line solutions, where an abstract address space is formed by a mesh topology. In it, each node represents a sensor in the production stage and the edge between two nodes represents the stage-by-stage relationship during the production. As an overall service architecture, European Telecommunications Standards Institute (ETSI) [5] proposed a solution interworking with 3GPP machine-type communication (MTC) stan-dard [6], to support machine-to-machine (M2M) communications when upgrading from traditional cellular networks where each device is with a unique international mobile sub-scriber identity (IMSI) and is IP addressable. Furthermore, as a service layer architecture, it defines a variety of service capabilities (SCs) including a network reachability, addressing, and repository (NRAR) SC. Our goal in this work is to work with any service platforms as a middleware at the back-end data center. Therefore, all these efforts pay attention only to a specific network or application domain, however, not applicable as a common platform managing different technologies and standards.

Figure 9.1 shows an overall architecture from the physical phenomenon all the way up to the data center, considered in this chapter. Devices such as sensors and actuators sense/

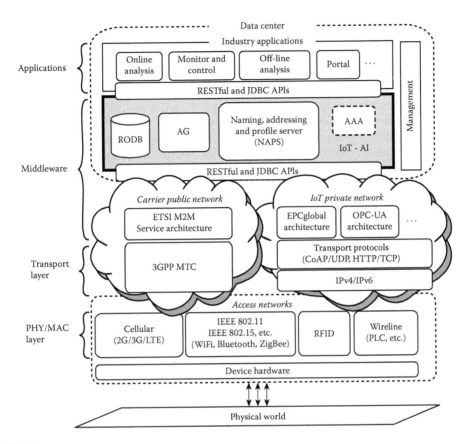

FIGURE 9.1
An overall architecture.

control the physical world, which are interconnected either wirelessly or wired through a variety of access network technologies. Some examples are cellular networks (2G/3G/LTE), IEEE 802.11/802.15 series of standards for Wi-Fi, ZigBee, and Bluetooth, RFID readers and tags, and wireline technologies like power-line communications (PLCs), etc. Usually there exists a gateway interconnecting the access networks and the backbone Internet. Data are then routed either through the carrier public network or IoT private network. For the former, standard like 3GPP MTC is defined to upgrade the existing backbone cellular network to manage M2M devices. For the latter, most service layer architectures, like EPCglobal RFID architecture or OPC-UA client–server model, leverage existing transport layer protocols such as CoAP [7] over UDP, and HTTP over TCP. In the service layer, ETSI M2M service architecture can interwork with 3GPP MTC via interworking function (IWF) that enables seamless integration of M2M SC layer with cellular MTC devices. That is, M2M SCs can invoke and leverage cellular MTC functions to optimize and support better M2M services. Meanwhile, cellular MTC functions can be enhanced for M2M SCs. Companies like InterDigital and Juniper Networks have this kind of solution [8,9].

Toward this end, there is a lack of a common platform interoperable with different platforms to hide this heterogeneity and provide a transparent naming service to applications. The key technical constituents of IoT-AI are application gateway (AG), NAPS and its service registration portal (Portal), and real-time operational database (RODB). AG coordinates the data filtering and processing and controls message delivery based on a uniform device naming and addressing convention in NAPS. The goal is to have applications access devices across different platforms without knowing their languages in detail, but focusing on the development logic only. The position of NAPS extends the functionality comparable to DNS in the Internet, to the profile services such as storage and query. We next present three assumptions of this work. First is service discovery. As the scope of NAPS is a middleware component at the back-end data center to hide the heterogeneous protocols and standards, here we assume that service discovery has already been successfully performed by each platform individually, and stored in our NAPS repository. Examples are service discovery server enhanced from ETSI M2M service architecture by Inter-Digital [8], discovery service set in OPC-UA standard, and protocols like Universal Plug and Play (UPnP) [10]. Second is the authentication, authorization, and accounting (AAA). Although it is not the focus of this work, the design can largely leverage the network security capability (NSEC) SC in ETSI M2M service architecture. It uses a key hierarchy, composed of root key, service key, and application keys. Root key is used to derive service keys through authentication, and key agreement between the device or gateway and the M2M SCs at the M2M Core. The application key, derived from service key, is unique as per M2M application.

Finally, we assume that wireless imperfection like packet errors and interference have been handled by the communication stack of each access networks. Solutions from PHY layer techniques (e.g., antenna techniques, modulation, and coding) and MAC/network layer protocols (e.g., scheduling and routing) are a few examples. Therefore, any wireless issues are completely transparent to the service layer operations, or the NAPS middleware considered in this paper.

EU FP7 project IoT-Architecture (IoT-A) extensively discussed the existing architectures, protocols, and platforms [11]. Besides, Castellani et al. [12] propose a framework to interconnect sensors running 6LoWPAN [13], where IEEE 802.15.4 and IPv6 were considered to connect wireless devices to the Internet. Silverajan and Harju [14] provide an XML schema to encode device profile information including its local name. Web-of-things (WoT [15]) make use of popular Web languages for building applications involving smart things and users.

As for industrial standards, based on DNS, the pure IP solution [16] is favored due to the recent development of IPv6 to connect "things" for IoT. EPCglobal [3] specializes in use of RFID in the information-rich and trading networks, especially for logistics. In it, similar to DNS, an ONS is designed to translate a 96-bit binary sequence to a uniform resource identifier (URI), which directs the query to a (set of) database(s) called EPC information service for information retrieval and update. OPC-UA [4] defines an address space where devices are interconnected to form a mesh topology. The connectivity represents the production line sequence, where the directional edge called "reference" links the next stage of behavior. ETSI M2M service architecture [5] assumes each M2M device is IP addressable. In 3GPP MTC [6], they propose to use the IMSI as the internal identifier for signaling and charging, while providing external identifiers to include domain information under the control of a network operator, and flexibly allocated descriptions as the customer-friendly local identifier. On the other hand, naming and addressing for wireless sensor networks have been extensively investigated [17–20]. The first kind of approaches rely on the efficient address allocation among nodes, where in [21,22] the assigned addresses are reused spatially and represented by variable length of code words. This was later extended by Kronewitter [23] who used the prefix-free Huffman encoding of node addresses based on the energy map, where nodes with little battery life left will have the advantage of a short address and check period. Elson and Estrin [24] proposed the "attribute-based naming." In the work by Heidemann et al. [25], clients use an intentional name to request a service without explicitly listing the end-node that ultimately serves the request. This level of indirection allows applications to seamlessly continue communicating with end-nodes even though the mapping from name to end-node addresses may change during the session. Finally, unique identifier generation problem is investigated by Shen et al. [26], where they designed a correlated lookup scheme with a distributed hash table to improve the performance of identifier management.

However, none of these schemes are motivated from the service layer as part of the middleware at the back-end data center, nor they tackle the fundamental problem of providing a homogenous, both human and machine understandable, and unique naming and addressing convention across different platforms. Furthermore, none of them support profile services and legacy system integration.

9.2 Name Service

With respect to the IoT, a name service helps in the following situation (Figure 9.2). We have read an identifier, regarded as a name, from an RFID tag attached to an object. Then we need to find the corresponding information sources on the Internet. That is, an IoT name service resolves object identifiers to information service addresses. In the main reference architecture by EPCglobal two categories of name services are used. The first is the ONS to locate the item manufacturer and uses—as of today—only EPC Manager and Object Class fields of an Serialized Global Trade Identification (SGTIN) EPC. The second category comprises EPCIS Discovery Services, which shall offer lookup of multiple information sources related to fully serialized EPCs.

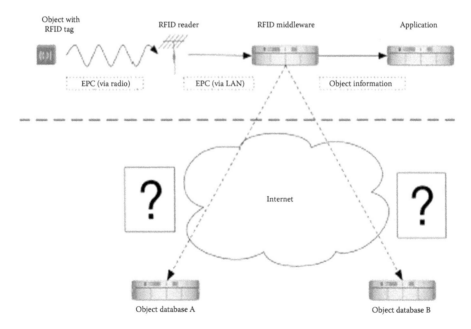

FIGURE 9.2
Function of an IoT name service.

9.2.1 Requirements

In this section we discuss the functional and performance requirements of IoT name services. Functional requirement means the functionality and services that a system should provide. Nonfunctional requirements include the constraints on the system functionality such as performance, quality, safety, etc. The following high-level functional, scalability, performance, and robustness requirements of an IoT name service S can be identified.

Function roles of IOT Name Service are shown in Table 9.1.

9.2.1.1 Functional Requirements for an IoT Name Service

1. *System Membership and Authorization Procedure*: A set of membership definition and authorization procedures for all publishers and clients of name service S shall be provided. Those procedures shall define:
   a. Which publishers shall be authorized to publish information about what kind of object identifiers (OIDs).
   b. Which clients shall be granted authorized access to what object information service (OIS) address information and to which actual object information.
   c. Which parties are allowed or even obliged to insert machine components into S and who may run sub-services of S, for example, security services in the form of Certification Authorities.

   The first and third procedures should be a global convention between all IoT users; the second set of procedures can be delegated to the authorized information providers. In theory, membership could be free to everyone, so that everyone may be able to publish and retrieve information about any object, but in practice there will be constraints by the information provider's economic and security interests.
2. *Flexible OID Support*: S should be flexible in its support for different OID schemes.
3. *Publishing*: An information provider shall be able to input address documents into S for OIDs for which he is authorized to publish information. These documents shall include addresses of OIS servers providing information about objects carrying those OIDs.

TABLE 9.1

IOTNS Function Roles

Functional Role	Stakeholder	Example
Object information service (OIS)	Information provider (Publisher)	Manufacturer EPCIS
OIS resolver	Client	Shop, smart home IT
IOT central node	IOT infrastructure provider	EPCglobal core service
IOTNS node	Node provider	ONS server
IOTNS special node	Node provider	ONS root server
OIS discovery service	Discovery service provider	EPCIS discovery service
Router	Internet service provider	Local or backbone router

4. *Multiple Publishers (Independent)*: Multiple independent but authorized publishers should be able to provide information for an OID by storing corresponding address data in S, without possible mutual interference, like censorship.

5. *Querying*: On input of OID e by a client, S shall output a current list of servers offering information about the object corresponding to e.

6. *Updating*: Authorized publishers shall be able to update the data records they published at will.

7. *Deleting*: Authorized publishers shall be able to delete the data records they published at will. A time-to-live (TTL) value should be provided for each document to indicate old data and to reduce overhead for deletion.

8. *Class-Level Addresses*: If the OID is structured into a class-level and serial-level part, S shall be able to work with partial OIDs at the class-level; for example, a partial SGTIN EPC consisting of EPC Manager and Object Class.

9. *Serial-Level Addresses*: If the OID is structured into a class-level and serial-level part, S should be able to work with fully serialized OIDs, for example a complete SGTIN EPC consisting of EPC Manager, Object Class, and Serial Number (see Figure 9.3). Most EPCs have a similar structure as shown in Figure 9.3.

10. *Object Information* (optional): S should itself be able to store and return (small amounts) of object information about OIDs to reduce query overhead, for example, directly indicating if an object's official lifetime has expired.

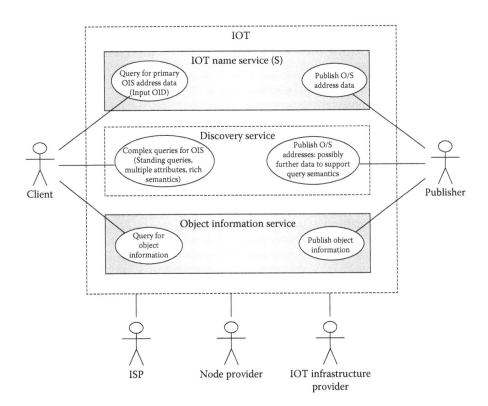

FIGURE 9.3
SGTIN-96 EPC.

9.2.1.1.1 Scalability

The system must be able to work on a global scale. Because it is used for the IOT, it is probable that S—in the long run—must cope with much more traffic than the usage of DNS for URL name resolution performed today.

1. *High Node Count*: S should work with a very large number of participating nodes (servers).
2. *High Client Count*: S should work with a very large number of participating clients.
3. *Scalability to Medium IoT Adoption*: S shall work in scenarios with a medium-level adoption of the IOT across businesses.
4. *Scalability to Large IoT Adoption*: Class-Level Lookups: S should work in scenarios with a high-level adoption of the IOT across business and society, serving class-level queries.
5. *Scalability to Large IoT Adoption*: Serial-Level Lookups: S should work in scenarios with a high-level adoption of the IOT, serving also serial-level queries.

9.2.1.1.2 Performance

The IOT name service S must be able to deliver a performance that is suitable for global use in very heterogeneous applications. This includes:

1. *Fast Update Propagation*: Information changed by authorized information providers should be propagated fast throughout the system, to avoid stale data.
2. *Low Latency*: The waiting time for an answer by S to a query shall be short, below one minute to enable nearly real-time operations.
3. *Ultra-Low Latency* (optional): The waiting time for an answer to a query should be very short, e.g., below a few seconds, to enable real-time or interactive applications with human beings who deem longer waiting times unacceptable.
4. *Acceptable Load* (Average Node): The network, storage, and processing load of an average node (server) of S must not be too high, to guarantee its correct and fast execution of tasks.
5. *Acceptable Load* (Special Nodes, Root): The network, storage, and processing load of all special or root nodes (servers) of S must not be too high, to guarantee their correct and fast execution of tasks.

9.2.1.1.3 Robustness

S should perform reliably in the face of apparently random errors and attacks common on the Internet. Again, most of the high-level requirements are not precise. Depending on particular application scenarios, each high-level requirement should be refined and mapped to several more exact metrics and corresponding tolerance intervals, so that their fulfillment can be verified. A similar refinement process can be described in a mathematically rigorous way for security, especially confidentiality requirements. During additional process iterations, additional time and domain-specific requirements—arising from specific application domains—should be combined and reconciled, and again compared to design options.

9.2.1.2 Security Requirements

Someone who places value into an information asset and wants it to be protected is called a stakeholder of that asset. This definition allows for a generalization of classical security requirements engineering to the multiple stakeholders of multilateral security [27]. In classic security requirements engineering, for example in parts of the Common Criteria, only one stakeholder is considered, i.e., the owner of a target of evaluation (ToE) [28].

In the following, we use the classical triad of protection goals, which a stakeholder may have with respect to an information asset: availability, integrity [29], and confidentiality [30]. We subsume, for example, anonymity under confidentiality of identity, and authenticity under integrity.

- *Availability*: The property of being accessible and usable upon demand by an authorized entity [31]. Alternatively, formulated by avoidance; the prevention of unauthorized withholding of information or resources.

- *Integrity*: The property of safeguarding the accuracy and completeness of assets. That means the prevention of unauthorized modification of information.

- *Confidentiality*: The property that information is not made available or disclosed to unauthorized individuals, entities, or processes. That means the prevention of unauthorized disclosure of information. Extending classical security engineering, the concept of multilateral security emphasizes the importance of taking the security goals of most or ideally all stakeholders into account before designing and building a system, not only of system owners or investors. This becomes especially important for confidentiality requirements of system users.

9.2.1.2.1 Availability

The system S, i.e., the IOTNS (Figure 9.4), should be constructed, and its data should be available to authorized users any time they need to access it. We assume this to be a requirement shared by all stakeholders. In particular, S should offer robustness to targeted (Distributed) Denial-of-Service (DDoS) attacks; the system should avoid single points of failure, and be able to adjust itself to failures of single components (servers or nodes).

9.2.1.2.2 Multipolarity

A specific case of availability concerns the anticipated future role of the IoT as critical IT infrastructure in many countries. Considered as stakeholders of S, those countries will

Header	Filter value	Partition	Company prefix (EPC manager)	Item reference (object class)	Serial number
8 bits	3 bits	3 bits	20–40 bits Total length: 44 bits	4–24 bits	38 bits
00110000 "SGTiN-96"	001 "retail"	101 "24:20 bits"	4012345 (decimal)	734 (dec.)	2 (dec.)

FIGURE 9.4
IOTNS function in context.

have a high interest that no single one of them controls access to S, or could prevent it from working. This requirement will be discussed in detail in the next chapters.

9.2.1.2.3 Integrity

S shall offer data integrity, including authenticity of data origin. All unauthorized changes to the data stored in S should be detectable by a client via means integrated into S. S should also prevent spamming and pharming attacks, which aim to add arbitrary, non-authorized data entries to S. All of those also will be assumed common requirements of all stakeholders.

In addition, there are special cases where data integrity may need to be enforced by system integrity in lower-level design steps; for example, to deliver authentic messages on non existence of records or during the publishing phase, where the system node that is contacted for publishing needs to be authentic. Data integrity should also include a measure to assess the age of data, as well as provide nonrepudiation, i.e., the fact that a provider published exactly these data should be provable to third parties, for example, for auditing or legal purposes.

9.2.1.2.4 Confidentiality

In classical security engineering, confidentiality requirements usually have been considered only for the information provider or server side of an Internet service, such as confidentiality enforced by access control for the data offered by a Web server. This provider perspective also applies to ubiquitous computing environments and the IoT, but is far from complete. Clients of IoT services are also stakeholders whose security requirements need to be accounted for.

Stakeholder confidentiality requirements on the client side of S, however, usually do not only refer to data processed in a system, but also to high-level information (e.g., turnover or lifestyle) inferable from using the system, and multiple entities (persons, organizations, competitors, criminals, the public) from whom that information must be kept confidential (counter-stakeholders).

As an example, consider the use of RFID, IoT, and the name service S in a smart home owned by an individual, Bob Concerned, and on a shop floor (Figure 9.5). Bob Concerned practices a lifestyle he wants to keep confidential from others (Figure 9.5a). These others are the counter-stakeholders of his confidentiality requirements, including neighbors, marketing companies, and governments, or other entities who adopt functional roles in the IoT (Table 9.1). The shop has confidentiality goals that have a similar structure to Bob's goals (Figure 9.5b). For example, the shop produces turnover that it wants to keep confidential from competing shops.

Many of those high-level information assets may be inferable—simply from the observation of queries to S, by using query data analysis and mining from content, location, time, frequency, clusters of queries, and changes over time. For example, lifestyle can be inferred by analyzing which item brands are in regular use at Bob's home and are creating periodic query patterns to the IoT, or how often new and potentially expensive or cheap items are detected by his RFID readers, while EPCs are resolved to retrieve item information for smart home services. Similarly, the shop's turnover can be inferred by observing periodic inventory queries to the IOT, watching for specific brands, items missing, returns, or new arrivals. However, Bob Concerned or the shop may not even be aware of the data traces that they are producing in their smart environments equipped with RFID readers, and across IOT name and information servers, simply by querying and retrieving item-related information. Therefore, it will become very difficult for security engineering

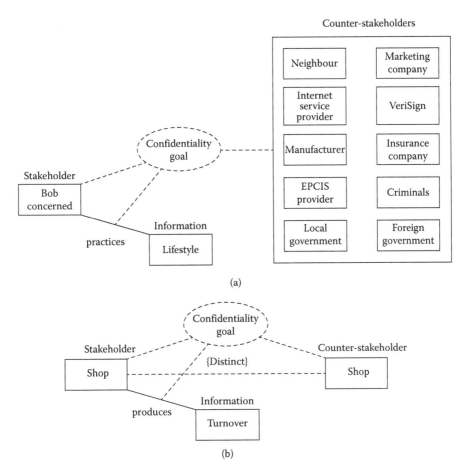

FIGURE 9.5
(a, b) Examples of stakeholders and confidentiality goals.

to state confidentiality requirements for Ubiquitous Computing (UC) in a rigorous way. Not only does the what to protect become harder to specify the more concrete a system becomes in the development process, also the against whom becomes difficult to state precisely for omnipresent, globally connected machines and environments operated by multiple stakeholders, some of which may not even be known in advance. On the other hand, protection against all potential adversaries seems to be impossible to achieve in practice.

Furthermore, mappings between counter-stakeholders and functional system roles are not always clear, in addition to the ever-present possibility of external attacks. Counter-stakeholders and adversaries usually also differ in the degree of background knowledge they have available for data analysis, which could abstractly be described by general conditional probability distributions, but seems nearly impossible to quantify for all adversaries during run-time, or in earlier development stages. To cope with those problems, we can use a pragmatically oriented approach and apply a rule-of-thumb guideline: low-level data in S should be as hard to collect or analyze as possible.

Consequently, while using the IoT, there will be many situations when the EPC belonging to an RFID-tagged item should be regarded as sensitive information—be it in a private context, where people fear to be tracked or have their belongings read by strangers, or in a business context, where product flows constitute valuable business intelligence. The combination of an EPC company identifier and item reference is usually enough to determine the exact kind of object to which it belongs. This information can be used to identify assets of an individual or an organization. If someone happens to wear a rare item or a rare combination of belongings, one could track that person even without knowing the actual serial numbers. In addition to this supply side, there are also many entities that have a certain or at least potential demand for EPC traces, which will be illustrated in the following section.

9.3 Object Naming Service

For ONS, a hierarchical, tree-like architecture has been proposed by EPCglobal. The ONS protocol is identical to the protocol used by the Domain Name System (DNS). The ONS root is the central root of this tree. Further delegation works as in DNS, and information providers itself will deploy authoritative ONS servers—for their EPC ranges—that point to their actual EPCIS. This architecture and protocol choice will have a deep impact on the reliability, security, and privacy of the involved stakeholders and their business processes, especially for information clients, as will be discussed after the technical inheritance of DNS has been described in the next section.

9.3.1 ONS Foundation: DNS

From a technical point of view, ONS is a subsystem of the DNS, and are codified in many requests-for-comments (RFCs). The main design idea of ONS is to first encode the EPC into a syntactically correct domain name, then to use the existing DNS infrastructure to query for additional information. This procedure makes use of the Naming Authority Pointer (NAPTR) DNS record, which is also used with other Internet applications, for example the Session Initiation Protocol (SIP) for Voice-over-IP (VoIP) to map phone numbers into corresponding URIs.

For a discussion of the DNS security heritage to ONS later in this chapter, in the following sections a short summary of the inner workings of DNS is given, discussing names, architecture, and protocol.

9.3.1.1 DNS Names and Architecture

The basic function of the DNS is that of an Internet name service: the resolution of human-memorable, alpha-numerical hostnames into the corresponding purely numerical IP addresses used for datagram routing. At an early stage of the Internet, the Advanced Research Projects Agency Network (ARPANET), name resolution was performed by referring to a flat text file that stored mappings between the hostnames and the IP addresses (hosts file). Obviously, maintaining and synchronizing copies of the hosts file on all computers connected to ARPANET were extremely inefficient. To address this issue, the name resolution protocol was updated to introduce a central distribution of the master hosts file via an online service maintained by the Network Information Center. This architecture worked successfully for about a decade. However, the rapid growth of the Internet rendered this centralized approach impractical. The increasing number of changes introduced to the hosts file and its growing size required hosts to regularly download large volumes of data and often led to propagation of network-wide errors. As a reaction, shortly after deployment of TCP/IP, the new DNS was introduced. This DNS still serves as the foundation of the Internet name resolution system today. A hostname now has a compound structure and consists of a number of labels separated by dots, e.g., www.example. com.—the final dot is often omitted. The labels specify corresponding domains: the empty string next to the rightmost dot corresponds to the root domain, the next label to the left to the top-level domain (TLD), followed by the second-level domain (SLD), and so forth.

The resolution of the hostname into the corresponding IP address is carried out by a tree-like hierarchy of DNS name servers. Each node of the hierarchy consists of DNS name servers that store a list of resource records (RRs) mapping domain names into IP addresses of Internet sites belonging to a zone for which the DNS servers are authoritative. Alternatively, in case of zone delegation, IP addresses of DNS servers located at the lower levels of the hierarchy are returned. The resolution of a hostname is performed by subsequently resolving domains of the hostname from right to left, thereby traversing the hierarchy of the DNS name servers until the corresponding IP address is obtained.

In addition to name-to-IP resolution used by nearly every Internet application today, there are several other established and future uses of DNS, such as inverse queries (IP-to-name), queries for mail server addresses (using MX RRs), DNS use for storing VoIP phone numbers, key distribution, and even for stating communication security requirements. ONS will place additional DNS burden on top of the load created by all those applications.

9.3.1.2 DNS Protocol

The DNS protocol is part of the application layer of the TCP/IP hierarchy. In general, it uses the User Datagram Protocol (UDP) with server port 53 as transport layer protocol for queries and responses. DNS uses the Transmission Control Protocol (TCP) for responses larger than 512 bytes, as well as for higher reliability of zone transfers between DNS servers. An exception is the use of the so-called extension mechanisms for DNS, which allow larger DNS payloads to be transported via UDP, and is important for transferring signatures for DNS security extensions (DNSSEC). To match incoming responses with previous queries, DNS uses a 16-bit query identifier located in the DNS header. In addition, the header carries multiple status bits indicating query or response, authoritative answer, response truncation, and the desire for—respectively, availability of—recursive query tasks for the name server. The actual query or answer DNS RRs, as well as possible additional information, follow after the header in specific section of a DNS packet. To reduce the message size due to the classical 512-byte limit, a compression and pointer scheme is used to avoid the repetition of names, which however increases the parsing complexity for human eye and DNS software, and has been the cause of implementation errors in the past.

In the following, the inner workings of the ONS resolution process and its use of DNS are described.

9.3.2 ONS Resolution Process

The ONS resolution process is described in the work of Mealling [32]. For a schematic view of the communication procedure, see Figure 9.6. After an RFID reader has received an EPC in binary form, it forwards it to some local middleware system. To retrieve the list of relevant EPCIS servers for this particular object, the middleware system converts the EPC to its URI form (e.g., urn:epc:id:sgtin:809453.1734.108265). Then this is handed over to the local ONS resolver, which in turn translates the URI form into a domain name

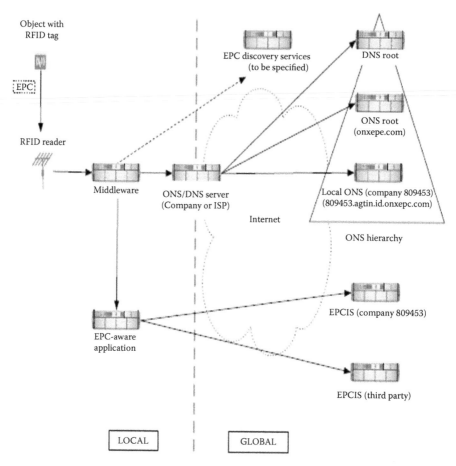

FIGURE 9.6
ONS resolution.

(e.g., 1734.809453.sgtin.id.onsepc.com) by following a well-defined procedure. This name belongs to a subdomain of the domain onsepc.com, which is reserved for ONS use.

The current ONS specification states that the serial part (item level, in the example: 108265) of the EPC, which differentiates between objects of the same kind and brand, should not be encoded as of now, but it leaves room for such a possibility: The ability to specify an ONS query at the serial number level of granularity as well as the architectural and economic impacts of that capability is an open issue that will be addressed in subsequent versions of this document. Its lack of mention here should not be construed as making that behavior legal or illegal. This newly created domain name is now queried for by using the common DNS protocol, possibly involving a recursive query to a local DNS or service provider. DNS server that then queries iteratively across the Internet.

In addition to this primary ONS use of the DNS, note a secondary dependency on the existing DNS hierarchy: The names stored in ONS records and returned by the ONS query process will again have to be resolved into IP addresses—using the standard DNS hierarchy—to receive the IP addresses of the EPCIS servers, cf. (3a, b) in Figure 9.6.

We turn now to the actual discussion of ONS security issues, especially with respect to the DNS heritage, which becomes critical in this new application domain.

9.3.3 ONS Security Analysis

DNS is an old and central Internet service with a long history of security and configuration issues in the protocol itself and in particular implementations. Various vulnerabilities and attacks can be listed by consulting established security sites as CERT, SecurityFocus, and the SANS Institute's top 20 list of internet security vulnerabilities.

A corresponding request-for-comments, RFC 3833 Threat Analysis of the DNS, was published quite late after two decades of DNS use, though many of its security problems have been identified before. Some of the main threats discussed are packet interception, i.e., manipulating IP packets carrying DNS information, query prediction by manipulating the query and answer schemes of the DNS protocol, cache poisoning by injecting manipulated information into DNS caches, betrayal by trusted servers controlled by an attacker, and denial of service, a threat to every Internet service—but DNS itself might be used as an amplifier to attack third parties.

Besides bugs in the code, the fundamental reason for most of these vulnerabilities is the fact that even though DNS is a central and highly exposed service by definition, it has—in its original and widely deployed form—no way of authenticating a client, the server, nor the information that is provided. In addition, DNS uses a clear text protocol, as do most of the early IPs.

These DNS weaknesses directly transfer to ONS. In the following sections, a discussion on ONS availability, integrity, and confidentiality risks is given.

9.3.3.1 ONS Availability

ONS will constitute a service highly exposed to attacks from the Internet, if only due to its necessary widespread accessibility. A particular threat is denial of service (DoS), which abuses system and network resources to make the service unavailable or unusably slow for legitimate users. This could include DDoS attacks overwhelming a particular server or its network connection by issuing countless and intense queries, e.g., by the use of zombie networks, Botnets or the so-called Puppetnets, i.e., hosts controlled by browser-based malware. DoS attacks can also use more sophisticated methods, e.g., targeted exploits that shut down the DNS server software or the operating system. Though distributed, DNS suffers from limited redundancy in practical implementations. Authoritative name servers for any given zone should be redundant according to RFC 1034. Recent studies on real implementations, however, show that for a non-insignificant part of the global name-space this requirement does not hold. Name servers storing the same information for a given zone are often few and not redundantly placed with respect to geographical location and IP subnets, and often reside inside of the same autonomous system (AS). There are many servers that have single distinct routing bottlenecks on paths to reach them—from every place in the world.

The small number of servers for a given zone information, and their limited redundancy creates single points or small areas of failure. Those are also attractive targets for DDoS attacks—not only at the DNS root, which is currently run by fewer than 150 servers and has been attacked with some, but so far moderate, success before. Failure of the root, though, would after some time to account for caching imply failure of the whole system, not only of some of its subtrees. Root and TLD servers, as well as name servers for domains that rise in popularity (flash crowds, for example the famous Slashdot effect), suffer from strong load imbalance induced by the architecture. Omnipresent DNS caching, on the other hand, reduces flexibility and the speed of update propagation. Studies also show the significance of human configuration errors that slow down the resolution process or even cause it to fail. Part of the problem is the complexity of the DNS delegation process, which is based on cooperation across different organizations. Therefore, an integration of the EPCglobal Network with ONS as proposed into core business processes could leave even formerly non-IT related companies dependable on the availability of Internet services. This will most probably increase overall business risk.

9.3.4 Named Data Networking

Shang et al. [33] discussed how named data networking (NDN) addresses the root causes of the various challenges in IoT and can help achieve the IoT vision in a more secure, straightforward, and innovation-friendly manner. The first challenge of IoT is how to enable all different types of digital devices that provide IoT functionality to communicate locally and globally. The second is how to consistently and securely communicate the data associated with the things themselves, once connectivity is achieved.

NDN fundamentally shifts the network communication model from host-centric to data-centric. Instead of sending packets between source and destination devices identified by numeric IP addresses, NDN disseminates named data at the network level, forwarding directly on names that carry application semantics. Moreover, each packet in NDN is secured at the time of production, allowing data replication anywhere in the network and preserving security properties of the data over its lifetime. Amadeo et al. [34] proposed a baseline NDN framework for the support of reliable retrieval of data from different wireless producers which can answer to the same Interest packet (e.g., a monitoring application collecting environmental data from sensors in a target area). The solution is evaluated through simulations in NDN SIM and achieved results show that, by leveraging the concept of excluded field and ad hoc defined schemes for data suppression and collision avoidance, it leads to improved performance in terms of data collection time and network overhead.

The research community is currently exploring cutting-edge approaches to transform the Internet, as we know it today, into a system more capable of and tailored for effective content distribution, according to today's (and tomorrow's) needs. Information-centric networking (ICN) has been recently proposed for this purpose and is inspiring the design of the future Internet architecture. Unlike the IP-address-centric networking of the current Internet, in ICN every piece of content has a unique, persistent, location-independent name, which is directly used by applications for accessing data. This revolutionary paradigm also provides content-based security regardless of the distribution channel and enables in-network data caching. Amadeo et al. [35] discussed the challenges and opportunities in ICN.

9.4 Efficient Naming, Addressing, and Profile Services in IoT Sensory Environments

In this section we discuss the naming, addressing, and profile services in IoT introduced by Li et al. [36]. They introduced a NAPS server to interconnect different platforms in the IoT environment. NAPS serves as the key module at the back-end data center to aid the efficient upstream sensory data collection, content-based data filtering, and matching and downstream efficient control by applications. The main difference between the existing approach and the Li's approach is the introduction of a middleware component. The earlier approaches focused on a specific protocol, while the approach by Liu et al. [36] gives more focus on the middleware component, which serves dynamic application needs, sensors/actuators deployment, and configurations across different platforms.

In the following sections, we discuss the design of NAPS, including its key functionalities, system flows, interfaces, and individual module design. We discuss also a unique device naming and addressing convention and show its applicability to a few widely used standards and protocols. We deal with also an efficient identifier generation scheme.

In this section, our contributions are summarized fourfold. First, we propose a complete and detailed design of NAPS, including its key system flows, interfaces, and individual module designs. Second, we propose a unique device naming and addressing convention interworking with different platforms, and we show its applicability to a few widely used standards and protocols. Third, we propose an efficient identifier generation scheme, not only used during data transportation, but also to facilitate the data filtering and matching. Fourth, we provide CURD operations on device, device type, and device group profiles, in the RESTful design style [37] over HTTP at runtime.

9.4.1 System Flows

To support the high scalability requirement, Li et al. further decomposed AG into four modules, front-end processor (FEP), command controller (CC), back-end processor (BEP), and message queue (MQ). FEP is used for data collection and format transformation, CC for application command parsing and translation from NAPS, BEP for rule-based data filtering and matching, and MQ for publish-subscribe-based topic services [38].

Then, users can identify its bottleneck and scale up/out the corresponding component, for example, to deploy MQs for large number of applications in a cloud. The list of components that interact with NAPS either off-line or at runtime are AG-FEP, AG-CC, AG-BEP, applications, portal, and RODB, as shown in Figure 9.4 for system context and associated interfaces. It is worth noting that the overall IoT-AI platform is only used as an example to demonstrate the system flow of NAPS, whereas its applicability can extend to any external component with similar interfaces.

We next present system flows, service registration and configurations, upstream data collection, and downstream command delivery. In all aspects, an AAA server interacts with NAPS for security authentication and authorization.

9.4.1.1 Device Registration and Configurations

Service discovery is performed at individual platform beneath the data center service layer, and NAPS only provides a set of interfaces to facilitate the device registration, either automatic or off-line. The registered capabilities include the ones offered by devices, device types, and device groups, and thus the repository stores the corresponding profile information. The provided interfaces are based on the RESTful design style, where standard HTTP request/response is used to transport the data. It is worth nothing that before the response is returned to the client, we generate a unique device identifier, or "devID." It contains key elements of the device metadata. This devID generation process is also applicable when device type and device group are registered.

9.4.1.2 Upstream Data Collection

As shown in Figure 9.5a, when the raw data are received from an IoT platform, AG-BEP translates the devID to the corresponding device name (devName) as more friendly to the application. Meanwhile, AG-BEP makes use of devID to perform efficient content-based data filtering and matching. For example, one application configures a topic on MQ to aggregate and average the room humidity data in a smart building environment. Then, as the designed devID contains key elements of device metadata such as associated domain information and device type, it helps to categorize, filter, and select the exact set of raw data from the massive data pool. This can be achieved by masking certain encoded profile parameters. To save the overhead, we can use devID as the unique identifier in the transport, e.g., through the carrier public network and IoT private network as shown in Figure 9.1, while using devName containing a set of human-readable properties for applications.

9.4.1.3 Downstream Command Delivery

Figure 9.5b shows the downstream system flow when control messages are initialized by the application to specific device groups. First, device (group) name is passed to the AG-CC, and the latter retrieves the list of devID(s) from NAPS over the RESTful interface. Then, AG-FEP translates the devID(s) to the corresponding device address(es) from NAPS, in which it specifies how to address the command(s) to the exact device or group of devices. In this way, similar to the functionality of DNS in the Internet, NAPS performs the name-to-address resolution. Here the main aim is to define a thin layer for address resolution, as a uniform convention to unify and cooperate different platforms. In our addressing convention (see Section 4.2), it specifies the way to route the command to the Internet gateway, which is IP addressable, such as the M2M gateway in ETSI M2M service architecture, OPC-UA server, or WiMax base station. In a hierarchy, these gateways further route the command to the next level gateway (e.g., the ZigBee coordinator), which maintains its own addressing mechanism to the device.

9.4.1.4 Application Query

Application developers may be same as, or different from, the device owners. Therefore, their development logic entirely relies on identifying the right set of devices from the physical world, which may eventually belong to different network domains/platforms. Toward this end, as the device, device type, and device group profiles are registered and stored in NAPS repository, we allow search services to retrieve a list of devices and device groups with certain geographical, domain, and device name information. Furthermore, this process can be coupled with downstream command delivery procedure where a retrieved list of device and device group names are used to issue commands to the physical world.

9.4.1.5 Integration with Different IoT Platforms

Nearly all third-party device vendors and platform operators have their own naming mechanism, and it is likely that they also have already developed their own applications. NAPS provides the translation between our uniform naming convention and the legacy naming, offering shared services to different vendors and applications. We thus provide two RESTful interfaces, getDevOldNamebyDevName() and getDevNamebyDevOldName(), where the former is used when the upstream data are received by the application attached with our device name, and thus to be translated to the legacy name, and the latter is used when applications issue commands to the actuators by translating the legacy naming to the proposed one. In this way, existing naming platforms like 3GPP MTC and EPCglobal can be seamlessly integrate with this NAPS middleware.

9.4.2 System Designs and Implementations

Here, we discuss the naming and address convention and the generation of devID.

9.4.2.1 Naming and Addressing Convention

In this section we discuss a novel naming convention for devices and device groups across different platforms in a form of:

dev://domain-series/devtype/legacy-name,

grp://domain-series/target/policy/grp-name,

where the prefix distinguishes its category (as for devices or device groups), followed by a series of device domain information. The device domain is organized in a tree structure, written in the above naming convention back trace from the leaf to the root node of the domain tree. After the domain series, for devices we use the device type information to further categorize all devices associated with a domain node on the leaf of the tree, and finally the legacy naming (e.g., serial number) from the production phase. Meanwhile, device groups use the monitoring target (e.g., the room temperature) and grouping policy for detailed classification.

Device domain information refers to either their deployed geographical information, or logical organization of these devices. In NAPS, we can store multiple device domains, and each domain indexed r is a tree structure with depth d_r and width of each level as $\{w_r^i, \forall i = 1, ..., d_r\}$. In other words, domain r is composed of total $N_r = \sum_{i=1}^{d_r} w_r^i$ domain nodes. We call the "partition" of a domain tree by parameters $\left(d_r, \{w_r^i\}\right)$ as the "domain rule," and the corresponding data structure representation as "rule assignment." For example, the rule assignment parameters are the number of bits to store the domain nodes in each level of the tree. The metadata of each node include the name, its parents and children, and other properties (see Figure 9.6). In practice, project managers of an IoT system will carefully plan the device deployment at a site of interest, where the first step is to plan an overall device domain, for example, in a smart building environment, how many temperature and humidity sensors of what device type should be deployed at which location, and this deployment stage eventually specifies a domain tree structure, and is stored in NAPS repository.

As for device addressing convention, we propose to use the format:

$$\underbrace{address-1@protocol-1/.../address-n@protocol-n}_{non-IP\ networks}/IP-address$$

to accommodate heterogenous protocols and standards in use across different platforms. To allow the granularity of addressing the device in a hierarchy, we repeat the element "address@protocol" one after the other from the device to the Internet gateway, which is IP addressable (as the last part of the convention).

TABLE 9.2

Bit allocation for devID

ID Category	Device Type	Control Flag	r/w Flag	Mobile Flag	Domain Rule	Domain Series	Sequence
5 bits	6 bits	1 bit	2 bits	1 bit	5 bits	34 bits	10 bits

9.4.2.2 Generating the devID

As mentioned earlier, when device profile information is registered either manually or automatically from each IoT platform, a devID is automatically generated. We use a 64-bit-long integer. It is used when the data and control messages transfer between the access network, carrier public network, and IoT private network, to save the communication overhead over any character-based naming. However, one cannot randomly generate this sequence to avoid potential collision when a cluster of NAPS are deployed in a cloud environment. Meanwhile, the allocation of these 64 bits should have to support other components. For instance, AG will deploy its policy-based rule engine like MQ topic according to the device domain information, where one example is that applications may efficiently query all devices with mobility for their connectivity and presence at runtime by simply masking a portion of this devID. Toward this end, we propose the following design, as shown in Table 9.2.

The category field identifies the type of sequence in the database (for devID this field equals to 1), followed by the device type, controllable flag, read/write access flag, and mobility indicator. The last part is a series of domain information, starting from the domain rule sequence r and back trace from the leaf to the root domain nodes of that domain rule.

9.5 Conclusions

In this chapter, we briefly discussed the role of naming service and the high-level requirements of the name service. Also we discussed NAPS, which is a middleware to support device naming, application addressing, and profile storage and lookup services in IoT environments. Earlier approaches were only focusing on specific standard/protocol, but the NAPS works in any platform. Here we briefly discussed its complete design including its functionalities, system flows, interfaces, etc.

References

1. K. Gama, L. Touseau, D. Donsez. Combining heterogeneous service technologies for building an internet of things middleware. *Computer Communications* 2012; 35(4): 405–417.
2. L. Roalter, M. Kranz, A. Moller. A middleware for intelligent environments and the internet of things. In: *ACM UIC'10*, Springer, Berlin, 2010.
3. EPCglobal. www.epcglobalinc.org/ (accessed January 3, 2017).
4. OPC Unified Architecture (UA). www.opcfoundation.org/ua/ (accessed January 5, 2017).

5. ETSI. Machine-to-Machine Communications (M2M); FunctionalArchitecture. Draft. ETSI TS 102 690 v0.10.4 (2011-01).
6. Service Requirements for Machine-Type Communications, 3GPP TS 22.368 v11.0.0.
7. CoAP. http://tools.ietf.org/html/draft-ietf-core-coap-04 (accessed January 5, 2017).
8. Standardized Machine-to-Machine (M2M) Software Development Platform, White Paper, InterDigital, Inc.
9. Machine-to-Machine (M2M)—The Rise of the Machines, White Paper, Juniper Networks, Inc.
10. UPnP. http://www.upnp.org (accessed January 4, 2017).
11. Internet-of-things architecture (IOT-A) project deliverable d3.1—Initial m2m api analysis.
12. A.P. Castellani, N. Bui, P. Casari, M. Rossi, Z. Shelby, M. Zorzi. Architecture and protocols for the internet of things: A case study. In: *IEEE PerCom Workshops'10*, pp. 678–683, 2010.
13. 6LoWPAN. http://datatracker.ietf.org/wg/6lowpan/charter/ (accessed January 4, 2017).
14. B. Silverajan, J. Harju. Developing network software and communications protocols towards the internet of things. In: *IEEE COMSWARE '09*, pp. 9:1–9:8, 2009.
15. D. Guinard, V. Trifa, E. Wilde. A resource oriented architecture for the web of things. In: *IEEE IoT'10*, 2010.
16. N. Kong, N. Crespi, G. Lee, J. Park. Internet-draft: The internet of things—Concept and problem statement, *IEEE*, 2010.
17. M.Y. Uddin, M.M. Akbar. Addressing techniques in wireless sensor networks: A short survey. In: *IEEE ICECE'06*, pp. 581–584, 2006.
18. M. Ali, Z. Uzmi. An energy-efficient node address naming scheme for wireless sensor networks. In: *IEEE INCC'04*, pp. 25–30, 2004.
19. F. Ye, R. Pan. A survey of addressing algorithms for wireless sensor networks. In: *IEEE WiCom'09*, pp. 1–7, 2009.
20. Z. Du, C. Zhang, Y. Su, D. Qian, Y. Liu. Two-tier dynamic address assignment in wireless sensor networks. In: *IEEE TENCON'09*, Singapore. pp. 1–6, 2009.
21. C. Schurgers, G. Kulkarni, M.B. Srivastava. Distributed assignment of encoded mac addresses in sensor networks. In *ACM MobiHoc'01*, New York, pp. 295–298, 2001.
22. R. Fonseca, S. Ratnasamy, J. Zhao, C.T. Ee, D. Culler, S. Shenker, et al. Beacon vector routing: Scalable point-to-point routing in wireless sensornets. In: *USENIX NSDI'05*, New York, 2005.
23. F.D. Kronewitter. Dynamic huffman addressing in wireless sensor networks based on the energy map. In: *IEEE MILCOM'08*, pp. 1–6, 2008.
24. J. Elson, D. Estrin. Random, ephemeral transaction identifiers in dynamic sensor networks. In: *IEEEE ICDCS'01*, pp. 459–468, 2001.
25. J. Heidemann, F. Silva, C. Intanagonwiwat, R. Govindan, D. Estrin, D. Ganesan. Building efficient wireless sensor networks with low-level naming. In: *ACM Symposium on Operating Systems Principles (SOSP'01)*, pp. 146–159, 2001.
26. Q. Shen, Y. Liu, Z. Zhao, S. Ci, H. Tang. Distributed hash table based id management optimization for internet of things. In: *IEEE IWCMC'10*, pp. 686–690, 2010.
27. N. Borisov. Anonymity in structured peer-to-peer networks. PhD thesis, UC, Berkeley, CA, 2005.
28. N. Borisov, J. Waddle. *Anonymity in structured peer-to-peer networks*. Technical Report UCB/CSD-05-1390. EECS Department, UC, Berkeley, CA, 2005.
29. BRIDGE. *BRIDGE WP02—Requirements document of serial level lookup service for various industries*, August 2007. http://www.bridge-project.eu/ (accessed January 6, 2017).
30. E. Brickell, J. Camenisch, L. Chen. Direct anonymous attestation. In: *Proceedings of the 11th ACM Conference on Computer and Communications Security (CCS '04)*, pp. 132–145, ACM Press, 2004.
31. H.J. Bullinger, M. ten Hompel, editors. *Internet der Dinge.* Springer-Verlag, Berlin, 2007.
32. M. Mealling. EPCglobal Object Naming Service (ONS) 1.0, 2005. http://www.epcglobalinc.org/standards/ (accessed January 8, 2017).
33. W. Shang, A. Bannis, T. Liang, Z. Wang, Y. di Yu, A. Afanasyev, et al. Named data networking of things. In: *Proceedings of the 1st IEEE International Conference on Internet-of-Things Design and Implementation*, April 4–8, Berlin, Germany, 2015.

34. M. Amadeo, C. Campolo, A. Molinaro. Multi-source data retrieval in IoT via named data networking. In: *Proceedings of the 1st ACM Conference on Information-Centric Networking*, pp. 67–76, Paris, France, 2014.

35. M. Amadeo, C. Campolo, J. Quevedo. Information-centric networking for the internet of things: Challenges and opportunities. *IEEE Network* 2016; 30(2): 92–100.

36. C.H. Liu, B. Yang, T. Liu. Efficient naming, addressing and profile services in internet-of-things sensory environments. *Ad Hoc Network* 2013; 18: 85–101.

37. R.T. Fielding. Architectural styles and the design of network-based software architectures, Ph.D. Thesis, Dept. of Information and Computer Science, University of California, Irvine, 2000.

38. IBM Websphere MQ. www.ibm.com/software/integration/wmq/ (accessed January 5, 2017).

10

Review on Communication Security Issues in IoT Medical Devices

Somasundaram R. and Mythili Thirugnanam

VIT University

Vellore, India

CONTENTS

10.1 Introduction

Internet of Things (IoT) is an emerging concept for the sensor-enabled things with internet protocol (IP) address. It can connect with the Internet and collect sensor data, analyze those data, and automatically take a decision. IoT is a network of devices (things) connected to improve the performance of day-to-day life and also incorporates different kinds of resource-constrained devices like implantable (RFID tags) (Fan et al. 2009), wearables (smart watches), and external devices (smart phones, thermostats, smart fridges). Medical research surveys indicate that about 80% of the people older than 65 years suffer from at least one chronic disease. Many aged people have difficulty in taking care of themselves (Dohr et al. 2010; Weinstein 2005). The report "IoT Healthcare Market of Components, Application, End-User—Global Forecast to 2020" (Anonymous 2016a) says that worldwide the IoT medical information market was worth $32.47 billion in 2015. Also, this market value is estimated to increase by $163.24 billion in 2020. The reason for this huge expectation is IoT health care's timely treatment and smart communication between the doctor and the patient. By using IoT medical devices, a doctor can remotely track a patient's health, the working of biomedical devices, and the treatment procedure for the patient. This real-time monitoring of the patient's physiological parameters are useful to prevent health problems in the initial stage. On the contrary, these medical devices are vulnerable to wireless attack when they are associated with Internet through IPv6, in spite of the devices being secure with cryptographic mechanism. When these devices are compromised by attackers, sensitive medical data will be exposed, which in turn threatens the patient's life. In case, security breaches are not addressed then it is difficult for the technology development and lead to significant financial losses. From the security perspective, it is a challenging task to develop a secure protocol for devices with low battery power, low processing capability, and low memory capacity. Therefore, this chapter focuses on reviewing security challenges in IoT medical devices during communication across the devices. The conclusion of the review analysis will provide the limitations and future scope of the IoT health care.

The chapter is encapsulated as follows: Section 2 describes the study on existing security issues in IoT communication protocol and solutions; Section 3 elaborates recent incidents in medical device security; Section 4 focuses on technology used for securing IoT; Recommended positive measures for existing security issues are discussed in Section 5; and Section 6 concludes with limitations and scope for future work.

10.2 Background

IoT is the buzzword for the wireless devices (things) with electronic sensor-based networking. A wide variety of cheap sensors (wearable, implanted, and environmental) have the possibility to create IoT health care. Also, these resource-constrained medical devices and insecure communication protocols in IoT become vulnerable to numerous security attacks. Therefore, this section concentrates on the study of the secure authentication scheme, protocol standardization, and optimizing communication security of medical devices (Olivier et al. 2015).

10.2.1 Highlights on Secure Authentication Scheme over IoT Communication

Moosavi et al. (2014) developed a secure mutual authentication scheme for a radio frequency identification (RFID) implant system. This system relies on elliptic curve cryptography (ECC) and the D-Quark lightweight hash technique. Performance of this work is quite desirable in order to implement resource-constrained medical devices. Raza et al. (2013) evaluated the system to find out intrusion in IoT network called SVELTE. In order to evaluate SVELTE, Raza tested it with routing attacks such as spoofing, sinkhole, and selective-forwarding. SVELTE was implemented in the Contiki OS. It detects attacks in simulated scenarios. Nieminen et al. (2014) proposed an anonymous verification scheme, which can defend some prominent properties, for example, sensor obscurity, cloning issue, and sensor untraceable. This work uncovered that the proposed authentication scheme will be appropriate in numerous IoT applications where the protection of the sensor development is assured. Qiu and Ma (2015) and Gope and Hwang (2015) proposed an enhanced authentication scheme for 6LoWPAN in machine-to-machine (M2M) communications. Rahman and Shah (2016) concentrated on a basic part of IoT connected with internet protocols. In IoT communication, constrained application protocols (CoAP) bind datagram transport layer security (DTLS) as the security operator. In spite of this, there are some areas where DTLS is as yet missing and can be considered a threat to the protocol.

The work carried out at present is focused only on the problem of secure authentication. Past research in this field was focused on security attacks. In-depth, major challenges in medical devices are protocol standardization and optimizing communication security of medical devices. The standardization and optimization of IoT protocol are discussed in the following sections.

10.2.2 Highlights on Standardization and Optimization of IoT Protocol

Sheng et al. (2013) conducted a survey on the Internet Engineering Task Force (IETF) protocol suite with regard to the IoT. Considering each layer in the IETF protocol suite, the author argued that the technical challenges such as constrained protocols in IoT and research opportunities still have more scope. Sye Loong Keoh et al. (2014) elaborated about the efforts in the IETF standardization IoT system. The author reviewed the communication security solutions for IoT, particularly security protocols, to be utilized as part of the CoAP.

Al-kashoash et al. (2016) proposed a new routing metric in routing protocol (RPL) which reduces the number of packet drops when congestion occurs. Additionally, Al-kashoash proposed a new RPL objective function called congestion-aware objective function (CA-OF). This function powerfully performs when congestion occurs by low congestion route selection technique. Raza et al. (2012) proposed 6LoWPAN header compression for DTLS. This compression technique considerably decreases the amount of extra security bits. Based on the review, the protocol constraints such as header compression, buffer overflow, routing selection, congestion, less throughput, lost packets, energy consumption, and packet delivery issues are identified.

10.2.2.1 Standards

Standards are essential certification for IoT medical devices and communication protocols. Many standards associated with medical devices are discussed in Table 10.1. Standards provide concrete support to the good security of the medical devices.

TABLE 10.1

International Safety Standards for Health Softwares

Standards	Objective
ISO/IEC 27032:2012	Provides guidelines for cyber security standards. It includes addressing cyber security problems associated with vulnerability in cyber security.
IEC 62304:2006	Provides guidance for software used in medical devices. It incorporates software requirements and life cycle management. This particular standard is presently under revision and management with ISO 82304.
IEC/ISO CD 82304	Provides general requirements for product safety and is a standard for the safety of health software, and an evolution of IEC 62304. This standard also provides safety requirements of health software devices. Two standards 82304 and 62304 concentrate on the work of device design and software evaluation.
ISO/IEC 80001	This standardization provides elaborate support to risk management in the software development for IoT medical devices.
ISO/DTR 80002–2	This standardization provides embedded software validation in terms of directed process in the medical device software view.
IEC/TR 80002–1:2009	This standard provides the risk management consultant advice to satisfy the requirements of ISO 14971 and is used as the principal standard for risk management regulation. Also provides guidance on the application of ISO 14971 to medical device software.
IEC/TR 80002–3:2014	This standard gives the report of the software life cycle methods and the associated safety class definitions, derived from IEC 62304.This also provides process reference model of medical device software life cycle processes (IEC 62304).

These standards provide fundamental protection to use medical devices in the IoT environment. These standards will support robust security solutions for both manufactures and software developers. Similarly, security in the device communication relies on standardization of different IoT protocols (Figure 10.1).

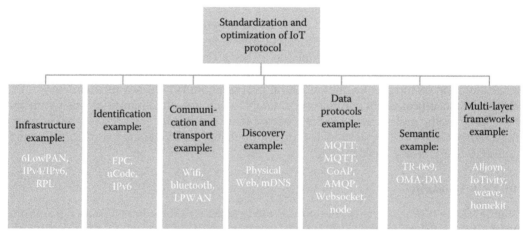

FIGURE 10.1

Different types of IoT protocols.

10.2.2.2 Infrastructure

Infrastructure-based IoT protocols are the basic blocks of connecting devices and routing the data packets.

1. **6LoWPAN**

   Expansion of 6LoWPAN is IPv6 over low power, wireless personal area networks. In IEEE802.15.4, links of 6LoWPAN perform as an alteration layer on behalf of the IPv6. Sensor usage and communication capability in wireless network is provided by EEE 802.15.4 standard. 6LoWPAN protocol is capable of transferring 250 kilo bytes and 2.4GHz is the operating frequency.

2. **IPv4/IPv6**

   Usage of IPv6 with the existing IPv4 over transition technology hinders security management. Weak IPv6 security strategies are an immediate consequence of the present shortage in IPv6. To take care of this issue, the IP network threat ontology presented to gather and analyze different network security issues on IPv4 and IPv6 are discussed by Tseng et al. (2014).

3. **RPL**

   Routing protocol(RPL) is used to overcome the issues of RFC5867, RFC5826,RFC5673 and RFC5548 standards. Winter and Thubert 2012 were examined RPL in low control and lossy networks(LLNS) to resolve the inefficiency routing, poor memory usage and lack of durability in battery capacity in network node).

10.2.2.3 Identification

Identification-based IoT protocols are an interface to identify the tag/device using a reader.

1. **EPC**

   EPC air interface protocol characterizes the physical and coherent prerequisites for an RFID framework. This protocol is capable of to work with in the range of 860 MHz to 960 MHz ultra high frequency (UHF). In the last 10 years, EPC Generation 2 has perceived itself as the standard for UHF executions over numerous portions, and is at the heart of more RFID usage (Anonymous 2016d).

2. **uCode**

   ucode [UCODE] is an identifier to be stored in many types of tags. The embedding of ucode in many types of tags like RFID tags, infrared markers, optical code, and even sound source is also identified by the Ubiquitous ID Center. The length of ucode is 128 bits; these details of ucode are stated by Ishikawa (2012).

10.2.2.4 Discovery

Discovery-based IoT protocols are used to search for a set of factors in devices to communicate.

1. **Physical Web**

   The Physical Web allows to view a set of URLS that are communicated by any devices which is surrounded by bluetooth low energy (BLE) signal area. Bluetooth beacons can be used in hospitals to make sure workers maintain certain standards. For instance, a beacon may be deployed at a hand sanitizer dispenser in a hospital; the beacons can help make sure employees are using the station often. Some of the beacon protocols are iBeacon, AltBeacon, URIBeacon, and Eddystone.

2. **Multicast Domain Name System**

   Cheshire and Krochmal (2013) explained about Multicast DNS (mDNS). mDNS is supports DNS operations on local network in order to improve the accessibility to establish connection in any traditional unicast DNS server. The essential advantages of mDNS names are that:

   a. It requires no configuration to set them up.
   b. It works when no infrastructure is available.
   c. It works while infrastructure fails.

10.2.2.5 Data Protocols

Data protocols support to ensure successful data transmission between devices.

1. **Message Queuing Telemetry Transport**

   Message Queuing Telemetry Transport (MQTT) is an ISO standard (ISO/IEC PRF 20922) publish-subscribe-based lightweight messaging protocol. MQTT is helpful for associations with remote areas where a little code footmark is required and arranged transfer speed is from an optimistic standpoint.

2. **Message Queuing Telemetry Transport-Sensor Network**

   An open and lightweight subscribe protocol designed, especially for mobile applications. MQTT-SN implements Mosquitto and IBM message sight applications..

3. **Constrained Application Protocol**

   Constrained Application Protocol (CoAP) is proposed for use in resource-constrained devices. CoAP is an application layer protocol. RESTful protocol design minimization, low header overhead, uniform resource identifier and content-type support, and support for the discovery of resources are optimization features for CoAP.

4. **Advanced Message Queuing Protocol**

   Advanced Message Queuing Protocol (AMQP) is an open standard application layer protocol. The important features of AMQP are queuing, message orientation, routing, privacy, and security.

10.2.2.6 Communication/Transport Layer

The main purpose of communication and transport layer IoT protocols is to connect and transmit data from one device to another device successfully.

1. **Wireless Hart**

   Wireless HART technology gives a concrete wireless network protocol for the complete range of control and process management applications.

2. **Near-Field Communication**

Near-field communication (NFC) follows the standard ISO/IEC 18092:2004, using the frequency of 13.56 MHz with 424 kbps.

3. **ANT**

ANT (Network) is a proprietary Wireless Sensor Networking technology highlighting a wireless communications protocol stack. Which empowers frequency for Industrial, Scientific and Medical is 2.4 GHz. sharing the Radio Frequency spectrum to communicate by creating concrete rules for signaling, data representation, authentication and error detection. ANT secured with AES-128 and 64-bit key.

4. **Bluetooth**

Bluetooth uses a frequency hopping technique, and the working frequency of Bluetooth is 2.4 GHz. Data rate is 3 Mbps and coverage area is 100 m.

5. **ZigBee**

The ZigBee protocol utilizes the 802.15.4 standard and works in the frequency range of 2.4 GHz. The transmission capacity of ZigBee is 250 Kbps. ZigBee can cover a distance of 200 m and accommodate 1024 nodes in the network. AES 128 bit encryption is used to secure ZigBee.

6. **WiMax**

It depends on the standard IEEE 802.16 developed for wireless metropolitan area networks. Coverage range of WiMax is 50 km, and frequency range is 2.5 GHz to 5.8 GHz. WiMax can transfer data in 40 Mbps.

7. **Multi-layer Frameworks**

Alljoyn is an open source software system that provides an environment for distributed applications running across different device classes with an emphasis on mobility, security, and dynamic configuration. IoTivity is a similar software like Alljoyn with same functionality which was developed by Linux foundation, and supported by the OIC. IEEE P2413, is the standard for an architectural framework for the IoT.

8. **IPSO Application Framework**

This framework designed with rest interface, that helps to connect the smart object in it is coverage in the easiest way.

9. **OMA LightweightM2M v1.0**

LightweightM2M is intended to build a quick deployable client server intended to give machine-to-machine benefit. LightweightM2M is a device administration protocol, but is not limited to device administration, LightweightM2M ought to be capable of exchanging application information.

10. **Weave**

Weave is an IoT device communications stand that empowers device setup between mobile and IoT device, and IoT device and the cloud environment.

10.3 Recent Incidents in Medical Device Security

The year 2016 appeared to be mostly free of industry grasping electronic Protected Health Information (ePHI) breaches. However, database containing over 9 million health insurance records appeared for sale on a darknet peer-to-peer marketplace. Frauds that merit a mention are: potential access and theft of over 2.2 million patient data files from the cancer treatment supplier, 21st Century Oncology; over 600,000 patient ePHI records compromised through stolen laptops; and an estimated 265,000 patient records illegally hacked via malware at Bizmatics, the electronic medical records management company. Up to 2016, 142 health care breaches were reported at the Department of Health and Human Services. Similarly, in 2015, 143 health care records were stolen (Anonymous 2016e; Maia et al. 2014).

10.3.1 Various Security Incidents Occurred in the Wireless Medical Devices

Wireless medical device usage in hospitals and other medical application areas have become common place. In this digital era, medical devices and RFID tags connected with Internet escalate new security vulnerabilities. In this section, we discuss the various incidents that have occurred with wireless medical devices in recent times. In the United States of America, there was a recommendation from Food and Drug Administration (FDA) (Alexandra 2015) that medical device manufacturers and health care facilities should ensure security against different kinds of attacks. The major reason behind these recommendations is the security vulnerability presented when using insulin pumps in the medical field. Insulin pumps are used to control and monitor the insulin level in patients. In order to operate the insulin pump, the doctor uses wireless devices like smart phone, smart watches, and personal digital assistant (PDAs). In case, an attacker observed the serial number of an insulin pump, then it is possible that the attacker can hack the control of the device from a few feet away. This scenario becomes worse when these kinds of devices are connected to the Internet. A few years back, Barnaby Jack a security researcher from McAfee (Dan 2011) hacked Medtronic insulin pumps on 25 October 2011 using a special software and antenna designed by Jack. This permitted him to find and seize overall control of any device around 300 feet of the antenna without knowing the serial number. In this context, the FDA has been working closely with other government organizations and device manufacturers to mitigate vulnerabilities and incidents. These incidents help create awareness in the secure usage of medical devices. In the next section, recent case studies are discussed.

Hackers are targeting the health care industry, especially when compared with the industrial, banking, and retail sectors. In health care, investing for security is negligible, and as the value of individual protected health information in the black market continues to increase, even the largest companies in the health care industry have become a victim to medical data breaches. The following case studies are examples of security data breaches in the health care industry, as a consequence of which millions of private records have been compromised (Anonymous 2016e; Rodika 2015).

Table 10.2 shows the recent health care data breaches that occurred in the United States; these breaches highlight the importance of ensuring the security of medical device usage in health care IoT. In the next section, different technologies used for security IoT are discussed.

10.4 Technology Used for Securing IoT

Simulation is a cost-sparing technique for creating and overseeing system network. Before deploying into the real world, users can recognize the basic performance of network protocols,

TABLE 10.2

Recent Health Care Data Breaches

Data Breach	Year	Consequences
Hackers breach Mass General vendor and compromise 4, **300 health care records**	2016	The Mass General Hospital data hacking of 4,300 patient dental records, medical id numbers, social security numbers, and other identifying information was stolen and/or compromised. This breach demonstrated that even utilizing the services of a third-party vendor hired to assist with management of patient data information leaves the organization open to data breaches and theft of patient information. Mass General Hospital (MGH) contracted with Patterson Dental Supply, Inc. (PDSI), specifically for the purpose of safely and securely managing their patients' data (Akanksha 2016).
US vice-president on hacking into his pacemaker	2013	Former US vice-president Dick Cheney told the truth in a meeting to CBS' "an hour," uncovering that when he had an implanted device to manage his heart pulse in 2007, he had his specialists turned off its remote capacities to protect against possible assassination attempts (Dan 2013).
Anthem health care industry affected by hacking, almost 80 million records were disclosed	2014	Anthem was the victim of the largest data breach in the health care industry. In the cyber attack that occurred in December 2014, Anthem found that hackers might have stolen the names, social security numbers, addresses, income data, and health care identification numbers of nearly 80 million customers. In addition, and perhaps equally as concerning, was the fact that Anthem believed, but could not confirm, that medical records or credit cards of customers were compromised (Charlie 2015).
Banner Health identifies cyber-attack	2016	A data breach at Banner Health compromised at least 3.7 million people. Compromised information included the names of patients and their physicians, Social Security numbers, and health insurance information. The first times ever, information gathered from purchases made at vending machines, including cell phone and payment data, was also compromised. Banner Health is yet to determine how hackers were able to infiltrate the organization's servers and computer systems (Anonymous 2016c).
Premera Blue Cross reveals cyber-attack that affected 11 million customers	2015	In 2015, over 11 million customer data records were compromised as a result of illicit access to Premera Blue Cross' networks by an unknown hacker. While information compromised was very similar to the Anthem data breach, Premera announced that this data breach might have also compromised customers' banking information and detailed insurance claims of customers dating as far back as 2002 (Huddleston 2015).
Hospital network hacked and nearly 4.5 million records compromised	2014	Community health care systems in the United States which operates 206 hospitals, fell victim to a cyber attack resulting from hackers exploiting heart bleed, a known SSL vulnerability. As a result, 4.5 million patients had their names, date of birth, and social security numbers potentially stolen in a cyber attack that may have been connected in some way to the Anthem data breach (Jose 2014).

and test the properties that are probably going to work. QualNet creates a platform for designing a protocol, picturing network scenarios, and analyzing the output. To simulate and analyze IoT network, there is another lightweight operating system called ContikiOS which gives Cooja network simulator. Cooja permits vast and small networks of Contiki motes to be simulated. Also, MONOSEK and TinyOS platforms are used for arranging, testing, and practicing a tool that "mimics" the performance of a real transaction between network nodes.

10.4.1 Monosek

Monosek can effectively utilize top of the line network processing cards like Netronome's NFE-3240 which has 40 smaller scale engines. Packet analysis and packet inspection are done by Monosek software. Monosek will be connected with either parallel network or serial network. It can adopt either local area network or wireless network. Monosek is also useful to understand various protocol traffic patterns. Packets are taken care with suitable flags (Anonymous 2006).

10.4.2 Qualnet

QualNet is a communication simulation platform. QualNet provides access to the users. Using Qualnet, an user can develop a new protocol model as well as analyze and improve the model. Qualnet also gives access to design large wireless and wired networks by utilizing the user-designed models (Anonymous 2016b).

10.4.3 TinyOS

TinyOS is an open-source operating system specially designed for energy-constrained wireless devices. Particularly utilized as a part of sensor system and personal area networks. It supports a complete 6lowpan/RPL IPv6 stack. Average download of TinyOs is 35,000 per year (Anonymous 2013).

10.4.4 Contiki versus TinyOS

ContikiOS is an open-source operating system for IoT and WSN. Performance of Contiki is good as it fulfills the requirements like efficiently using resource-constrained hardware platform, concurrency of the operating system, flexibility among different applications, and energy conservation compared to TinyOS. Comparison between Contiki and Tiny OS is shown in Table 10.3.

Contiki and TinyOS are both operating systems (OSs) that run on microcontrollers with very restricted resources. In any case, because of the higher complexity of the Contiki operating system kernel, TinyOS for the most part brings low-resource requirements. Both OSs are adaptable to handle applications.

ContikiSec and TinySec tools are provide the service of security during the node communication in both operating systems. ContikiSec is a connection layer security answer for Contiki OS. Thus, TinySec is the security layer for TinyOS. Design of ContikiSec is incredibly impacted by TinySec. ContikiSec gives secrecy, trustworthiness, and verification. ContikiSec has three methods of operation: ContikiSec-Enc, ContikiSec-Auth, and ContikiSec-AE. ContikiSec-Enc gives just secrecy; however, ContikiSec-Auth gives integrity and validation. ContikiSec-AE gives secrecy, confirmation, and integrity. Like TinySec, classification is accomplished utilizing block cipher as a part of CBCCS mode.

TABLE 10.3

Comparison between ContikiOS and TinyOS

	Contiki	TinyOS
Kernel architecture	Modular	Monolithic
programming model	Proto threads and event driven	Primarily event-driven support for TOS threads has been added
CPU scheduling	Current events will be dismissed when higher priority interrupts occur.	FIFO
Resource sharing	Serialized access	Virtualization and completion of events
Memory management and protection	Dynamic memory management no process address space protection	Static memory management with memory protection
File system support	Coffee file system	Single-level file system
Communication protocol	uIP and Rime	Active message
Communication security	ContikiSec	TinySec
Simulator	Cooja	TOSSIM
Programming language	C	nesC

Contiki utilizes AES rather than Skipjack, and ends up being more secure, despite the fact that it calls for more calculation (Kamalov 2009).

Overall, comparison of both the OSs shows ContikiOS is more compatible to use in resource-constrained devices, using secure cryptographic mechanism, Flexibility and easy updating of applications in ContiSec is more preferable for testing secure communication protocols. In the next section, solutions and recommendations of existing communication security issues are discussed.

10.5 Solutions and Recommendations

This section focuses on providing solutions based on public key cryptography to communication security issues, especially authentication issue between IoT medical devices.

10.5.1 Asymmetric Key Cryptography

Asymmetric Key Cryptographic algorithms like RSA, ECC, and DSA are more challenging to implement in wireless sensor networks because of less processing power and constrained memory usage. Compared to other algorithms, ECC is most suitable for IoT environments (Eisenbarth et al. 2007). Comparatively, ECC uses less processing power and storage than other algorithms (Lopez 2006). Explanation and comparison of RSA and ECC are presented in the following sections.

10.5.1.1 RSA

In 1977, Rivest, Shamir, and Adleman presented a public key cryptographic algorithm named RSA. This algorithm is accepted to be secure and is generally utilized for e-business and transactions. The general belief about RSA's security stems from the difficulty in factorization of vast numbers. RSA algorithm has three parts, specific encryption part, public key generation part, and decryption part. In the key generation part, two pairs of keys are generated, namely private key and public key. Whereas the private key is a secret key which is maintained by the sender, the public key can be shared with the receiver. Time taken to generate a key is very slow, it is also said to be a disadvantage of the RSA algorithm.

There are five major steps involved in RSA key generation (Anonymous 2012).

1. Initially, generate two large prime numbers m and n. Make sure that m is not equal to n.
2. Calculating product of generated prime number m and n, $q = mn$. The calculated number q should be a larger number so that the generated number p and q cannot be identified.
3. Calculate phi $\Phi = (m-1)(n-1)$.
4. Select a public key exponent e such that $1 < e < \Phi$ and ensure that gcd $(e, \Phi) = 1$.
5. To end, find the decryption key d by calculating $d \bmod t = 1$, where t is the plaint text.

Finally, generated public key {q, e} and private key {d} are used to encrypt the secure plain text into cypher text as well as decrypt the cypher into plain text. RSA encipher is constantly done by utilization of public key. Any individual who needs to send message to the receiver utilizing RSA encipher should first get its public key. It should be possible by direct exchange of keys or by utilization of any accreditation power called "CA." If the message is larger than q then, split the message into two block of messages. Then encrypt the plaint text "t" into scrambled text called cypher text "c" and generate public key{q,e}. Unscrambling or deciphering process is carried out by the private key generated by the RSA. Then the plain text can be obtained from the scrambled text. Using RSA algorithm, public key digital signatures are generated. This signature can be verified later by any receiver with the knowledge of generated public key (Qamar and Hassan 2010).

RSA based signature is always computed over the hash of the original message. In order to make a signature, the sender node will produce a hash of the message and then raise it to the power of "d mod q", and then attach this to the original message. Receiver gets the digitally signed message from the sender, and the same hash function is utilized to compute the hash of the plain text. By verifying the signature, the receiver can ensure the confidentiality of the message.

10.5.1.2 Elliptic Curve Cryptography

Elliptic curve cryptography (ECC) is a public key cryptography developed in 1985 by Victor Miller and Niel Koblitz. ECC is a solid cryptographic algorithm compared to other public key cryptographic algorithms like RSA. ECC is a mathematical structure of elliptic curves over limited fields. Compared to the RSA public key algorithm, ECC is very secure. It is difficult to discover huge prime numbers in RSA, whereas ECC is more secure and furnishes equal security while generating little key size. ECC is thought to be exceptionally valuable for wearable devices, smart phones, and IoT implantable medical devices (IMDs). ECC's public key brings about data transfer capacity utilization and quick computation capability. Algebraic formula of ECC is as shown below.

$$d^2 = b^3 + ab + c$$

In this formula, b, d, a, and c are real numbers. To generate different elliptic curves, we can change the values of "a" and "c." Public key is generated by scalar multiplication with consistent base point G in the elliptic curve, whereas private key is generated by random number generations. Little key size in ECC is a greatly favored option. Comparing with 1024 key size in RSA algorithm, ECCs of 160 bit key size is equal.

Two major mathematical operations are done over elliptic curve. One is point addition and the other is point doubling. Point addition is shown in Figure 10.2, where addition of two points P and Q in elliptic curve $y^2 = x^3 - 3x + 5$ gives another point R. Example: R = P + Q.

In point doubling, addition of a P point to itself to get another point Q, that is, Q = 2P. Point doubling operation is shown in Figure 10.3, where addition of the same point P to itself in elliptic curve $y^2 = x^3 - 3x + 5$ gives another point Q. Example: Q = 2P.

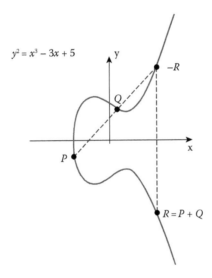

FIGURE 10.2
Figure point addition.

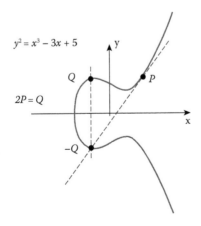

FIGURE 10.3
Point doubling.

The algorithm is beneficial when users can utilize private and public keys to secure the sensitive private data to transmit over the public medium.

10.5.1.3 Elliptic Curve Digital Signature Algorithm

Implementation of Elliptic Curve Digital Signature Algorithm (ECDSA) is based on the algorithm proposed. Signature generation and verification algorithms can be seen in the given Algorithm 1 and Algorithm 2. As seen in the algorithm, the implementation of signature generation requires less computational effort than the verification. Especially, the generation operation requires only one scalar multiplication, whereas the verification requires two scalar multiplications (Khalique et al. 2010).

Algorithm 1: Implemented EC signature generation algorithm. It consists of a modular inverse.

Operation: an addition and one scalar multiplication.

Input: Message to be signed m, EC parameters (generator point G, order o), secret key a

Output: (m, R, w) forming the signed message

1. Select a random integer n with $1 \leq n \leq o$
2. Compute R = kG = (x, y)
3. Compute s = $k^{-1}1$(m + ax) (mod o)

Algorithm 2: Implemented EC signature verification algorithm. It consists of a modular inverse operation, multiplications, and two scalar multiplications.

Input: Signed message to be verified (m; R; w), EC parameters (generator point G, order o), sender public key Q

Output: Validating the signature for the message m

1. Compute d1 = w^{-1} m (mod o) and d2 = w^{-1} (mod o)
2. Compute V = d1G + d2Q
3. If V = R then
4. Signature is valid
5. Else
6. Signature is invalid

10.5.2 Implementation

To incorporate public key cryptography with any sensor-based OS, numerous critical components should be considered before arrangement. This section discusses the possibility of adapting cryptographic mechanism in Cooja simulator in Contiki OS. Cooja simulator has a special kind of mote called Cooja mote, which helps to collect various types of information from nodes.

1. **Starting Cooja Simulator in ContikiOS**
   Instant Contiki development environment is a single download file. Contiki is an Ubuntu Linux virtual machine that keeps running in VMware player.

Instant Contiki supports developers to use all in-build services to understand the process in easy way.

2. **Downloading Instant Contiki**

   Contiki is a large file, greater than 1 gigabyte. Once downloaded, unzip the file, and place the unzipped directory on the desktop. Instant Contiki can be downloaded from the bellow link.

   http://sourceforge.net/projects/contiki/files/Instant%20Contiki/

3. **Installing VMware Player**

   Download and install VMware player. It is free to download but requires a registration. It might require a reboot of your computer, which is unfortunate but needed to get the network working.

   Download from the link https://download3.vmware.com/software/player/file/VMware-player-12.5.0-4352439.exe for Windows.

   Download from the link https://download3.vmware.com/software/player/file/VMware-Player-12.5.0-4352439.x86_64.bundle for Linux.

   Run VMware player and Contiki OS. Run Cooja simulator using the command "ant run" in the terminal. Once command is executed, Snapshot 10.1 Cooja simulator home window will open.

4. **Creating New Simulator**

   Create a new simulator by giving simulator name, radio medium, Mote startup delay, random speed, and new random speed on reload. The Snapshot 10.2 shows the creation of a new simulator in Cooja.

5. **Add New Motes**

   Once the simulator is created, we can add motes using motes menu (Snapshot 10.3).

6. **Add Cooja Motes**

   Add Cooja mote with the help of creating a new mote type option as shown in the Snapshot 10.4.

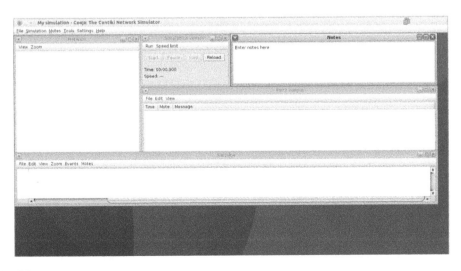

SNAPSHOT 10.1
Cooja simulator window.

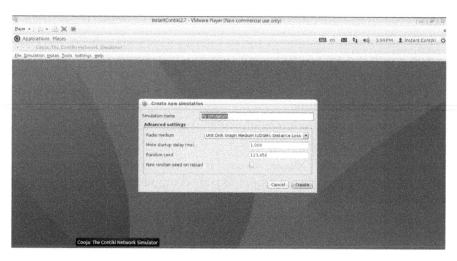

SNAPSHOT 10.2
Creating a new simulator.

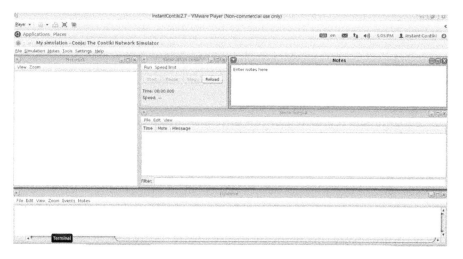

SNAPSHOT 10.3
Adding motes in new simulator.

7. **Selecting Contiki process file**

After creating the Cooja mote, select the process file by browsing the location. The Snapshot 10.5 shows the selection of Test_ecdh.c file. This file is an elliptic curve signature generation and verification file to authenticate the sensor data transmission. In order to compile this process file, we need to include the ecc.h header file.

8. **Add new motes**

Add new motes by giving values like number of motes, positioning, and position interval. Two Cooja motes are added in Snapshot 10.6, considered to be sender and receiver of sensing data.

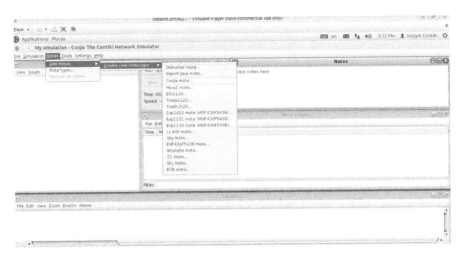

SNAPSHOT 10.4
Adding Cooja motes.

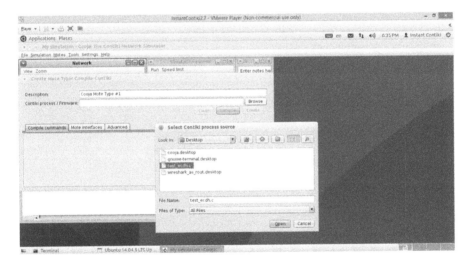

SNAPSHOT 10.5
Selecting Contiki process/firmware file.

9. **Start simulator**

   Start simulation and reading data transmission between the nodes in the encrypted form. We can use the run and speed limit window to start, pause, stop, and reload (Snapshot 10.7).

10. **Collect simulated results**

    Experimental results of ECC implementation in Cooja simulator.

    Snapshot 10.8 shows connections between two notes and respective readings of data packet transmission, temperature of node, beacon interval, delay, processing speed, and lost packets. Securing this data transmission using most powerful public key cryptographics like elliptic curve cryptographic is challenging.

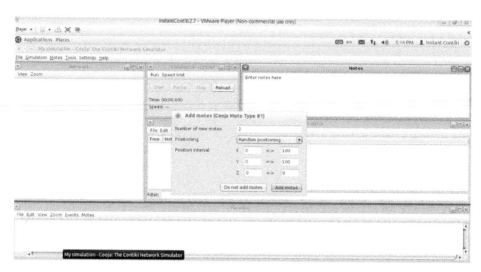

SNAPSHOT 10.6
Adding new motes.

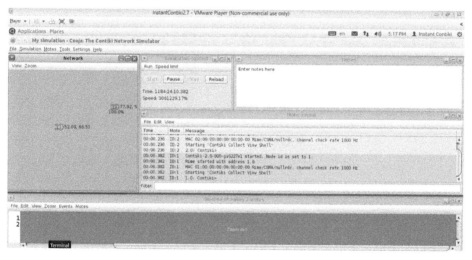

SNAPSHOT 10.7
Staring simulator.

10.5.3 Analysis of Collected Result

Collected data was annualized in terms of memory occupancy, battery consumption, and processing capacity in IoT health care devices.

After adopting the principle of ECC - based authentication in Contiki Cooja simulator, the battery consumption of node will be normalized. The sample battery voltage consumption of particular node is shown in Figure 10.4. This scenario moderately differs in real time, but this result guarantees that experiment on simulated way of implementation using ECC-based authentication system is more adoptable for lightweight IoT medical devices.

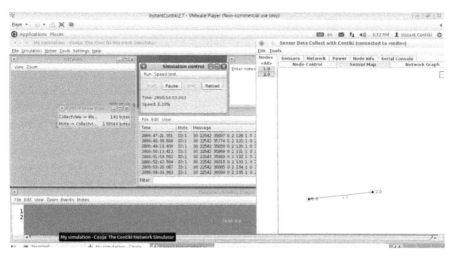

SNAPSHOT 10.8
Sensor data collection using Cooja simulator.

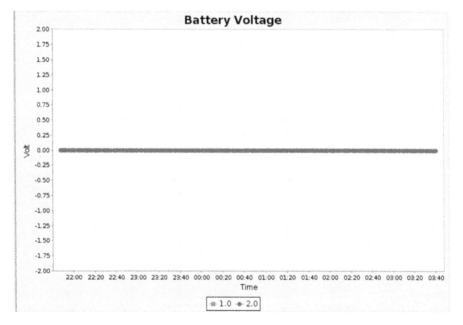

FIGURE 10.4
Battery voltage.

10.5.4 Recommended Positive

In general, the major security authentication issue in medical IoT is an insecure communication between IoT medical sensors/tags and sinks nodes/readers, which can be mitigated using the strong public key algorithms like ECC and hash function in a lightweight manner to adopt less memory, less battery, and less processing capacity in IoT health care devices.

10.5.5 Scope of Future Research

Most of the research work carried out in this field is focused on simulator-based approach rather than real-time environment. Implementation of EEC-based lightweight cryptography has proved that this approach is the most suitable approach for simulation environment and no work is concentrated on real-time environment. In future, the same work can be extended in real-time environment by utilizing the optimized resources, that is, speed, memory to mitigate the unauthorized access, and provision of strong security in IoT authentication system.

10.6 Conclusion

The overall aim of this study is to provide new security solutions as a recommendation to strengthen communication security. This chapter reviewed several security issues like mutual authentication issue, sinkhole attack, third-party intrusion, target routing, and spoofing, as well as device constraints such as less battery power, less memory capacity, and less processing capacity. In IoT medical device communication, all the security issues identified can be vulnerable to trustworthiness, or lack thereof, of the system. This chapter gives new security solutions as recommendations that can be used to provide reasonable guidance to IoT device communication development in future.

References

Akanksha, J., 2016. *Hackers breach Mass General vendor, compromise 4,300 records* [online]. Available from: http://www.beckershospitalreview.com/healthcare-information-technology/hackers-breach-mass-general-vendor-compromise-4-300-records.html (accessed September 1, 2016).

Alexandra, O., 2015. *FDA issues warning about hackable medical devices* [online]. Available from: http://www.popsci.com/fda-issues-warning-cyber-security-risks-medical-devices (accessed September 1, 2016).

Al-Kashoash, H.A., Al-Nidawi, Y. and Kemp, A.H., 2016. Congestion-aware RPL for 6L0WPAN networks. In *2016 Wireless Telecommunications Symposium (WTS)*, pp. 1–6, April, IEEE.

Anonymous, 2006. *Nihon communications solutions* [online]. Available from: http://www.ncs-in.com/index.php?option=com_content&view=article&id=78&Itemid=485 (accessed September 1, 2016).

Anonymous, 2012. *RSA Algorithm* [online]. Available from: http://pajhome.org.uk/crypt/rsa/contrib/RSA_Project.pdf (accessed September 1, 2016).

Anonymous, 2013. *TinyOS open source operating system* [online]. Available from: http://webs.cs.berkeley.edu/tos/ (accessed September 1, 2016).

Anonymous, 2016a. *IoT healthcare market worth 163.24 billion USD by 2020* [online]. Available from: http://www.marketsandmarkets.com/PressReleases/iot-healthcare.asp (accessed September 1, 2016).

Anonymous, 2016b. *Qualnet communications simulation platform* [online]. Available from: http://web.scalable-networks.com/qualnet (accessed September 1, 2016).

Anonymous, 2016c. *Banner health identifies cyber attack* [online]. Available from: https://www.bannerhealth.com/news/2016/08/banner-health-identifies-cyber-attack# (accessed September 1, 2016).

Anonymous, 2016d. *EPC UHF Gen2 air interface protocol* [online]. Available from: http://www.gs1.org/epcrfid/epc-rfid-uhf-air-interface-protocol/2-0-1 (accessed September 1, 2016).

Anonymous, 2016e. *Healthcare attacks statistics and cases studies* [online]. Available from: http://resources.infosecinstitute.com/category/healthcare-information-security/healthcare-attack-statistics-and-case-studies/ (accessed September 1, 2016).

Charlie, O., 2015. *Health insurer Anthem hit by hackers, up to 80 million records exposed* [online]. Available from: http://www.zdnet.com/article/health-insurer-anthem-hit-by-hackers-up-to-80-million-records-exposed/ (accessed September 1, 2016).

Cheshire, S., Krochmal, M., 2013. *Multicast DNS* [online]. Available from: https://tools.ietf.org/html/rfc6762 (accessed September 1, 2016).

Dan, G., 2011. *Insulin pump hack delivers fatal dosage over the air* [online]. Available from: http://www.theregister.co.uk/2011/10/27/fatal_insulin_pump_attack/ (accessed September 1, 2016).

Dan, K., 2013. *Dick Cheney feared assassination via medical device hacking: I was aware of the danger* [online]. Available from: http://abcnews.go.com/US/vice-president-dick-cheney-feared-pacemaker-hacking/story?id=20621434 (accessed September 1, 2016).

Dohr, A., Modre-Osprian, R., Drobics, M., Hayn, D., Schreier, G. 2010. The Internet of things for ambient assisted living. *ITNG* 10: 804–809.

Eisenbarth, T., Kumar, S., Paar, C., Poschmann, A., Uhsadel, L. 2007. A survey of lightweight-cryptography implementations. *IEEE Design & Test of Computers* 24(6): 522–533.

Fan, J., Batina, L., Verbauwhede, I. 2009. Light-weight implementation options for curve-based cryptography: HECC is also ready for RFID. In *Internet Technology and Secured Transactions, 2009. ICITST 2009. International Conference for 2009*, pp. 1–6, 9th November, IEEE.

Gope, P., Hwang, T. 2015. Untraceable sensor movement in distributed IoT infrastructure. *IEEE Sensors Journal* 15(9): 5340–5348.

Huddleston, T. 2015. *Premera blue cross reveals cyber-attack that affected 11 million customers* [online]. Available from: http://fortune.com/2015/03/17/premera-blue-cross-hacking-breach/ (accessed September 1, 2016).

Ishikawa, C. 2012. *A URN namespace for ucode* [online]. Available from: https://tools.ietf.org/html/rfc6588 (accessed September 1, 2016).

Jose, P. 2014. *Hospital network hacked, 4.5 million records stolen* [online]. Available from: http://money.cnn.com/2014/08/18/technology/security/hospital-chs-hack (accessed September 1, 2016).

Kamalov, M. 2009. *Security in wireless sensor networks: An overview*.

Keoh, S.H., Kumar, S.S., Tschofening, H. 2014. Securing the Internet of things: A standardization perspective. *IEEE* 1(3): 265–275.

Khalique, A., Singh, K., Sood, S. 2010. Implementation of elliptic curve digital signature algorithm. *International Journal of Computer Applications* 2(2): 21–27.

Lopez, J. 2006. Unleashing public-key cryptography in wireless sensor networks. *Journal of Computer Security* 14(5): 469–482.

Maia, P., Batista, T., Cavalcante, E., Baffa, A., Delicato, F.C., Pires, P.F., et al. (2014). A web platform for interconnecting body sensors and improving health care. *Procedia Computer Science* 40: 135–142.

Moosavi, S.R., Nigussie, E., Virtanen, S., Isoaho, J. 2014. An elliptic curve-based mutual authentication scheme for RFID implant systems. *Procedia Computer Science* 32: 198–206.

Nieminen, J., Gomez, C., Isomaki, M., Savolainen, T., Patil, B., Shelby, Z., et al. 2014. Networking solutions for connecting bluetooth low energy enabled machines to the internet of things. *IEEE Network* 28(6): 83–90.

Olivier, F., Carlos, G., Florent, N. 2015. New security architecture for IoT network. *Procedia Computer Science* 52: 1028–1033.

Qamar, T., Hassan, R. 2010. *Asymmetric key cryptography for Contiki* [online]. Sweden, Chalmers University of Technology. Available from: http://publications.lib.chalmers.se/records/fulltext/129176.pdf (accessed September 1, 2016)

Qiu, Y., Ma, M. 2015. An authentication and key establishment scheme to enhance security for M2M in 6LoWPANs. In *2015 IEEE International Conference on Communication Workshop, (ICCW)*, pp. 2671–2676, IEEE.

Rahman, R.A., Shah, B. 2016. Security analysis of IoT protocols: A focus in CoAP. In *2016 3rd MEC International Conference on Big Data and Smart City (ICBDSC)*, pp. 1–7, March, IEEE.

Raza, S., Trabalza, D., Voigt, T. 2012. 6LoWPAN compressed DTLS for CoAP. In *2012 IEEE 8th International Conference on Distributed Computing in Sensor Systems*, pp. 287–289, May, IEEE.

Raza, S., Wallgren, L., Voigt, T. 2013. SVELTE: Real-time intrusion detection in the Internet of things. *Ad Hoc Networks* 11(8): 2661–2674.

Rodika, T. Third Certainty. 2015. *Healthcare data at risk: Internet of things facilitates healthcare data breaches* [online]. Available from: http://thirdcertainty.com/news-analysis/internet-things-facilitates-healthcare-data-breaches/(accessed September 1, 2016).

Sheng, Z., Yang, S., Yu, Y., Vasilakos, A.V., McCann, J.A., Leung, K.K. 2013. A survey on the IETF protocol suite for the internet of things: Standards, challenges, and opportunities. *IEEE Wireless Communications* 20(6): 91–98.

Tseng, S.S., Weng, J.F., Hu, L.L., Wen, H.N. 2014. Ontology-based anti-threat decision support system for IPV4/IPV6. In *2014 Tenth International Conference Intelligent Information Hiding and Multimedia Signal Processing (IIH-MSP)*, pp. 61–64, August, IEEE.

Weinstein, R. 2005. RFID: A technical overview and its application to the enterprise. *IT Professional* 7(3): 27–33.

Winter, T., Thubert, P. 2012. *RPL: IPv6 Routing Protocol for Low-Power and Lossy Networks* [online]. Available from: https://tools.ietf.org/pdf/rfc6550.pdf (accessed September 1, 2016).

11

IoT Challenges: Security

Neha Golani and Rajkumar Rajasekaran

VIT University

Vellore, India

CONTENTS

11.1 Introduction

The IoT devices and services, which are not well secured, can cater the potential entry of cyberattacks and leave the data streams poorly protected as a result of exposing the user data to unauthorized access. Due to the presence of interconnectivity feature of IoT devices, also the devices that are not securely connected online affect the global security and capacity to recover quickly from difficulties. As shown in Figure 11.1a and b, an IoT schematic has its applications in various fields.

Looking into the ethics and principles, the developers and users of IoT systems and devices bear a collective obligation to ensure that IoT systems and devices do not make the Internet and other users vulnerable to any potential harm. This calls for an interconnected approach to ensure the security in order to come up with pertinent and effectual IoT solutions. Being frequent users of the Internet, a certain level of trust over the Internet and the connected devices is highly required, that is, because if people don't believe the information exchange and their connections are safe and secure from harm, it would result in their reluctance in the use of the Internet.

Many IoT devices would be used in the circumstances that make it very difficult to upgrade or reconfigure them. The reasons might be their deployment with longer anticipated service durability than what they are originally associated with or their upgrade process is cumbersome or impractical. At times, they have to be abandoned as they are left with no means of support in long terms, due to which they might outlive the company that designed them. The aforementioned scenarios indicate that the mechanisms adopted for security that is sufficient at the time of deployment of the device or system might not be sufficient to check the evolving and upcoming security threats for its full lifespan. This could lead to the long-term and persistent vulnerabilities. Whenever the manufacturer provides an update, there could be a change in functions as well without any notification, which leaves the user vulnerable to whatever changes the manufacturer makes. In this way, the management and support of IoT is a prevailing security challenge in longer terms.

In some cases, the IoT devices hold very little or no visibility into the internal function of the device or the data streams they produce. This makes the device vulnerable in such a way that it leads the user to believe an IoT device is performing certain functions when it might be undergoing undesirable data collection which the user does not intend to provide.

Previous models of IoT would be the products of large private or public technology enterprises, but in the future, the technologies with customized features and the ones built by the users themselves would be more prevalent as mentioned by *Raspberry Pi60* and *Arduino* developer communities. These technologies may or may not take the industry security standards into consideration.

The technology that we consume today is going to create a major disruption where people absolutely connect things and are becoming connected; we are going to see some fundamental changes in how we interact with the world, for example, *People will be moving from the Internet to Consumption of things through Internet.* The way the IoT is currently set up where huge content on the Internet is generated, people request for data over the Internet from uploading videos, writing, or whatever we are doing. There is a lot of really interesting second- and third-order effects to this—for one, it means that the Internet does not have any kind of objective view of the world—at best, it has an amalgamated set of subjective views, from which it can tease out what the real world is like.

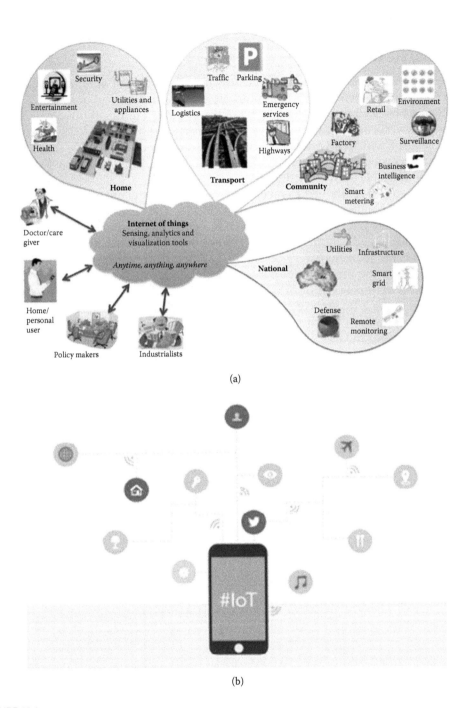

FIGURE 11.1

(a) An IoT schematic with its applications in various fields. (b) Applications in which IoT is used. (Available at http://internetofthingsagenda.techtarget.com/photostory/4500253982/Top-Internet-of-Things-privacy-and-security-concerns/5/Privacy-by-design-and-other-IoT-security-best-practices.)

11.2 Literature Review

If we consider the terms of the Internet facility and advancement of today, it was not difficult to make different computers communicate with each other and make them perform simple operations such as sending and receiving e-mails. We can think about interconnecting people, information, and systems. The emergence of miniaturized systems that are equally powerful, which can act independently and smartly due to the advanced software, empowers the concept that in many of the scenarios there might be implanted or linked processors responsible for its functioning.

In the IoT, resource-restrained elements are connected to the untrustworthy and undependable Internet via IPv6 and 6LoWPAN networks [1]. The wireless intrusions from both the Internet and the internal part of 6LoWPAN network are what make these susceptible to both external and internal attacks. The requirement of intrusion detection systems (IDSs) is considered to be necessary due to the probability of occurrence of these attacks.

Presently, due to the limitation of IDS approaches, which are being designed only for wireless sensor networks (WSNs) or traditional Internet, there is no IDS that can live up to the necessities of the IoT connected to IPv6.

In the work of Raza et al. [1], an innovative IDS for the IoT called SVELTE has been blueprinted, enacted, and assessed. In the execution and evaluation, the authors have fundamentally focused on attacks on the routing like imitation or alteration of information, sinkhole, and selective forwarding. However, they claim to extend the applicability of their designed system to detect other intrusions. They implement SVELTE in the Contiki OS and assess it profoundly. Their assessment depicts that in the simulated cases SVELTE spots all harmful nodes that initiate the executed sinkhole and/or forwarding intrusions with an accuracy rate of <100%, because of certain false notifications during the spotting. Figure 11.2 represents a general summary of the IDS, one of whose prime goals is that IDS should be lightweight and compatible with the processing abilities of the nodes. They also have enacted the idea of a distributed mini-firewall to secure the 6LoWPAN networks from the unauthorized users at a global level. They implement SVELTE in the Contiki operating system.

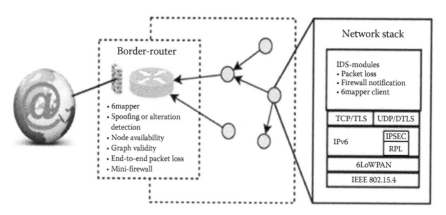

FIGURE 11.2
Illustration of IoT where IDS modules are kept in 6BR and in the separate end points.

11.3 Authentication Using OTP Validation

The IoT apps require the establishment of more steps to recognize the relevant appliances. Keeping this in mind, the illustrated work has been proposed for applying the concept of using one time password (OTP) as a validation criterion so that only the users who are eligible to or authorized for logging into the system can do so.

With the help of this procedure, only the users who are authorized to do so would be able to initiate the transmission and reception of the information after booting of the system. The description is illustrated in Figure 11.3 [2].

The system accesses the IoT broker application by sending OTP request after booting with the help of regular Message Queue Telemetry Transport (MQTT) messaging.

The IoT app creates an OTP, delivers that to the system admin individually, and notifies it to the system. Once the OTP is entered into the system, it is sent to the broker application. The broker application then authenticates the OTP intimated by the system and responds with a notification of the successful or failed message (usually invalid OTP or session time-out) back to the system. There is an allowance of trying the OTP validation again which depends on the number in the configuration of the retry counter. This further generates two cases:

1. *Successful OTP authentication even on retrying*: This results in the shutting down of the application.

2. *The inability of OTP validation*: This results in skipping of the OTP validation after initiation.

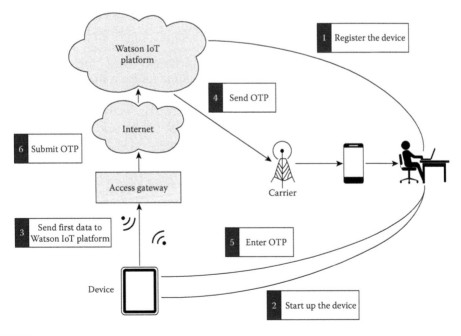

FIGURE 11.3
The switching on and switching off of OTP authentication with the help of system property.

11.4 How Security Threats Affect the Users

As PCs and devices that work on the Internet have become smaller in size, faster in functioning, and more connected, there has been an evolution in the threats in terms of security that has direct and indirect effects on the users.

With the assistance of developing cryptography and several protocols of security, we have also exploited that power to overcome those threats. There is a fight between the recently introduced threats and the recently introduced preventions and measures that counteract threats.

One of the factors that have a fundamental contribution to the security issues has been the Internet and its connectivity. When people had computers in a local setup that are usually accessed by a dial-up, the measures like passwords were enough to ensure the security. With the increasing outreach and coverage of the Internet, the chances of threat have increased and eventually the response demanded by that. Passive tags such as barcodes, quick response (QR) codes, and radio-frequency identification (RFID) are comparatively safer options as they are not prone to viral or malware-related attacks that degrade the functioning of the computer systems. The prime threat is in terms of accessibility of the information, that is, if a file can be read, then any user might be able to read that data and that can be used in any way by the user. The repercussions of this are determined by how convenient it is to obtain that information and how much value the information holds in terms of usefulness [1].

11.5 Security Requirements for the Internet of Secure Things

Some of the factors that the solution of security for embedded devices should ensure are that the device firmware has not been altered, damaged, or misused by malicious users and the device is secure from cyberattacks.

We should design the device in such a way that it includes security measures in its initial phases of modeling and prototyping. There is no definite formula that can solve the security issues of all the embedded devices.

The security specifications should take into account the security execution charges, security failure expenditure, intrusion risks, and available intrusion vectors. The security features that should be taken into account are discussed in Table 11.1 [3].

11.6 Toward the Secure Solutions

Concepts representing the risks that increase the probability of intrusions in the IoT devices are discussed subsequently [4].

11.6.1 Emerging Hazards and Limitations

With the evolution of IoT technologies, certain techniques replaced the ones with more difficult implementation.

TABLE 11.1

Security Features and Their Respective Applications in the Embedded Appliances

Security	Application in Embedded Appliances
Safe booting	Obtained from the producer with the help of a cryptographically signed code as well as with the hardware support to validate the code. This ensures that there has not been any changes made in the firmware.
Secure code revision	A technique of revisions of safe code that ensures that the code in the appliance can be revised for fixing the errors, safety patches, etc.
Information security	Avoids access without an official permission or approval to the appliances. Prevents the transmission of encrypted data and retention of the same.
Authentication	All the transmission of data with the appliance must be validated through passwords of good strength.
Safe transmission and reception	The transmission to and reception from the appliance require a secure encrypted protocol in order to prevent the usage of unsafe encryption algorithms; care should be taken.
Safeguarding against cyber intrusions	This involves introduction of the embedded firewalls for preventing intrusions. A firewall has the ability to include only the known, trusted users, thereby preventing the hackers from even initiating the intrusions.

Source: Alan Grau. (2014). *What is Really Needed to Secure the Internet of Things?*. Icon Labs Whitepaper. https://www.automation.com/pdf_articles/Internet_of_Secure_Things.pdf.

11.6.1.1 *Blacklisting versus Whitelisting*

Whitelisting took the place of blacklisting as it is easier to allow only what is necessary and deny the rest of the requests. It was better than the conventional antivirus techniques that would scan for malware in each and every input file. Whitelisting, on the contrary, only allows the authentic applications blocking every other one (http://searchsecurity.techtarget.com/answer/Application-whitelisting-vs-blacklisting-Which-is-the-way-forward). Figure 11.4 is an illustration of blacklisting and whitelisting.

The blacklisting method is becoming more inefficient because of the increasing saturation of malware. In order to update a blacklist, the services based on a cloud which accumulates the data from several nodes because of which it becomes a less burdensome task to keep a whitelist.

However, there are certain functional constraints in maintaining an application whitelist when the conventional blacklist methodologies are proved to be less efficient. While whitelist holds the capability of blocking the unauthorized files and data from entering the system it is installed in, it can restrain the initial acceptance of the recent and more advanced methodologies and technologies.

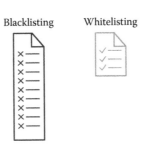

FIGURE 11.4
Representation of blacklisting and whitelisting.

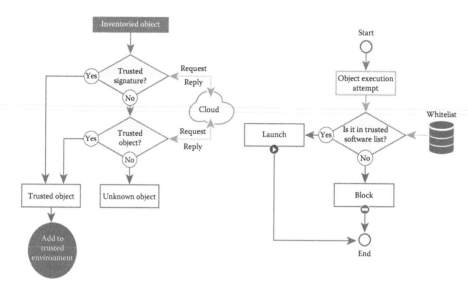

FIGURE 11.5
Depiction of the flow of procedure in blacklisting and whitelisting.

Whitelisting is the pre-acceptance of the favorable and source instead of the denial of a pernicious one. Although there are some applications are derived from some new applications which that give rise to new applications which make it impracticable to detect by the whitelisting due to the lack of knowledge of dealing with it and its absence in the whitelisting database. In that case, the whitelisting would trust the application if it was originated and promoted by a trustworthy and reliable source, and it would be categorized as trusted. This technique is termed as "Trust Chain" (https://blog.kaspersky.com/wonders-of-whitelisting/6367/). There are certain variations in the algorithms blacklisting and whitelisting, which are shown in Figure 11.5.

11.6.2 Towards Safer Solutions

In order to overcome the limitations of the device functionalities, there is a need of a significant reconstruction and redesigning applying several evolved techniques of safety in IoT. For instance, a lot of storage space in RAM is needed for backlisting to fulfill the IoT applications. Figure 11.6 shows the way in which constituent parts of IoT are interrelated or arranged. The designing of embedded gadgets usually comes with constraints in terms of interconnection because of the silicon form factor not being large enough.

The vast diversity in IoT applications puts the security factor in jeopardy. The examples include:

A smart energy meter: The one that reports the energy consumption statistics to the functional operator should be able to secure it from the revelation to an unauthenticated node. This report could be proved to be harmful of the house owner if it is disclosed to the ones with malicious intent as it would depict the house is vacant and would make it vulnerable to burglary.

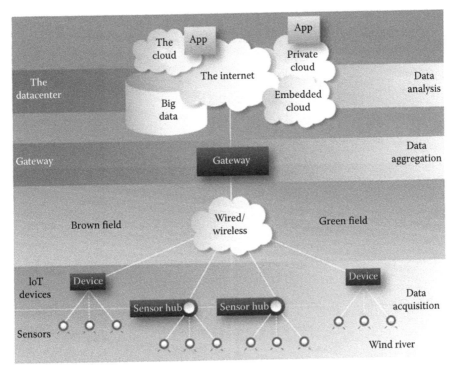

FIGURE 11.6
Depiction of ways in which constituent parts of an IoT are interrelated or arranged. A quintessential IoT deployment would include the appliances facilitated by sensors on the physically connected or unconnected network.

11.7 The Domino Effect of IoT Safety

Several safety issues are originated in terms of IoT execution and its traits. Users are required to keep a certain level of faith and reliance with the confidence that the appliances that use IoT are safe from the intruders to reduce the system's information insurance, peculiarly because this application has become more ubiquitous and well incorporated in our daily activities. The IoT appliances in which the security is compromised can cater various intrusions by attacking any of the components shown in Figure 11.7 and unlicensed login. This increases the chances of data invasion by exposing the data networks with insufficient safeguarding.

The interlinked fashion of IoT services denotes that every appliance that is linked online has the capability to influence the security and the ability to withstand external disturbances and internal collapses of the Internet worldwide. The factors such as software or hardware implementation of uniform IoT services or appliances, the autoconnection features of certain IoT appliances with other appliances, and the probability of employing or running them in environments that are not well protected tend to enhance the potential vulnerability. Figure 11.8 lists the potential challenges in terms of IoT security.

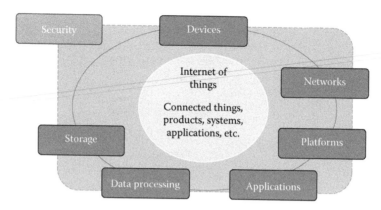

FIGURE 11.7
Illustration of the essential components of IoT security.

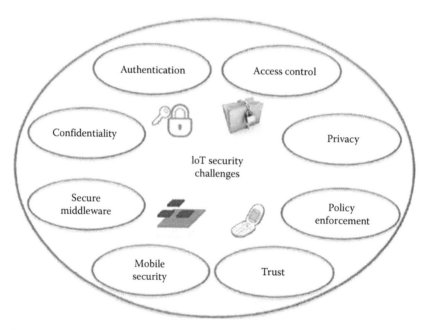

FIGURE 11.8
Listing of potential challenges in terms of IoT security.

The following are some of the challenges:

- The bringing of several IoT services into an effective action would include an assemblage of similar or sometimes indistinguishable appliances. This similarity amplifies the influence of any possibility of the devices getting intruded by outside devices having similar traits and functionalities.

  For instance, the likelihood of system of rules for transmission of information easily getting invaded of a certain company's brand of CFLs operated by the

Internet may expand to each design and prototype of the device that considers an identical set of rules of design or equivalent traits or functionalities of construction.

- The functioning of various IoT devices involves very scanty transparency level into its interior functions or the exact flow of information chain they generate. Due to this, the chances of the system security are likely or liable to be influenced or harmed, as the lack of transparency might lead the user to think that IoT is functioning in a regular manner when it might be indulged in undesirable activities or accessing more information than what is required by the user. When the producer of the product incorporates an update in the product's features, the functioning of the product can also undergo variations without any awareness increases in the user's likelihood of the susceptibility in terms of security to any further variations introduced by the producer.

- One of the crucial factors to be considered when the IoT products/services are implemented or brought into action is physical security at all the times. The intruders might have chances to get their hands on the complete IoT system or one of its components directly. Anti-interfering characteristics and transformation to update the modeling and prototyping are required and must be to be taken into account in order to maintain security in such situations.

- Certain IoT services and products, smoke detectors, burglar alarms, temperature and humidity sensors, door switches, etc., are generated with an aim to be integrated into the environment. Here, the user usually does not supervise the system in a continuous manner to check its functioning. Moreover, these systems may not have an explicit or direct technique to alarm the users whenever an issue regarding the safety is created. This increases the inconvenience for the user to be aware of any kind of intrusions arising.

- Due to both the versatility of the IoT devices and dynamic nature of the lifestyle of the users, there is no restriction on the number of IoT services. There is a need to establish an agreement of the extent of shared data and opted privacy settings to the services or systems using IoT, which can be created by building an efficacious technique that can withstand the enormous amount of IoT appliances having controls that are elucidated by the users.

- Usually, a more conventional privacy design has been followed in which the users are supposed to choose the privacy options and settings by making inferences based on the description given on the screen of the app. When we download the apps from Play Store, the user is immediately asked to "accept and install." A list of objects show up, which can be accessed by this particular app to perform the functions. IoT appliances generally do not offer the liberty of configuring privacy choices and settings because of the absence of the user interface. In several configurations of IoT, there is very little or no regulation or any kind of influence by the users over the method of accumulating and utilizing the information.

11.7.1 The Complications of a Security Hazard

Figure 11.9 presents the factors that pose a threat to the IoT security. These include Unprotected Network services, Cloud, mobile, an absence of security configuration, etc.

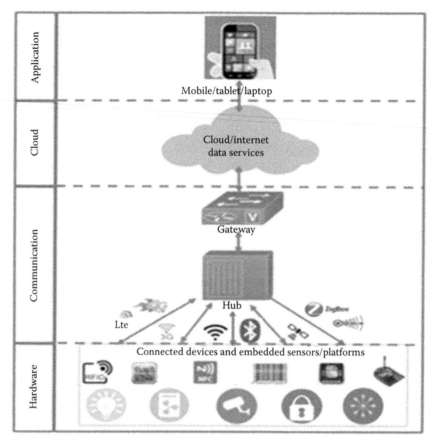

FIGURE 11.9
Visual representation of certain initial nodes or phrases which serve as potential challenges for security when the systems are interconnected.

11.8 Bottom-Up Approach for Ensuring Security

The security of a device must be ensured throughout its course of development, from the early prototyping to the publishing phase and even further [4].

11.8.1 Security While Booting

When a system is started and put into a state of readiness for operation, the software on the gadget is validated on the basis of its credibility and authenticity with the help of cryptography by generating signatures or authentication log-in systems in a digital form.

Even though authentication has been made, there is still a requirement of securing the devices from intrusion during the period when a computer program is being executed.

11.8.2 Control over the Accessibility

This involves restricting the liberty of the devices and its constituents to only the relevant receivers by designing the OS access controls, which are data retrieval controllers based on the roles integrated into the OS. The aim is to let these applications approach only the resources relevant to finish their tasks. In the case of intrusion of one element of the system, this feature of control over the accessibility ensures that the attacker does not have any further access to the system or the access is kept to a minimum. Even if the attacker somehow gets a hand to the classified information of a certain company or institution by obtaining the login details of, say, an employee or an internal staff member, the information leaked should be restrained to only the parts of the network that are given official permission for or approval to only those login details. This above concept illustrates that the effect of an intrusion can be kept to a bare minimum only if sufficient details needed to accomplish a task are permitted.

11.8.3 Certification/Validation of Appliances

Every time an appliance is connected to a network, it should be able to check the validation on its own before any type of exchange of information. The integrated systems usually do not have manual functioning with the provision of users entering the details needed to login into the network. This makes it difficult to confirm recognition rightly before sanctioning authority. The verification process of user identity trying to access the system with the help of machines provides access to an appliance to a network on the basis of an equivalent set of login details allocated in a safe space.

This is exactly like how user authorization by comparing their credentials with the provided database on the local OS or within the corporate server has a provision of allowing users to acquire a network of an institute with the required authentication credentials.

11.9 Secure Framework of the IoT Related to Perceptual Layer [5]

The IoT possesses traits that involve certain risks: innate/built-in exposure, diversity, and exposure to the possibility of being attacked or harmed at the end nodes. For this reason, a novel framework needs to be introduced to check the security breach. This framework should focus on enhancing the efficacy, dependability, accountability, and regulation in the overall structure that checks the security issues. This must also address almost all technologies relevant to security and confirm effective compatibility of several mechanisms of safety. The following proposed work addresses all potential threats to the IoT system and suggests relevant security methodologies. This proposal contains a blueprint of worldwide two-dimensional (2D) secure framework of security embedded with relevant safety technologies. The IoT is a more compendious association of the nexus which is more effective, witty, and strong, rendering IoT the benefits over other applications used for sensing which was prevailed in the past. There is a lot to develop in the field of IoT implementation as the exploration of security challenges regarding IoT is still insufficient.

However, the complications of the IoT segregate the associated facets of the exploration in this field. Besides, the secure path, recognition of attacks on the system, the incognito privacy, the established process of trust and credence, and other features and aspects are disparate. This is why there is a dire need for a long-term introduction and acceptance of a novel, more advanced framework that could address most of the technology to ascertain the functioning of several security processes together without having to make alterations for successful working and effectively enhancing the resilience, regulation, dependability, and accountability of the whole system. Taking into account that the biggest gap between IoT and the web remains in the ability to interpret or become aware of something through senses, the secure framework with a bottom-up approach is presented for consideration with features which are aligned to the perceptual layer. A set of application requirements is presumed to be fulfilled by this framework.

11.9.1 Challenges Posed in IoT Security

The application threats existing in the IoT are closely associated with its application domain. A comprehensive argument is given below.

11.9.1.1 Perceptual Layer Security Threats

In the perceptual activity layer, sensory activity nodes sometimes build an ad hoc network with a dynamic distribution. Given restricted node resources, dynamic modification in configuration, and dispersed systematic structure, the most threats that come from the sensory activity layer are the following:

a. *Manual annexure*: Many nodes are steadily installed in the place and can simply be seized by intruders and thus are physically jeopardized.

b. *Brute force invasion*: The capability of storage of related amenities along with the computation of the perceptual node is restricted and is highly probable to face brute force invasion.

Number of combinations = Possible character (password string length)

11.9.1.1.1 How Brute Force Intrusions Function

A brute force intrusion follows the cryptography algorithm. The attackers who hack into the system have knowledge of the password and user name saved in an organized repository of the system. So when the users try to sign in and the request of their page is transmitted from the server side to the client system, these intruders are more operational to hack into the account. These attackers know that there exists an encrypted key through which the password and other details required to login can be made intelligible. So they try to access the full functionality or data of the system by obtaining the password or another form of authentication from the coded information. They achieve this by trying all the combinations that are possible to login into the system. The hackers use the set of coded instructions on the computer for the automatic performance of the task to retrieve the password and other confidential details. They take help of the highly efficient, quick computer systems that are programmed to calculate lengthy problems in very short time. This brute force intrusion (Figure 11.10) functions with digits, numerals, special symbols, and characters and combines them to match with the combination required to login and unlock the account.

11.9.1.2 Blocking of Brute Force Attack

A brute force invasion is proved to be dangerous for the operators at the nodes. This intrusion takes up a lot of space allocation, time, and resources; however, if we manage

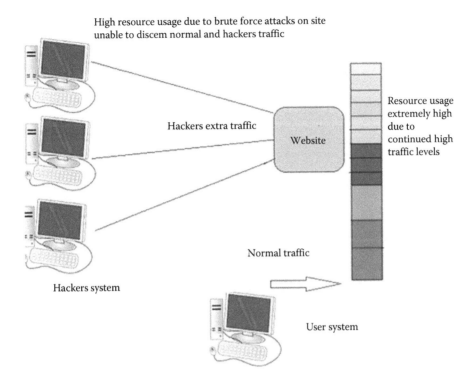

High resource usage due to brute force attacks on site unable to discern normal and hackers traffic

Hackers extra traffic

Website

Resource usage extremely high due to continued high traffic levels

Normal traffic

Hackers system

User system

FIGURE 11.10
Depiction of how brute force intrusion works and leads to indistinguishable normal and hacker's traffic. (Available at http://www.c-sharpcorner.com/UploadFile/66489a/brute-force-attack-and-how-we-block-it/.)

to execute certain security measures, then it would make it less susceptible for attackers to use this force to hack into the user's system. Some of the precautionary measures would be:

- Having a complex (e.g., alphanumeric or a mix of special symbols) or lengthy password.
- Restrict the number of attempts to sign into the account. And in the case of failure to log in within the given limits, the account should be provisionally inaccessible. For instance, when one enters the incorrect password multiple times in a Gmail login, there is a CAPTCHA code automatically sent along with a request to enter the text in the image sent which is system generated (as shown in the screenshot in Figure 11.11). This is done to check whether the password is being entered on its own or is being entered by the human robotically, which Gmail is unaware of.

11.9.1.3 Clone Node

The hardware framework of various perceptual end points of an equipment, such as a computer or peripheral, attached to a network is not very complicated and, therefore, is susceptible to be duplicated by the hackers.

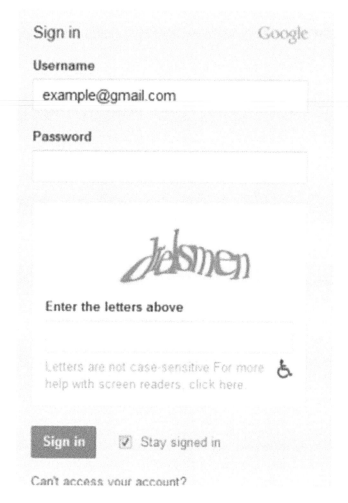

FIGURE 11.11
A screenshot with CAPTCHA code sent to be filled out in a Gmail login.

11.9.1.4 Impersonation of the User to Access the Account

The authorization in the dispersed setting or conditions in which a certain proceeding is being occurred is veritably tough to achieve for the perceptual end point. This facilitates the intruders at the nodes with malevolent intent to attack using the identification details imitated with fraud means.

11.9.1.5 Routing Attack

Data forwarding and transmission of information or a message occurs during the accumulation of perceptual information. This is what gives a chance to the information transmitted through the medial points to be intruded while it is being redirected further ahead.

11.9.1.6 Denial-of-service (DoS) Attack [6]

It is the kind of attack where the access to a system or a certain network is denied by the users and the culprit misuses this unauthorized access. It could be for the purpose of terminating an ongoing or upcoming process of a host linked with the Internet for an unlimited or unspecified period of time. This condition of DoS can be achieved by inundating the aimed system or device with surplus and redundant requests in order to overload it and block the permissible and authorized requests from being acknowledged and entertained. The analogy drawn could be a scenario where the entrance to a particular door to a shop is jammed by the crowd and does not allow the authorized people who are supposed to get into the shop, thus interrupting the regular and smooth functioning of the shop. Figure 11.12 illustrates the distributed denial of service (DDoS) intrusion framework.

DoS intrusions are featured by a distinct effort by the hackers in order to cause hindrance to the authorized users for using the particular system. There are two kinds of DoS

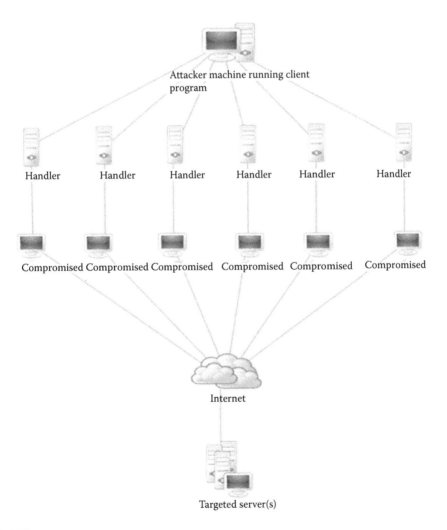

FIGURE 11.12
DDoS intrusion framework.

intrusions: one that makes the system processes fail and one that inundates the system processes. The distributed attacks are the most damaging ones [6].

11.9.1.7 Node Privacy Leak

Hackers can directly or indirectly steal compromising data in the end point. The devices at home are situated at particular places and, instead of attacking them, the hackers try to immorally obtain the information stored in those devices. They use a guerrilla approach to take over the minor parts of someone's system by eventually influencing the functioning of the network [7].

Furthermore, the hardware framework for certain perceptual end points is very straightforward and uncomplicated, which gives a chance for the hackers to duplicate it. Because of the restriction of the resources at the nodes, the dispersed framework, this perceptual layer results in several damages, including annexing it manually, being jeopardized by annexation.

11.10 An Innovative Framework for IoT Security [8]

Keeping the security challenges in mind, we suggest an interesting framework for the IoT security after the analysis of the secure framework needs. The secure framework suggested consists of four layers (see Figure 11.13).

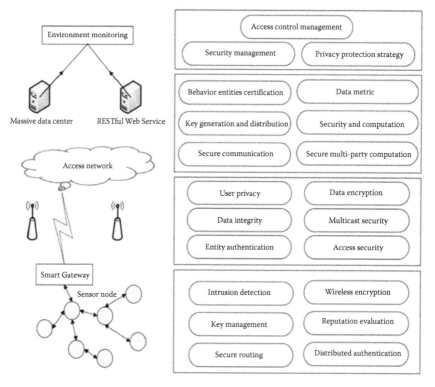

FIGURE 11.13
A secure framework for IoT.

1. Data Perception Layer

   A major problem is to come up with a prototype which is not very heavy for managing this layer in the IoT. Some of the safety safeguards incorporated in this particular layer are listed in the following:

   a. Safe routing
   b. Supervision and command of the key
   c. Recognition when the attack occurs
   d. Conversion of information or data into a code to prevent unauthorized access

2. Varied and Divergent Network Access Layer

   A means or process of ensuring the authenticity of the user accessing the data must be integrated into all kinds of devices irrespective of their complexity level. The precautions included in this layer mainly consist of:

   a. Security and seclusion of user data
   b. Integrity of the information
   c. Authorization for the users while logging in
   d. Safety measures while accessing
   e. Encryption of the confidential information

3. Information Management Layer

   To deal with an enormous amount of data and the decision-making taking into account the nature of the connections, firmly setting up the techniques for supervision and assessment of trust and achieving efficacious mining have to be scrutinized for the information security techniques. Keeping this in mind, this layer would consist of the resistance mechanisms as follows:

   a. Information metric
   b. Behavior elements authorization
   c. Key production and dispensation
   d. Calculation of security factors and values

4. Smart Service Layer

   To yield an appropriate security facility for a particular user on the basis of his or her requirements, suitable accuracy should be established for various applications.

   a. Regulation of access and its organization.
   b. Organization of security.
   c. Methodologies to security of confidentiality.
   d. Figure 11.13 shows the IoT framework and factors that should be taken into account for the security.

11.11 Security Authentication and Confirmation System

In this section, we discuss the structuring of a security authentication system in the field of IoT.

11.11.1 System Framework

The nodes at the endpoints of the 3G network: RESTful Web Services that are the web services based on REST framework, that is, if surpassed the user verification procedure. All the resources work as a source in the REST framework. Figure 11.14 illustrates the functioning of the REST approach between the client and the server.

The security authentication and confirmation system consists of four layers:

- **Perception Layer Appliances (Figure 11.15):**

  This layer would include the sensors of factors such as temperature, the component of a television signal which carries information on the brightness of the image. Two AA rechargeable batteries are used for the source of power.

- **Heterogeneous Network Access Layer Appliances:**

  The smart Gateway with two AA rechargeable batteries or a USB connecting wire is used for the source of power.

- **Data Management Layer Appliances:**

  Intel(R) Core (Trademark) i5 with 4 GB RAM is used

- **Intelligent Service Layer Devices:**

  Web server using RESTful approach, Intel(R) Core (Trademark) i5, with 2 GB RAM.

  Figure 11.16 shows the Perception Layer applications and smart Gateway.

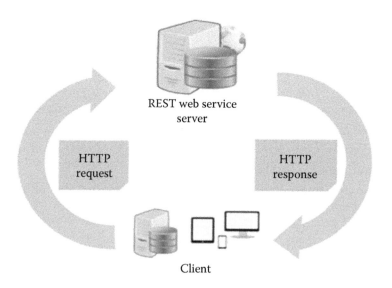

FIGURE 11.14
Functioning of REST approach. (Available at http://www.cousinsinfotech.com/restful-services/.)

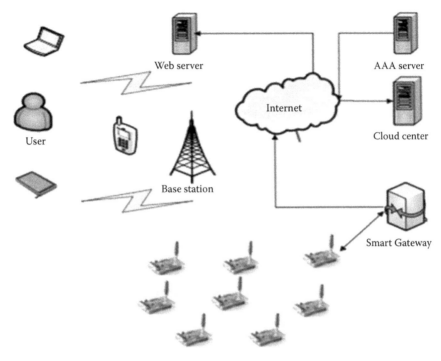

FIGURE 11.15
Appliances in the perception Layer.

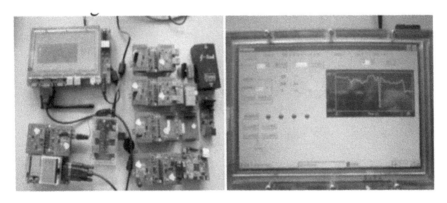

FIGURE 11.16
Perception Layer applications and smart Gateway (web services based on REST approach).

11.12 Conclusion

As we increasingly connect the devices to the Internet, it gives new opportunities to the development of cyberattacks by letting the malicious users to make the device malfunction or reprogram the devices. Sometimes, due to the cost being competitive or the existence of certain technical constraints, it becomes challenging for the manufacturers to include the features which would be adequate to check the probable security breaches.

Along with potential security design deficiencies, the increase in the number of IoT devices could invite the opportunities to attack the systems. Every device with compromised security that is connected online has a potential to affect the resilience of Internet on a global level in case of its high interconnection.

Many IoT devices are designed in such a way that they could be deployed in a huge amount in the market, far more than the traditional Internet-connected devices. It results in the creation of unparalleled interconnected links in terms of quantity. This would call for a new consideration of the existing methods and tools as many of these Internet-connected devices would become capable of establishing communication with other devices they are connected to in an unpredictable and undesirable manner. Moreover, the homogeneous nature of IoT deployments resulting from a collection of similar functioning devices possessing the same characteristics would amplify the effect of each one of the connected points or devices.

The effective and appropriate security solutions can be achieved in many ways.

One of the suitable and efficacious solutions to the security concern lies in applying a collaborative security model which serves as an effective approach. This collaborative model has served as an effective approach for all the authorities using IoT services in order to help them safeguard the cyberspace and the Internet, including the IoT. The model consists of effective tools and a variety of practices which include the sharing of information (i.e., bidirectional and consensual); preparedness from the incidents and cyber exercises; consenting to certain international behavioral norms, recognition, and improvement of international practices and standards; training; and awareness creation. The assistance of consistent efforts would eventually lead to the evolution of collaborative approaches based on the shared management of all the probable risks which would appropriately match the complexity and scale of the possible security challenges posed by the IoT devices in future.

IoT is going to give the Internet a more objective way of gathering data, meaning that the conclusions that will be drawn by artificial intelligence (AI) and machine learning algorithms are potentially going to be very different from what we expect. It is also means that the scale and quality of decisions being made by virtual agents are going to become an order of magnitude better than what it is today.

11.13 Future Scope

IoT is an emerging field add-on of the Internet. Thus, the design, security, and privacy issues on the Internet inherit the IoT.

- With the development of the IoT, different wireless communication technologies and network infrastructures are continuously integrated, including WSN, RFID systems, mobile vehicle network, 3G technology, WIMAX, PAN, etc. Even though with the emerging connecting technologies like IPV6 which has a mandatory Internet Protocol Security (IPSEC) with it, the security of IoT is still a matter due to various attacks. The nature of IoT, that is, dynamic (i.e., ability to connect any new object into the existing network and a variety of communication techniques) is also a reason for the security issues in IoT than in other fields such as WSN and mobile ad hoc.

- Since the corresponding communication network environment has become more complex and the security issues involved are more complex than any existing network infrastructures, the things in IoT should have inbuilt device authentication and encryption using the lightweight cryptographic protocols. Public key infrastructure (PKI) should be there in IoT environment and the damage to the devices should be also addressed.

- The security should be enabled in the technologies which should ensure that there is no tag collision problem. This chapter proposes a novel network secure framework that can be adapted to the future IoT and finally gives details for the security verification system design.

- We can introduce the sensors that are reliable enough for the users in case of emergencies, power failures, and other electricity issues. With this, we can move toward more healthy technological solutions which can take care of the patient's health.

- Issues such as sudden heart attacks, silent deaths, can be reduced if the real-time notification features are included in the IoT devices. For example, building the wearable body sensors that can notify the nearest hospitals in case of abnormalities in the patient's body can help in providing ambulance service and emergency aid to the patient.

- The new home automation systems can be advanced in the sense that the whole idea of physical switches can be eliminated. This can be achieved by directly connecting the home appliances with IoT appliances and sensors so that these appliances can function on their own without being dependent on us. This would decrease the chance of electric shocks and other related household accidents.

- In future, it is pretty clear exactly how IoT will look; the next 20 years is a known unknown to us.

References

1. Raza, S., Wallgren, L., & Voigt, T. (2013). SVELTE: Real-time intrusion detection in the Internet of Things. *Ad Hoc Networks*, 11(8), 2661–2674.
2. Gantait, A., Patra, J., & Mukherjee, A. (2016). *Design and build secure IoT solutions*, Part 1. Securing IoT devices and gateways. https://www.ibm.com/developerworks/library/iot-trs-secure-iot-solutions1/index.html.
3. Grau, A. (2014). *What is really needed to secure the Internet of things?*. Icon Labs Whitepaper. https://www.automation.com/pdf_articles/Internet_of_Secure_Things.pdf.
4. Shipley, A. J. (2013). Security in the internet of things, lessons from the past for the connected future. Security Solutions, Wind River, White Paper. http://www.windriver.com/whitepapers/security-in-the-internet-of-things/wr_security-in-the-internet-of-things.pdf.
5. Zhang, W., & Qu, B. (2013). Security architecture of the internet of things oriented to perceptual layer. *International Journal on Computer, Consumer and Control*, (IJ3C) 2(2), 37–45.
6. Taghavi Zargar, S. (November 2013). A survey of defense mechanisms against Distributed Denial of Service (DDoS) flooding attacks. *IEEE Communications Surveys & Tutorials*. 2046–2069.
7. Weber, R. H., & Weber, R. (2010). *Internet of Things*. Vol. 12. Springer, New York.

8. Chen, D., et al. A novel secure architecture for the Internet of things. Genetic and Evolutionary Computing (ICGEC), *2011 Fifth International Conference on. IEEE, 2011.* DOI: 10.1109/ICGEC.2011.77. https://people.umass.edu/~dongchen/secure.pdf.

Bibliography

1. Values and Principles. Principles. Internet Society, 2015. http://www.internetsociety.org/who-we-are/mission/values-and-principles
2. *Collaborative Security: An Approach to Tackling Internet Security Issues.* Internet Society, 2015.
3. http://www.internetsociety.org/collaborativesecurity
4. http://www3.ca.com/us/lpg/ca-technology-exchange/security-and-the-internet-of-things.aspx
5. https://inform.tmforum.org/internet-of-everything/2016/09/iot-isnt-secure-people-wont-use/#prettyPhoto
6. Belissent, Jacques E. Method and apparatus for preventing a denial of service (DOS) attack by selectively throttling TCP/IP requests. U.S. Patent No. 6,789,203. 7 Sep. 2004.
7. Shuler, R. L., & Smith B. G. (2017). Internet of things behavioral-economic security design, actors & cyber war. *Advances in Internet of Things*, 7(2), 25.
8. Alsaadi, E., & Tubaishat, A. (2015). Internet of things: Features, challenges, and vulnerabilities. *International Journal of Advanced Computer Science and Information Technology*, 4(1), 1–13.
9. Miorandi, D., Sicari, S., De Pellegrini, F., & Chlamtac, I. (2012). Internet of things: Vision, applications and research challenges. *Ad Hoc Networks*, 10(7), 1497–1516.
10. Barnaghi, P., Wang, W., Henson, C., & Taylor, K. (2012). Semantics for the Internet of things: Early progress and back to the future. *International Journal on Semantic Web and Information Systems (IJSWIS)*, 8(1), 1–21.
11. Xia, F., et al. (2012). Internet of things. *International Journal of Communication Systems*, 25(9), 1101.
12. Stankovic, J. A. (2014). Research directions for the internet of things. *IEEE Internet of Things Journal*, 1(1), 3–9.
13. Borgohain, T., Kumar, U., & Sanyal, S. (2015). *Survey of security and privacy issues of Internet of Things.* arXiv preprint arXiv:1501.02211.
14. Atzori, L., Iera, A., & Morabito, G. (2010). The internet of things: A survey. *Computer Networks*, 54(15), 2787–2805.
15. Granjal, J., Monteiro, E., & Silva, J. S. (2015). Security for the internet of things: A survey of existing protocols and open research issues. *IEEE Communications Surveys & Tutorials*, 17(3), 1294–1312.
16. Uckelmann, D., Harrison, M., & Michahelles, F. (2011). An architectural approach towards the future internet of things. In *Architecting the Internet of Things* (pp. 1–24). Springer, Berlin.
18. Zhang, Z. K., Cho, M. C. Y., Wang, C. W., Hsu, C. W., Chen, C. K., & Shieh, S. (2014). IoT security: Ongoing challenges and research opportunities. In *Service-Oriented Computing and Applications (SOCA), 2014 IEEE 7th International Conference on* (pp. 230–234), IEEE.
19. Mukhopadhyay, S. C., & Suryadevara, N. K. (2014). Internet of things: Challenges and opportunities. In *Internet of things* (pp. 1–17). Part of the Smart Sensors, Measurement and Instrumentation book series (SSMI, volume 9). Springer International Publishing.

12

Non-Physician Primary Healthcare Workers: Making Wearable Computational Devices and Electronic Health Records Work for Them

Katpadi Varadarajan Arulalan

AA Child Care Center

Vellore, India

Vijai Shankar Raja

CEO, HELYXON® Healthcare Solutions P Ltd

Chennai, India

CONTENTS

12.1 Introduction 236
    12.1.1 Occurrence of Diseases and Scope of the Internet of Things......................236
    12.1.2 Control of Communicable Diseases with Environment Manipulation
        and Anti-Microbials 237
    12.1.3 Matching Health Budget to Disease Levels—Neither Possible
        nor Necessary 237
    12.1.4 Transition from Symptom-Based Treatment to Personalized Treatment....238
    12.1.5 Placing the Responsibility with the Patient Himself/Herself—Primary
        Health Care..238
    12.1.6 Monitoring Vital Signs through Sensors..238
12.2 Objective..239
    12.2.1 Explanation of the Core Problem with the Very Common
        Example: Diabetes..239
    12.2.2 Relationship with Paramedical Workers, EHRs, and Gadgets.....................241
    12.2.3 IoT Is Capable of Providing Such a Solution, Which We Have
        Demonstrated in Chennai and Vellore 241
        12.2.3.1 AA Child Health Center, Vellore 241
        12.2.3.2 HELYXON® Health Care Solutions, Chennai 242
    12.2.4 Why Do We Need to Choose "Child Health"? 242
    12.2.5 Use of Wearable Computational Medical Devices by
        Paramedical Workers 242
    12.2.6 Role of Continuous Temperature Measurement in the Community by
        Paramedical Workers 243
    12.2.7 Depending upon the Anxiety and Communication Capability There
        Can Be Misguidance to Physician 245
    12.2.8 How About Making the Continuous Monitoring of Vital Parameters a
        Possibility at Pocket Level?..245

12.1 Introduction

As a society, we have come a long way since the World Wars and pre-independent India, when food for survival was itself a question. Millions died due to famine and lack of food. Mortality due to hunger has come down, except in a few countries. Human lifespan has increased by 30 years in the last century. One of the best inventions made by man after fire and wheel is TOILET. Our modern toilet dates back not more than 250 years. In 1775 Alexander Cummings invented the S-trap, which is still in use. Toilet discovery needs highlighting because it helped to reduce infection level in the community, even before antibiotics were discovered.

12.1.1 Occurrence of Diseases and Scope of the Internet of Things

Environment: There is no simple theory to explain how diseases occur; for our purposes the following paradigm is more than adequate. Interaction between the host (man), agent (living and nonliving), and the environment is called the epidemiological triad. Figure 12.1 shows this truth (Krieger 2001).

To raise a plant, seed and mud alone are not sufficient. Somebody has to water it. Similarly host–agent interaction alone will not cause disease. Facilitating environment is crucial. As a corollary it is just not enough to deal with host and agent alone to cure illness, but the environmental contribution has also to be corrected, which is precisely *not* done by physicians. To deal with environment, *intersectoral coordination*, by various segments of the society is urgently required. Without the intervention on the environmental factor, diseases are bound to persist leading to various conditions like drug resistance, increased mortality from diseases like tuberculosis, etc. Such events can be prevented by primary health care, which has three pillars, viz. equity, community participation, and intersectoral coordination. *Internet of things* (IoT) is one such intersectoral coordination.

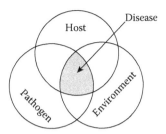

FIGURE 12.1
The epidemiological triad—environment is the most important contributor for disease occurrence.

12.1.2 Control of Communicable Diseases with Environment Manipulation and Anti-Microbials

The day we started to understand that the disease has a relationship with hygiene, toilets were invented, antibiotics followed, vaccines were developed, etc. New way of healthcare delivery became prominent over all the existing practices, which in many circumstances were not scientific. In 1824, first public toilets were introduced in Paris. Subsequent decades saw the birth of epidemiology, which proved the relationship of all diseases *is* with the environment.

Eradication of infectious diseases became easy as we understood the bacteria and virus more and more. Antibiotics and painkillers changed the very practice of medicine, by providing more ammunition in the physician's bag. Of course the science of medicine further progressed in the area of surgical treatment and issues related to child birth. All these contributed to the extension of human life by 30 years. But any further improvement in longevity and quality of human life is becoming a bigger question and certainly not possible in the near future. The life-span extension has been achieved by only curtailing the communicable diseases, and controlling the contribution by non-communicable diseases was virtually nil. Chronic diseases like diabetes, cardiac issues, neurological issues, psychology, oncology, etc., which are popularly called lifestyle diseases of the twentieth century, have to be controlled to improve the quality of life.

12.1.3 Matching Health Budget to Disease Levels—Neither Possible nor Necessary

In India there are around 14,00,000 hospital beds and we need 2.5 times more beds to meet world average and at least 20 times to meet Japan's standards. The average cost of creating a hospital bed is approximately Rs. 50 lakhs [24], including the infrastructure and optimal equipments inventory. That works out an investment of Rs 60,000 crores each year for the next 20 years [25]. Current healthcare budget by government is Rs 30,000 crores, which itself is few times more than previous years. This being the case, investment for constructing the hospital beds looks practically impossible for now. We need alternative technology in place [26]. Although we need physicians and hospitals in place immediately, we neither have time or money waiting for. Fortunately, we have 106 crore cell phone connections [27] for 74 crore adults (aged between 15 and 64 years) [28]. There are 46 crore internet users, and this is growing at 30 to 50% every year [29]. We have 1.4 cell phones per adult and 0.6 internet connection per adult. By 2020 India will have one internet connection per each adult [30]. All these happen without any investment from the health budget. Monitoring the health can be done with *right application, right platform, along with the right IoT sensors* connected to the experienced and sophisticated regional tertiary healthcare centers.

Also we can maximize the use of precise talents to get the correct solution even to the remotest location on time. This also helps to teach and train the next generation of physicians and paramedical staff more effectively.

12.1.4 Transition from Symptom-Based Treatment to Personalized Treatment

Today the solution toward health issues is just symptom based or based on blood tests or imaging studies. The causes of such diseases are yet to be established well with each individual, i.e., cause can only be generalized, and cannot be pinpointed as to why it has occurred in the individual. We need to move from symptom-based management to prevention and prediction, to bring out a cost-effective solution. This is called prognosis and has to be *individualized*. Thus we have a different set of issues to be solved, requiring different set of tools to capture, analyze, and prognosticate the outcome. We need to understand the personal needs of the patient, his/her genetic background, family surroundings, socio-economic challenges faced by the individual, the required behavioral change to prevent the illness, etc. Big data analysis is required for predicting the risk factors for each individual.

12.1.5 Placing the Responsibility with the Patient Himself/ Herself—Primary Health Care

In the decades of struggle, awareness to our own body is becoming better, but limited to few set of learned people. *The actual responsibility of health has to be with the patient himself which is one of the eight principles of primary health care.* It is time that he/she is empowered with such information in the right way to bring in the next big shift in healthcare. To achieve this, he/she needs an electronic health record (EHR) and gadgets to help him/her monitor his/her health indices and trained medical people to help him/her. These personnel need not be physicians. But they must capture the information with the specialist capabilities through EHR and gadgets. The healthcare system can take up the responsibility of continuously informing or predicting the healthcare status of each individual based on his/her responses or behaviors, affecting his/her health.

12.1.6 Monitoring Vital Signs through Sensors

There are many symptoms a person can feel and experience, but a few critical and vital parameters cannot be felt by a person; however, through IoT it can be monitored easily on a continuous basis. Well-documented examples include pulse, blood pressure, and body temperature. These measured parameters are remotely read and interpreted by an expert on a live basis and opinion on this reaches back to the person, while he continues to do his/her daily routine at a remote place. This is called healthcare anywhere, anytime. This will permanently change the landscape of healthcare system, for insurance companies, pharmaceutical industry, policymakers as well as for healthcare workers, above all patients themselves.

It is possible with the age of IoT. Various sensors are fixed to the body as wearable sensors, using low-energy radio frequency signals. The signals are sent to a nearby mobile device, through which the measured signal is sent to a remote cloud server through the Internet. There the required algorithms are applied and re-transmitted to the required people anywhere across the globe. An intelligent interpretation brings the report back to the person on whom the measurements were made. This completes the entire cycle.

It is almost like carrying an expert physician along anywhere, anytime. Today with the advancement of nanotechnology, information communication technology maturing, the Internet becoming pervasive, all the citizens having a mobile device with them, quality of life can certainly be improved.

In this paper we record our ultimate objective as *empowering the patient and at the same time amplifying the physician's strength*. The concept is further explored with diabetes as an example in which failure of current system and the proposed model are explained. Working of FEVERWATCH®, a wearable computational device, is explained and its role in the community. It can be used in more than 15 situations. Five essential phases for success of IoT in this country are enumerated. Three 'Is' (3I), information, intelligent, and instant decisions, are possible through sensors. EHRs will be the backbone for such changes using IoT devices in the country. The current problems in health information system in the country are explained. Complex process of development in EHR is described using HELYXON® software, which can also be used as a mobile application by the patients. The problems with traditional software and the advantages of INDIVO Open Source EHR are explained along with other technical details. The various features, such as how parents use it at home, paramedicals at front desk, and the physician in the consulting room, are explained with particular reference to symptom analysis, growth monitoring, immunization, and non-visit consultation with the physician. The scope for paramedical workers is further discussed with references. Finally the barriers, challenges, and the road ahead for use of wearable computational devices and EHRs are discussed and the solutions tried out in Chennai and Vellore are presented.

12.2 Objective

The main objective is to create an integrated and collaborative healthcare platform that empowers the poor and rich, and the urban and rural population alike. The solution addresses the core problem of severe shortage of physicians, specialists, and inadequate healthcare delivery mechanisms. The solution should be highly secured healthcare networking portal along with the IoT sensors and mobile application that allows registered consumers to create their profiles, upload electronic healthcare records (both live and offline), create their own care groups, and share selected information with care groups. It should help the end user to connect with healthcare providers and clinical laboratory personnel, pharmaceuticals, etc. Using smart sensors, smart hardware and software, network, and data science, all these are possible.

12.2.1 Explanation of the Core Problem with the Very Common Example: Diabetes

Complicated and expensive goods and services in their early days are beyond the reach of the poor. IoT in general, particularly in health care, has not developed widely, meaning it is yet to become a way of life for many people and things. Any industry in early stages is almost always appears to be complicated. As a corollary, it becomes expensive and on the other hand only the well-trained people can use it. Computers and photocopying machines in their early days are everyday examples to comprehend the meaning of "only the rich can afford it and the well-trained expert can use it." As the years pass by, the gadgets become cheaper, at the same time, becoming more versatile enabling even an ordinary person with a little training to get experiences that were never thought a generation ago.

India has the maximum number of diabetic patients, it appears at an earlier age, and its complication is far reaching when compared to western nations. In the early twentieth century, before the advent of insulin, diabetes was an acute illness, meaning once diagnosed they had a short lifespan and died of complications. Now diabetic acute complications are well managed, making diabetes a chronic disease. As a chronic case, management of diabetes by the medical world is poor, and far too many progress to kidney failure, develop heart attacks and strokes, and lose their sight and limbs due to the vascular complications.

We know the science of diabetes up to the molecular level. We have trained physicians to take this understanding to the patient. We have patients who can afford the drugs or as in the state of Tamil Nadu, government hospitals providing free insulin injections. We have laboratory facilities to do blood sugar even in villages at primary healthcare centers. Yet the disease is not controlled. Complication incidence (number of new ones appearing every year) is increasing. From where does the problem come from?

There are many dimensions, but the single most reason is that the disease has so many components, no single physician can document, let alone mange all of them. But a network of physicians and paramedical workers can handle it more appropriately. Enlarging the diabetes example, the first step in management of diabetes is the diagnosis of the disease. Diagnosis is made by the symptom or complication of diabetes e.g., foot ulcer—by the physician or by the laboratory test done for some other problem or as a screening test. After diagnosis, it is screening for complications: eye checkup, cardiac assessment, cholesterol level, and kidney function measurement. After that, based on the individual's body constitution and the complications present, drugs are chosen. If it is not controlled, insulin injection is given. Lifestyle modification in the form of exercise and avoidance of smoking and alcohol is stressed. An exhaustive diet charting is done including what to be avoided, what can be liberally taken, what can be taken in moderation. Patient education like care of the foot and careful execution of simple procedures like nail cutting are stressed.

In vast majority of health centers, diabetes is managed only by the physicians, who are at the top of healthcare hierarchy. They concentrate only on bio-medical aspects, and not stressing the other components, mainly because of lack of time availability. Suppose we train the paramedical workers to carry out the work missed by physicians, complications can be greatly reduced. Here comes the comprehensive health record and smart gadgets to assist them. Physicians can then delegate work, which can be done easily by paramedical workers. This is happening in exclusively diabetic hospitals, particularly in the city of Chennai.

This delegation has the following components:

1. Domains which can be delegated are basic general information, diet, exercise, and lifestyle modifications.

2. These "modules" must have evidence-based guidelines. The guidelines if adopted scientifically should give measurable outcomes, which can be validated.

3. All these have to be "caught" systematically in a health record, which can be transported along with the patient. The entry in them should be legible in a standard format. When the patient leaves the primary physician, he/she should not fumble at the second hospital.

4. Several checklists for the diabetic are once a year, eye check-up; once in 3 months, hemoglobin A_1c measurement; before retiring to the bed, wash the feet and look

for any injury, etc. A system has to be maintained for prompting and confirming that these are done.

5. The drugs have complications which have to be monitored regularly.

Needless to say that such a system is not possible, with current methods in healthcare delivery and hence complications are rising.

12.2.2 Relationship with Paramedical Workers, EHRs, and Gadgets

1. If a physician does not have time, it has been clearly proved beyond doubt that paramedical workers can share and complete their work. Several evidences are presented later.

2. Now the next step is to maintain a record, which should be simple and which can be used by paramedical workers and the patient themselves, with physician contribution limited to biomedical aspects, beyond the capacity of paramedical workers.

3. The third and final step is to create gadgets which can embed the physician experience, knowledge, and skill effectively and simply so that the paramedical worker and patient themselves can use it.

    Figure 12.2 explains the objective of this relationship between the patient, physician, and the paramedical worker.

12.2.3 IoT Is Capable of Providing Such a Solution, Which We Have Demonstrated in Chennai and Vellore

12.2.3.1 AA Child Health Center, Vellore

This center has been functioning in Vellore for the past 25 years, providing service to the needy, with the primary aim of preventing unnecessary hospital visits and reducing healthcare cost. Computer systems are in use since 1990.

FIGURE 12.2
Three people, parents, paramedical workers, and the physician, are always connected through the software, sharing the responsibilities.

12.2.3.2 HELYXON® Health Care Solutions, Chennai

The CEO of this center has been in the healthcare industry for more than 29 years. They are at the IIT-Madras Research Park in Chennai addressing medical gadgets and software needs of the country.

12.2.4 Why Do We Need to Choose "Child Health"?

Simple, everybody likes children. Not even a criminal would bear to see a child suffering. Beyond the emotional aspect, child health is complicated by several factors. And, in the first year of life, children are brought to the hospital four to seven times for immunization purposes alone. Moreover, children tend to fall sick often, from 6 to 18 months of life.

12.2.5 Use of Wearable Computational Medical Devices by Paramedical Workers

HELYXON® has built the capability to develop Bluetooth-based IoT product for monitoring body temperature continuously from remote, called FEVERWATCH®. Though called as FEVERWATCH®, it is opposite; hypothermia monitoring is vital in the new born period, and this instrument can save lives in those situations.

1. The instrument consists of an IoT sensor, cloud server, and a smart application, which works on any iOS/android mobile devices, like phone, pod, pad, etc.
2. The sensor is attached to the body of the baby using a plaster at the appropriate location and it can accurately measure body temperature and communicate the same to the application on the mobile device through Bluetooth (Bluetooth version 4.0/Bluetooth low energy: BLE).
3. The data are continuously recorded at the sensor and application, and data/alert is sent to a remotely monitoring person via cloud server. Figure 12.3 schematically represents this fact.

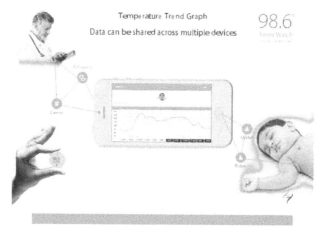

FIGURE 12.3
Wearable computational device FEVERWATCH®.

Features:

1. 24/7—continuous wireless remote monitoring.
2. Multipatient, simultaneous monitoring in hospital environment.
3. No loss of data, even when user with sensor moves away from the Bluetooth range.
4. Pre-settable alarm range for hyper-/hypothermia.
5. Directly computable data provide reliable temperature trend in graphical form.
6. Data and graph can be mailed, stored, and printed.
7. Shareable and reusable.

12.2.6 Role of Continuous Temperature Measurement in the Community by Paramedical Workers

1. There are exceedingly high rates of resistance of Gram-negative bacilli to almost all antibiotics (Giuliano KK 2006).

   Several reasons for this can be cited. In the community setting, fever without focus is treated with antibiotics by all and sundry, including parents. The single most reason cited is the *fever phobia*. "If something bad happens…, so we give antibiotics."

   This fear can be greatly reduced if temperature is monitored every second and the same is sent to both parents and the treating physician. The treating physician can fix his upper limit and can be messaged. The very thought that the child is remotely monitored by the medical team gives immense relief to the mother.

2. So is the case of febrile seizures, where every shake of the child makes the parents wide awake.

3. Last 20 years have seen the use of transplants and home management of leukemia going up.

   The single most predictor of early infection in this class is the rise in temperature or its opposite hypothermia. Febrile neutropenic patient numbers are also going up.

4. For want of beds, people, including children and the elderly, are discharged with an intravenous cannula in the forearm. They usually complete the antibiotic course at home. They visit the nearest healthcare facility once or twice daily. The one sure, noninvasive way of managing them is to record temperature.

5. India is becoming the preferred destiny for research and vaccine evaluation. The temperature monitoring in those cases expands the role of FEVERWATCH®.

6. Regarding temperature measurement in seizures, febrile seizure is differentiated from epilepsy on the basis of temperature. If temperature does not rise, within 24 hours of onset of fits, then we can conclude it to be afebrile seizure, possibly epilepsy.

7. Infrastructure-related issues like logistics management severely affect the healthcare outcome, especially vaccination/organ transplantation movements. Cold chain management as well as the refrigerators at each practitioner's place, where power supply is erratic, needs proper continuous monitoring.

8. Other community use is in hostels. During fever cases, question arises, to treat the inmates in hostel or at hospital or send them home, preventing the illness spreading to other children. Selectively, the students can be sent home or hospitalized or monitored in the hostel by paramedics.

9. Now that we have special ambulances for neonatal transport, FEVERWATCH® becomes inevitable in such situations, to monitor hypothermia (Kumar V et al. 2009).

10. Exercise among elderly patients will need temperature monitoring, along with pulse and blood pressure.

11. Infertility management by closely observing basal body temperature using FEVERWATCH® can be another compelling use requirement among youngsters, to find out ovulation time.

12. Remote monitoring of the postoperative patient after discharge, especially for medical tourism, patients who have come long way.

In the second phase, FEVERWATCH® can be expanded to monitor blood pressure, partial pressure of oxygen, pulse rates, which could then be transmitted along with temperature.

From Box 12.1, everything except the fourth point is already being done in an isolated way. We right now need a reliable solution to monitor the required vital parameters and essentials—temperature, pulse, blood pressure, and oxygen saturation. Based on this, if a physician continuously communicates to the patient, then the dependency of the patient to that particular physician becomes higher and higher. The mind share of the patient is fully occupied with that physician and the loyalty with that physician grows up. If the first phase of above-mentioned five steps doesn't work, then after certain telephonic follow-up we need to move on to phase 3, where a paramedic/a physician visit the home for the physical examination. Phase 3 asks the patient to come to a hospital for further examinations/disease management. As a final step, an exhaustive examination, diagnostic efforts, and reconfirmation are done before attempting the right therapy/surgery. In all these, the first key point is to ensure that the patient is resting all the while, and his/her systems are kept most comfortable to take on the fight against the disease more effectively. Everything else is put on second priority. If this process is going to be good only for the rich, how can we enable it for the poor and the needy? After all health and human life are invaluable to anyone, there is no rich or poor here.

BOX 12.1   FIVE ESSENTIAL PHASES TO OCCUR IN IoT TO SUCCEED IN HEALTHCARE INDUSTRY

1. A known physician/a set of physicians need to be made available all the time for a 24/7 any time connect. This brings both trust and confidence for the person to move forward in life with fewer worries.

2. Remote connect for easy conversation through video-conferencing using the readily available gadgets and freely available applications.

3. A sharable demography, past history, medical record, etc. in a repository under patient's control for any authorized physician to view it instantly.

4. All past and current vital and essential parameters available live to verify, track, or provide alert. This provides decision-making easier and symptoms-based approach.

5. Upon remote consultation, the medicines automatically reach the home through a pre-arranged system.

FEVERWATCH® is one of the first baby steps toward creating that new environment a real possibility. The best of the healthcare delivery is practiced at intensive care unit (ICU). Significant difference between ICU, the place of best care, and outside being the ability to continuously monitor vital parameters of the patient and intervene instantly.

Movie = *1,000 Pictures and each picture is 1,000 words = A movie is million words* *(1,000 × 1,000)*

Continuous monitoring = *multiple instant direct measurement is 1,000 times better than relying on* Patient's narration = *Continuous monitoring is million times better than patient's expression about symptoms.*

12.2.7 Depending upon the Anxiety and Communication Capability There Can Be Misguidance to Physician

Each individual is a complicated unique system and relying purely upon the patient's expressions can be misleading and can complicate things even more. There should be a matured system for physicians to rely upon to take decisions. Today's patients demand more than what can be delivered.

12.2.8 How About Making the Continuous Monitoring of Vital Parameters a Possibility at Pocket Level? (Fayyoumi et al. 2014; Mare and Kotz 2010; Zhu et al. 2015)

At the outset it looks like it is practically impossible for physician to go through all details to make a decision, even if we are going to invent new kind of sensors and gadgets to generate stream of measured data. This overwhelming amount of data is going to throw the entire current system out of gear; no one has time to review such a huge data for making decisions.

12.2.9 The 3I (Figure 12.4)

Consider we have an algorithm and if it can help physician in making *Instant/Informed/ Intelligent decisions*. The 3I. Physicians need to focus on handling the patient and not going through the piles of report of the past and current issues. As the technology advances more and more fragmented information about the patients keeps pouring in. The chances of losing piles of information are higher. Being the case he/she needs a new tool; we call it as 3I. Physician gets more time to build relationship with patient and such a decision-making tool is the need of the hour. It is like autopilot feature assisting an expert pilot. This helps the physician to move out of current practice of treating the patient through perceived symptoms to more scientific, evidence-based decisions. This for sure will lead to a new kind of sigh and relief to the patient as well as the physician and also build a path for the trust between them.

Figure 12.5 illustrates that there is no physical entry of data; *all data are directly computable* with high security and integrity.

Patient, practitioner, and payer all are in single platform. Using this readily available technological platform, we defy the current issues around the requirement of huge investment toward creating more hospital beds and more physicians. Recently Sweden's health minister announced that their country would eventually stop building any additional hospital beds.

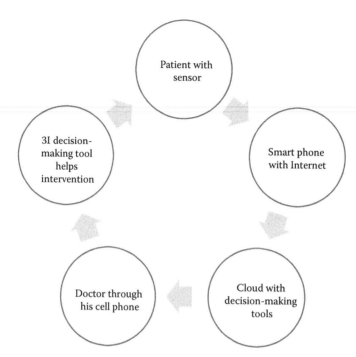

FIGURE 12.4
Wearable computational devise for 3I—informed, intelligent, and instant decisions.

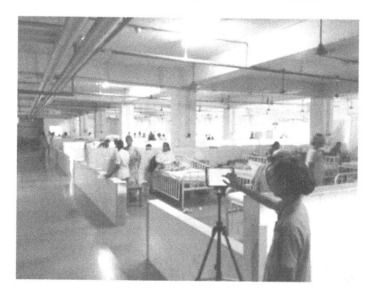

FIGURE 12.5
Use of FEVERWATCH® connected to an iPad in a postnatal ward.

12.2.10 Bringing Delight to All

Healthcare is a service industry. Any good-quality service is defined by delivery in time, any time and consistency at all times. For a patient, a physician available over a call from his/her cell phone anytime is *Comfort*.

FIGURE 12.6
Olden days approach: physician carrying all the information in his head: no longer acceptable or possible.

For a physician who can advise knowing the current status through simplified algorithm is *Confidence.* The closed loop complete cycle of experience brings *Trust* to the patient. Consistently improving results bring *Delight* to all. Extensive studies have been conducted in Tertiary care hospitals of Chennai with FEVERWATCH® and will be soon published.

12.2.11 Need for Electronic Health Record

Just paramedical workers and gadgets alone are not enough. They have to be supported by a health record which can capture what the paramedical workers do and the gadgets record. This calls for an EHR. Figure 12.6 shows a snapshot of olden days. It is not just enough to treat the patient with paper and pen.

12.3 Problems in Health Information System in the Country

1. The cause of death statistics obtained from the civil registration system are often incomplete and of poor quality (Pandey et al. 2010).

2. There is little coordination between the agencies managing health information and little integration and reconciliation of diverse data sources.

3. Data gathering is incomplete, and the non-inclusion of the private sector excludes the major provider of health care in India.

4. Although data collection on some indicators is duplicated, vast gaps exist on others. For example, information on health determinants, adult mortality and cause of death, adult morbidity, and the coverage and costs of many interventions are poorly recorded.

5. Use of data is limited by an inadequate focus on outputs and outcomes when making decisions for allocation of funds and a shortage of skilled managers who can analyze and use the data for decision-making (Mishra et al. 2012).

6. Although no systematic data exist on the use and costs of medical technologies (e.g., cardiac stents and knee implants), their irrational use has been widely reported (Times of India).

7. The single most important impediment to a holistic approach to health governance in the country is probably the inadequate convergence between various departments within the Ministry of Health and Family Welfare that deals with medical education, health services, family welfare, and a multitude of vertically implemented national programs, and other ministries related to health, such as those that deal with water and sanitation.

8. Additionally, Village Health Sanitation and Nutrition Committees and Patient Welfare Committees have been formed in public health facilities to instill citizen accountability of the health system; however, their overall effect is limited by a range of challenges related to selection procedures in place to ensure proper representation of marginalized populations and difficulties in effecting transfer and use of funds (Ganesh et al. 2013).

If the above deficiencies are corrected with a proper electronic record, vast improvements can be made. That should be possible, if the components of hospital visits are enumerated and the health record is designed according to that need.

12.3.1 Registration

The first step is registration of the patient. It has name of the patient, parent's name, address, telephone number, etc. for the first visit. For the subsequent visits, the appointment time and the token number is allotted to the patient. If the same is done using pen and paper, it is not going to be analyzed. As shown in Figure 12.7, this prevents the delay in developing EHRs.

12.3.2 Waiting Hall (see Figure 12.8)

Maximum amount of time is wasted when people are forced to wait. As sickness situations are emotionally charged, they cannot focus their minds on anything else and the

FIGURE 12.7
Paper entry of registration is not only labor intensive, but also prevents the development of portable electronic health record.

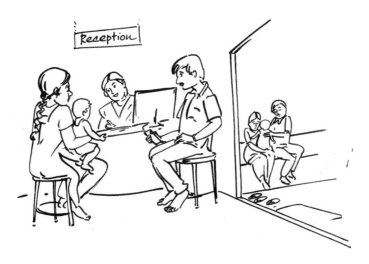

FIGURE 12.8
Using waiting hall to enter the doubts of parents, in EHR.

time is wasted. Paramedics enter those doubts and problems of patients into a system. As they enter the problems, they can tell whether the symptoms are mild, moderate, or severe.

For example in the neonatal period, the following issues do not require oral medications.

1. Noisy breathing
2. Straining
3. Cry before passing urine and motion
4. Motion: one per feed or once a week
5. Curd-like vomiting
6. Watering of eyes
7. Mild wetness around umbilicus
8. Several skin problems
9. Milk secretions from breast
10. Bleeding from the vagina

As these are entered into the system, the paramedical worker answers the parents that these are normal.

The same time is used to explain simple things like not blow the nose or apply oil in the nose while feeding. This act, still rampant in 2016, causes pneumonia.

As Box 12.2 and Figure 12.9 shows, such a comprehensive consultation can be an overload for the young mother.

The physician's consultation leads the patient to either pharmacy or if the illness demands to the laboratory.

12.3.3 Pharmacy

Each medical consultation ends up in physician writing a prescription, and each drug has at least 8 parameters to be addressed. 1. Formulation: Tablet or liquid. 2. Generic name.

BOX 12.2   CAPTURING THE PHYSICIAN'S
CONSULTATION FAITHFULLY INTO THE EHR

Components of a Consultation	Rationale
Complaints	The problem for which child has been brought today. Child may have several problems, but parents bring to the hospital for the problem which is distressing to them. For example, child may not be eating well, but would have been brought to the hospital particularly on that day for diarrhea.
History of present illness	Details of the presenting problem. For example for diarrhea, frequency, fluidity, and volume of the motion. Was there fever? Blood?
Past history	In case of diarrhea, the history of a similar problem in the past would indicate either unfavorable environment or a susceptible body constitution, both of which require evaluation.
Family history	Family history of wheeze, fits, and diabetes has an important bearing in treating the child. At present if there are any communicable problems in the family, like virus, fever is important to elicit.
Antenatal history	Hypertension, diabetes, and low thyroid problems can create havoc in the newborn period. When the child grows they can act as impediments.
Birth history	Type of delivery indicates the amount of suffering the child might have had; Usually prolonged labor results in operative deliveries.
Neonatal history	Seeds of problems like mental retardation are sown in the neonatal period. Many disorders have their beginning in the neonatal period.
Developmental history	Human brain grows only in the first 3 years and if mental delay is not picked up then there is high chance of the child becoming disabled. This is the status of several children.
Nutritional history	Food intake is an important component in the development of a child, which has both direct and indirect bearing on health.
Immunization history	Important in preventing many communicable illness.
Social and economic history	Economic and social circumstances make a child sick.
Other relevant issues	1. Separated parents. 2. Transport facilities not available for reaching the health center.

FIGURE 12.9
A comprehensive physician consultation can be an information overload for the already exhausted mother.

3. Trade name. 4. Strength: each 5 ml contains __mg or each tablet contains __mg. 5. Dose. 6. Frequency: How many times in a day or once in __hours in a day. 7. Duration: How long the physician has to be taken. 8. Relationship to food: before or after food. This process is to be repeated for every drug and if a patient receives 4 drugs means then there will be 32 data entries. Suppose a physician sees 50 patients in a day means then 1600 data entries for pharmacy alone. Needless to say errors do occur, and this can be solved only through electronic records and by paramedical workers.

12.3.4 Laboratory

This includes blood tests like blood counts, malaria, typhoid detection, and biochemical tests.

Imaging services: like X-rays and scans. If sound clinical principles are applied, they may not be required. EHR can display these results in a meaningful way.

Non-physician therapy, which includes other treatment modalities including counseling. Follow-up: The entire process has to be repeated for the second visit.

Because the healthcare visits involve so many components, the cost of health care is high. We need a good electronic record to capture the same.

The EHR was designed and developed jointly by A A Child Care Center in Vellore and HELYXON® *Health Solutions of IIT-M Research Park, Chennai.*

12.4 Problems with Traditional EHR Software

HELYXON® software has been designed so that the paramedics can enter most of the data. Physician's typing is less. This can be done before the patient is seen by the physician. As Figure 12.10 shows the waiting period is meaningfully utilized, using the software (Box 12.3).

FIGURE 12.10
Use of HELYXON® software by paramedical workers for data acquisition.

BOX 12.3   THE DRAWBACKS OF TRADITIONAL EHR SOFTWARE THAT PREVENTS THE EFFECTIVE USAGE

1. Patients are not allowed to control who can see, use, or disclose sensitive health data.
2. Lack of inter-operability with other EHR software and medical systems.
3. Hard to use by physicians.
4. Limited interaction with patients.
5. It is not patient-friendly.
6. Based on outdated technology and standards.

12.4.1 INDIVO Open Source EHR

INDIVO is the original personal health platform, enabling an individual to own and manage a complete, secure digital copy of his/her health and wellness information. INDIVO integrates health information across sites of care and over time. INDIVO is free and open-source uses open, unencumbered standards, including those from the SMART Platforms project and is actively deployed in diverse settings.

12.4.2 HELYXON®'s Vision for an EHR Targeted to Physicians of Indian Market

HELYXON® vision was to build an EHR system based on open standards and to empower the patient to control their data. The EHR should be built on loosely coupled building blocks which can be easily replaced. Physicians should also be able to easily customize the system to their needs. Multiple channels of access like mobile, tablets, etc. should be provided for parents and clinical support staff use.

12.4.2.1 Technical Approach

INDIVO was originally developed on Python 2.7 and Django 1.4. INDIVO provided Representational State Transfer—Application Programming Interface (REST API) which could be used by web, mobile apps, and other systems to interact with. INDIVO OAuth 1.0a provided the authentication system which could be used by other systems and end users. The INDIVO REST API provided facilities for user account creation and management, records management, storing medical facts, generating reports based on medical data, data management and sharing from patient login, care nets, and messaging.

Though INDIVO provided the basic API for common use cases for patients using EHR, the clinical management features and physician's services and functionality were missing. INDIVO was originally designed to be extensible and other apps could be easily built leveraging INDIVO REST API.

12.4.2.2 Following Events Handled by HELYXON® EHR System

The users of this system (Box 12.4) are parents, paramedical workers, and physicians.

Parents: When a child falls sick, even when they do not come to the hospitals, parents record the child's symptoms, in their smart phones using this software. This is called

BOX 12.4   SPECIAL FEATURES OF THE HELYXON® EHR SYSTEM

1. Migrated to the latest Python 3.4 and Django 1.8.

2. Added support for use by multiple physicians.

3. Enhanced account management and registration API to support registration of children and access of child data by parents.

4. Improved messaging by adding support for email and SMS integration. Developed scheduled jobs which will automatically message parents for appointment, vaccination reminders, inform parents of any communication from physicians, etc.

5. Developed an appointment system which supports both slot and token-based booking.

6. Developed a vaccination system for managing child vaccination as per IAP, reminders to parents, advisory information for parents, etc.

7. From the physician's perspective, inventory management for stocking vaccination, automated order processing for vaccines with the vendors/pharmacist chosen by physician, forecast to the physician for managing vaccination inventory.

8. Growth charts for track and monitor child growth as per Indian Academy of Pediatrics (IAP) standards.

life log and Figure 12.11 shows it further. Examples include recording of cold cough, fever, vomiting, etc. whenever they occur.

As depicted in Figure 12.12, during the next visit, the previous occurrences of illness recorded to the system through mobile will be discussed by the physician.

The second component is handled by the paramedical worker. When the child comes to the hospital, the complaints are further discussed in detail and recorded to the system, as shown in Figure 12.13.

This can save tremendous time for the physician. For example, we can consider a very common problem diarrhea. World Health Organization classifies diarrhea into three groups: Those requiring admission to hospitals—children who have not passed urine in the last 8 hours or sick looking children; children requiring antibiotics and children who need only oral rehydration solution. It will take latest 2 minutes to gather all the information and to take a decision. So for 50 patients in a day, it would mean 100 minutes. This can be easily done by the paramedical worker. The software discussed here helps them to do so. Moreover, even if data are entered wrongly by the worker, this can be always be corrected by the physician. This portion of the data can be stored in both the hospital computer and parent mobile.

After the basic complaints are entered, the growth of the child is recorded. One distinguishing feature of children is their growth. Weight recording tells about the immediate past events of the child (Figure 12.14).

For example if a child had weighed 9 kg last month and now it is 8 kg, then the cause should be found out.

Head circumference measurement is a quick guide to the brain development of the child. Any child delay in brain development has to be picked up in the first year of life.

FIGURE 12.11
Life log to record events at home by parents.

FIGURE 12.12
Life log data reviewed by physician during the next visit.

The easiest way, which can be done by anybody, is to check the head circumference and these are shown in Figure 12.15; it just takes less than a minute for a trained paramedical worker to do these.

The height measurements and its plotting in a graph can predict endocrine problems, constitutional growth delay, and stunting (Figure 12.16). This will lead the physician to take appropriate action.

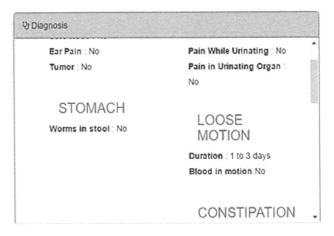

FIGURE 12.13
Paramedic expands the complaints, using HELYXON®.

FIGURE 12.14
Weight measurement says about recent past of the child.

Growth monitoring, no wonder, has been used as a tool for child survival (Figure 12.17). The greatest advantage here is the paramedic can do it and parents also can do it at home.

As the data travel along with the parents, whenever the child is taken to a new center, they are visible in the second place instantly. *These graphs are automatically generated once the data are entered by the paramedic. This eases the physician task and helps him to take quick decisions.*

Children become adolescents and then adults, and pediatrician will be replaced by the adult physician. What has been happening over the years can be brought to the screen instantly. The type of illness child has been undergoing, if displayed chronologically, can help in planning appropriate preventive measures as in the case of asthma.

Other advantages of using the EHRs:

1. Facility to handle multiple child profiles
2. Manage child vaccination: Vaccine reminders are automatically sent

FIGURE 12.15
Head circumference measurement is mandatory in the first year. Chest circumference between 1 and 5 years.

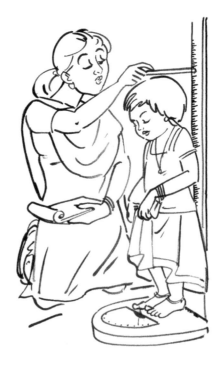

FIGURE 12.16
Height measurement can tell about the past health and predict the future problems of the child.

FIGURE 12.17
Children growth chart visible both to the physician on the computer and to the parents in their mobile phones.

3. Messaging management: for example by eight months a child should sit on its own. If there is a delay the physician should be consulted.

4. Control access and sharing information with other physicians

5. Schedule appointment with physicians

6. Chat with physician online

7. View advisory content from physician: For obesity, asthma, epilepsy, etc.

12.4.3 Role of Paramedical Workers in Various Health Programs

1. Paramedics can effectively take the burden from physicians has been proved in many studies. Even in the litigation-prone country like USA, non-physicians like registered nurses are allowed to practice (Clayton 2009).

2. Psychological interventions delivered by non-specialist health workers are effective for the treatment of prenatal depression in low- and middle-income countries (Chowdhary et al. 2014).

3. Structured and practical teaching and learning with minimal theory in resource-limited setting had a positive, short-term effect on the competence of individual staff to carry out an initial assessment and manage an acutely unwell patient (Stanley et al. 2015).

4. Home visits by community health workers to improve identification of serious illness and care seeking in newborns and young infants from low- and middle-income countries (Tripathy 2016).

5. There is a unique referral model in private eye care institutions linking primary- to tertiary-level eye care, at LV Prasad Eye Institute, Hyderabad (Rao et al. 2012).

6. In a Tanzanian study just four neurosurgeons could train paramedics in surgical procedures (Ellegala et al. 2014).

7. A descriptive study was performed to collect information on the feasibility of sub- stituting hospital care with primary care. This study concluded that it is possible to substitute hospital care with primary care and one important tool required is integrated information technology system (Van Hoof et al. 2016).

12.5 EHRs: Challenges and Solution to the Challenges

There exist barriers to adoption of e-health. Not all parents or physicians or paramedical workers accept the EHR. Following are some of the challenges faced during transition from conventional methods to electronic health care.

1. *Time:* Most of the physicians do not take sufficient time to become successful in EHR (Loomis et al. 2002).

2. *Cost:* For small- to medium-sized practices without large IT budgets, costs remain the biggest barrier to adoption (Rao et al. 2011).

3. *Absence of computer skill:* Lack of coordination between patient care and data entry by most users (Loomis et al. 2002).

4. *Concern about security and privacy:* Despite evidence to the contrary, non-users believe that there are more security and confidentiality risks involved with EHRs than paper records. Figures 12.18 and 12.19 convey these concerns in a lighter vein.

5. *Communication among users:* Positive EHR experience has to be shared, and social networks are very useful (Castillo et al. 2010).

6. *Interferes with physician–patient relationship:* Figure 12.20 captures this concern.

7. Some physicians reported that they sometimes stop using EHRs because hunting for menus and buttons disrupts the clinical encounter (Loomis et al. 2002).

8. *Lack of incentives:* This barrier could be easily corrected through financial rewards and appreciation for quality improvement (Meinert 2004).

9. *Complexity:* Physicians should spend sufficient time initially to understand the complex world of EHRs (ibid).

10. *Physical space barriers:* Space shortage, as the computers and accessories may take away large spaces (Ford et al. 2006).

11. Concern about the ability to select an effective EHR system. Physicians wonder if the software will suit their style of practice (Miller and Sim 2004).

12. *Technical support:* Particularly during off hours, holidays are a concern for many physicians (Ford et al. 2006).

13. *Interoperability as a determinant factor for adopting these systems:* Interoperability is important because it decreases the cost of EHRs and makes it feasible for an

FIGURE 12.18
Concern about data theft.

FIGURE 12.19
Concern about data privacy.

FIGURE 12.20
Wrong notion that computers affect physician–patient relationship.

individual or small group of physicians to acquire and adopt these systems (Castillo et al. 2010).

14. *Access to computers and computer literacy:* The low level of electronic medical records system use could be explained by a lack of available computers (Ross 2009).

15. *Vendor trust:* Physicians are concerned that vendors are not qualified to provide a proper service, or will go out of business and disappear from the market, leading to a lack of technical support and a large financial loss (Boonstra and Broekhuis 2010).

16. *Expert support:* Expert support refers to the assistance provided from a physician to another physician. This can be divided in two aspects: (1) a physician with experience in EHRs usage assists with information about how to use the system to another physician; (2) a physician has the knowledge to help another physician accomplish a medical task. Such assistance can be given through personal contact or via documents (Castillo et al. 2010).

17. *Concern about data entry:* Family physicians are sought after for their counseling abilities. Time taken to make data entry is the largest potential obstacle to the effective use of computers in family medicine (ibid).

18. *Reliability:* Reliability refers to the dependability of the technology systems that comprise the EHR. As more vendors enter the lucrative healthcare market, the number of competing systems will increase. Vendors will seek to differentiate themselves from competitors using quality and reliability of their EHR systems as evidence of their superiority (Rao et al. 2011).

19. *Inadequate data exchange:* Data are generated from physicians, pharmacy, lab, radiology, and referral to other hospitals. From patient point of view, policies should hasten the creation of community-wide data exchange systems that allow clinicians to view all of their patients' data, regardless of provider and care site (Meinert 2004).

20. *Concern about patient acceptance:* Physicians would spend more time interacting with the computer than the patient (Ajami et al. 2011). This need not be so always. With experience it will disappear.

21. *Formal training:* Physician style of learning and his/her clinical needs have to be taken into account in training (Ford et al. 2006).

12.5.1 Problems Faced with Wearable Devices in Health Care

At present, FEVERWATCH®, which is available in the market, measures only temperature. In an inpatient setting this may not be sufficient. The same sensor should be able to transmit blood pressure and respiratory rate, so that the paramedical worker can make intelligent decisions and help the physicians.

There are other issues to be supervised.

- *Legal requirements:* As this is a new field, legal requirements is not fully known to use in a hospital setting. However, pilot studies are going on in tertiary care centers in Chennai.
- *Cost:* Like all gadgets the initial cost will be high and it will come down only when the volume increases.

12.6 Best Practices for Usage of EHRs

We must remember the following points while moving toward EHR usage

1. Use open standards for architectural design and implementation to develop an EHR system. When developed using building blocks, it can be easily replaced with other plug-ins.
2. Develop user-friendly mobile and tablet apps to improve patient usage and interaction with the system.
3. Provide real-time and static-messaging facilities to make the patient constantly in touch with the physicians.
4. Provide reminders and notifications to parents on trigger of some health condition, which deserves attention, like asthma.
5. Life log facility to enable parent to store observations by parents and subsequent review by parents.
6. Care net feature which enables sharing data with near and dear ones.

12.7 Conclusions and Future Works

Despite the positive effects from using EHRs in medical practices, the adoption rate of such systems is still low, and they meet resistance from physicians. EHR usage requires the presence of trained users, system facilities, support from others, and numerous organizational and environment facilitators. In addition, difficulty of using EHRs and the non-use of specific functions is due to the presence of above-mentioned barriers. For the EHR systems to have a positive impact on patient safety, clinicians must be able to use these systems effectively with the provided system features. It has been shown that implementers can insulate the project from concerns by establishing strong leadership, using project management techniques, establishing standards and training their staff to ensure the reliability and superiority of implementation. Paramedics can be trained at medical college level itself for usage, maintenance, and upgradation of such as system with ease.

Continuous monitoring of critical parameters along with right information is going to reduce the anxiety of the patients. Car panel digital display of number of kilometers left before the next gas station, laptop displaying the remaining MB of data in downloads time, and cell phone showing the remaining battery charges in % are possible due to continuous monitoring. These kinds of warnings reduce the anxiety of the users. Why can't they be adapted in healthcare? People need to know their health status in a way they can understand and relate. The continuously monitored health data can be interpreted by the system in the same way for the physician to alert the patients before they could end up with a major health risk. To enable this, tie-up between the certified physicians within the vicinity, patients, and IoT service providers is essential to bring personalized health monitoring and to start first aid treatment on emergencies even before the ambulance arrives.

Awareness brings the higher quality need. It becomes easy to convince the patient on early actions required like vaccination and other precautionary medications. We need to exploit the technological capability to develop a well-informed patient and the physician through different communication media such as email, SMS, message chat, voice chat, and phone calls. Patients can be encouraged to start noting all the symptoms in the right sequence, so that the exaggerations or miss-communication by patient to physician can be avoided completely. We are heading toward that day when physician is able to provide best possible personalized care, which can provide from symptomatic remedy, to cause-based solution and prevention. Prognostication becomes a science instead of educated guess. Above all, these situations can be handled by a trained paramedical worker, provided suitable EHR and wearable computational devices are available.

References

Ajami, S., Ketabi, S., Saghaeian-Nejad, S., Heidari, A. Requirements and areas associated with readiness assessment of electronic health records implementation. *Journal of Health Administration* 2011; 14: 71–8.

Black, E. 2016. *Obamacare and the cost of gridlock.* Available from: http://data.worldbank.org/indicator/SH.MED.BEDS.ZA?year high desc=false (accessed January 11, 2017).

Boonstra, A., Broekhuis, M. Barriers to the acceptance of electronic medical records by physicians from systematic review to taxonomy and interventions. *BMC Health Services Research* 2010; 10: 231.

Castillo, V., Martinez Garcia, A., Pulido, J. A knowledge-based taxonomy of critical factors for adopting electronic health record systems by physicians: A systematic literature review. *BMC Medical Informatics and Decision Making* 2010; 10: 60.

Chowdhary, N., et al. The content and delivery of psychological interventions for perinatal depression by non-specialist health workers in low and middle income countries: A systematic review. *Best Practice & Research. Clinical Obstetrics & Gynaecology* 2014; 28(1): 113–33.

Clayton, M.C., Grossman, J.H., Hwang, J. 2009. *The innovator's prescription: A disruptive solution for health care.* New York: McGraw-hill.

Demographics of India. Available from: https://en.wikipedia.org/wiki/Demographics_of_India (accessed January 11, 2017).

Ellegala, D.B., et al. Neurosurgical capacity building in the developing world through focused training. *Journal of Neurosurgery* 2014; 121: 1526–32.

Fayyoumi, E., Idwan, S., Karss, A.A. Developing fever watcher-kit via bluetooth wireless technology. *International Journal of Business Information Systems* 2014; 16: 4.

Ford, E.W., Menachemi, N., Phillips, M.T. Predicting the adoption of electronic health records by physicians: When will health care be paperless? *Journal of the American Medical Informatics Association* 2006; 13: 106–12.

Ganesh, S., Kumar, S., Sarkar, S., Kar, S., Roy, G., Premarajan, K. Assessment of village water and sanitation committee in a district of Tamil Nadu. *Indian Journal of Public Health* 2013; 57: 43–6.

Giuliano, K.K. Continuous physiologic monitoring and the identification of sepsis: What is the evidence supporting current clinical practice? *AACN Advanced Critical Care* 2006; 17(2): 215–23.

India Internet Users. Available from: http://www.internetlivestats.com/internet-users/india/ (accessed January 11, 2017).

India's Internet users to double to 730 million by 2020 leaving US far behind. Available from: http://economictimes.indiatimes.com/tech/internet/indias-internet-users-to-double-to-730-millionby-2020-leaving-us-far-behind/articleshow/53736924.cms (accessed January 11, 2017).

Isaacs, D. Editorial The neonatal antibiotic crisis. *Indian Pediatrics* 2005; 42: 9–13.

Krieger, N. Theories for social epidemiology in the 21st century: An ecosocial perspective. *International Journal of Epidemiology* 2001; 30(4): 668–77.

Kumar, V., Shearer, J.C., Kumar, A., Darmstadt, G.L. Neonatal hypothermia in low resource settings: A review. *Journal of Perinatology* 2009; 29; 401–12.

Loomis, G.A., Ries, J.S., Saywell, R.M., Thakker, N.R. If electronic medical records are so great, why aren't family physicians using them? *Journal of Family Practice* 2002; 51: 636–41.

Mare, S., Kotz, D. *Is bluetooth the right technology for mHealth?* Position paper in USENIX Workshop on Health Security (HealthSec), 2010. Available from: http://www.cs.dartmouth.edu/~dfk/papers/mare-healthsec10.pdf (accessed January 17, 2017).

Meinert, D.B. Resistance to electronic medical records (EMRs): A barrier to improved quality of care. *Informing Science: International Journal of an Emerging Transdiscipline* 2004; 2: 493–504.

Miller, R.H., Sim, I., *Physicians' use of electronic medical records: barriers and solutions*, Heallth aff (Millwood). 2004; 23(2): 116–26.

Mishra, A., Vasisht, I., Kauser, A., Thiagarajan, S., and Mairembam, D. Determinants of Health Management Information Systems performance: Lessons from a district level assessment. *BMC Proceedings* 2012; 6: O17.

Pandey, A., Roy, N., Bhawsar, R., Mishra, R. Health information system in India: Issues of data availability and quality. *Demography India* 2010; 39: 111–28.

Pavithra, M., 2016. *Health budget figures tell a sick story.* Available from: https://thewire.in/24924/health-budget-figures-tell-a-sick-story/ (accessed January 11, 2017).

Rao, G.N., Khanna, R.C., Athota, S.M., Rajshekar, V., Rani, P.K. Integrated model of primary and secondary eye care for underserved rural areas: The L V Prasad Eye Institute experience. *Indian Journal of Ophthalmology* 2012; 60: 396–400.

Rao, S.R., DesRoches, C.M., Donelan, K., Campbell, E.G., Miralles, P.D. Electronic health records in small physician practices: Availability, use, and perceived benefits. *Journal of the American Medical Informatics Association.* 2011; 18: 271–75.

Ross, S. Results of a survey of an online physician community regarding use of electronic medical records in office practices. *The Journal of Medical Practice Management* 2009; 24: 254.

Stanley, L., et al. A tool to improve competence in the management of emergency patients by rural clinic health workers: A pilot assessment on the Thai-Myanmar border. *Conflict and Health* 2015; 9: 11.

Sukumar, S. 2012. *The average cost of building a hospital is Rs. 3,800/Sq.Ft.* Available from: http://www.constructionworld.in/News.aspx?nId=7hJUPJaHhkVfHvkKChsWWg==andNewsTy pe=The-average-cost-of-building-a-hospital-is-Rs-3,800/sq-ft (accessed January 11, 2017).

Telecommunications in India. Available from: https://en.wikipedia.org/wiki/Telecommunications_statistics_in_India (accessed January 11, 2017).

Tripathy, P., et al. Effect of a participatory intervention with women's groups on birth outcomes and maternal depression in Jharkhand and Orissa, India: A cluster-randomised controlled trial. *Lancet* 2010; 375(9721): 1182–92.

Van Hoof, S.J.M., et al. Substitution of hospital care with primary care: Defining the conditions of primary care plus. *International Journal of Integrated Care* 2016; 12: 1–11.

Zhu, Z., et al. Wearable sensor systems for infants. *Sensors* 2015; 15(2): 3721–49.

Video:

Use of FEVERWATCH®, 2015. https://www.youtube.com/watch?v=S32qnaGLiUQ

13

Semantic Technologies for IoT

Viswanathan Vadivel and Shridevi Subramanian

VIT University

Vellore, India

CONTENTS

13.1 Introduction

Semantic web technologies are becoming very popular and are adopted by companies such as Google and Yahoo. Websites use Schema.org to enhance the research results. Schema.org is a set of vocabularies called ontologies (Gruber 1995) to describe data on the web in an unified way such as Person and Organization. Google introduces the idea of the knowledge graph to structure and connect data with each other. Moreover, "Linked data" is very popular to share and reuse data to build and enhance rich web applications with little effort. The major challenge for the IoT would be to help developers in designing and developing interoperable inter-domain IoT applications. As devices are not interoperable with each other due to their proprietary formats, they do not use vocabulary to describe interoperable IoT data. One way to make them interoperable would be a common protocol used by all devices. Another solution would be to work on the interoperability of this data, since these devices are already deployed and data are already produced. Exploiting, combining, and enriching device data to build smarter interoperable applications is becoming a real challenge. The growing linked open data encourages to share the data on the web, including sensor data. To assist users or even machines in interpreting and combining the sensor data, there is a real need to explicitly describe sensor measurements according to the context, in a unified way and in a manner understandable by machines. For instance, temperature measurement does not mean the same always and its meaning varies according to the context (room temperature, body temperature, water temperature, or external temperature) and the machine will not infer the same knowledge (fever deduced with body temperature, abnormal temperature for room temperature). We also need to deal with implicit units (e.g., Fahrenheit, Celsius, Kelvin).

The second biggest challenge is combining domain-specific applications and data together to create innovative cross-domain IoT applications. Existing applications are specific to one domain such as smart home, smart health care, transportation, and smart garden. Examples of innovative cross-domain IoT applications: (1) suggesting food according to the weather forecast, (2) suggesting home remedies according to health measurements, and (3) suggesting safety equipment in a smart car according to the weather, etc. Future smart fridges will enable to purchase groceries online. In case of RFID tags embedded in food, it will be easy to recommend the menu for dinner or automatically order essential ingredients. If you are an athlete, the smart fridge will recommend you the perfect diet in case of compatibility with Apple Nike shoes. Finally, Google's car could automatically take you to the grocery store to grab the missing ingredients. Last but not the least, security issues should be considered when combining IoT data or designing IoT applications. For instance, health data are more sensitive than weather data and need to be secured.

Further, according to Barnaghi et al. (2012), semantics is required at different levels in IoT; it can be used to: (1) describe things and data, (2) reuse domain knowledge, (3) interpret IoT data, (4) provide smarter applications, and (5) provide security (Figure 13.1).

The existing challenges for the semantics-enabled IoT can be summarized as follows:

Generating interoperable cross-domain semantic-based IoT applications. This process should be flexible enough to be performed either on the cloud, constrained devices, or machine-to-machine (M2M) gateways. M2M gateways mean that processing is done automatically, without requiring human intervention.

Interpreting sensor data and inferring new knowledge by reusing domain knowledge expertise. Reusing domain knowledge (e.g., ontology) is highly recommended (Simperl 2009; Suárez-Figueroa 2010). The interoperability of domain knowledge enables building of cross-domain expertise.

Securing IoT applications when designing these applications. This challenge should be solved using the same approach as for the two previous challenges by combining security knowledge expertise to find the most suitable security mechanisms to secure IoT applications.

To deal with the challenges, we exploit semantic web technologies (Berners-Lee et al. 2001) for several reasons. First, semantics enables an explicit description of the meaning of sensor data in a structured way, so that machines could understand it. Second, it facilitates interoperability for data integration since heterogeneous IoT data is converted according to the same vocabulary. Third, semantic reasoning engines can be easily employed to deduce

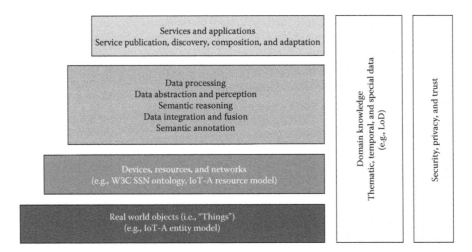

FIGURE 13.1
Semantics is required in the IoT. (From Barnaghi, P., et al., *International Journal on Semantic Web and Information Systems (IJSWIS)*, 8(1), 1–21, 2012. With permission.)

high-level abstractions from sensor data. Fourth, context-awareness could be implemented using semantic reasoning. Finally, in theory, semantics eases the knowledge sharing and reuse of domain knowledge expertise which should avoid the reinvention of the wheel. Indeed, each time a new domain-specific vocabulary is defined.

13.2 Semantic Technology

13.2.1 Resource Description Framework

The resource description framework (RDF) (Klyne and Carroll 2004) is a framework for graph-based information representation about resources and was intended for representing metadata about WWW resources. RDF gives interoperability between various applications that exchange information, which can be inferred by machine on the web. Various uses of RDF are given below:

- Providing standards for exchanging data between databases.
- By connecting RDF descriptions of people across multiple websites, distributed social networks can be built.
- Cross-dataset can be executed with the help of Simple Protocol and RDF Query Language (SPARQL) query if the datasets are interconnected in an organization.

13.2.1.1 RDF Data Model

An RDF is a graph-based data model, and subject–predicate–object expression is its basic structure. We can make a statement about resources using RDF. Each resource is associated with universal resource identifier (URI). An URI is structured like an URL (Uniform Resource Locator) or some other type of identifier. Special kinds of resource called properties or predicates describe relations between resources. For example: "hasLoca-tion," "hasType," "hasID," "deviceID." The relationship from subject to object is expressed in a directional way. Properties in RDF are also identified by URIs. Say, for example, "hasLocation" can be defined as: http://www.loanr.it/ontologies#hasLocation. RDF state-ments are called triples because they consist of three elements. An RDF statement has the following syntax:

<subject> <predicate> <object>

Example:
<"Sensor", Of_Type, "Temperature">

IRIs, literals, and blank nodes are three different types of nodes in an RDF graph. The positional constraints in the RDF are as follows: Predicate must be an URI; subject must be an URI or blank node; and object must be a URI, blank node, or literal. An RDF graph is a set of RDF triples. It can be visualized as a directed labeled graph. Each triple in an RDF graph is represented as a node-arc-node connectivity. Figure 13.2 shows the structure of RDF graph and Figure 13.3 shows the example of an RDF graph.

13.2.1.2 Internationalized Resource Identifier

An IRI is an Internet standard extended upon the existing URI, and it was published in RFC 3987 standard (Horrocks et al. 2004). The RDF resources are identified by IRI, which can appear in the subject, predicate, and object positions of a triple. IRI is in the URL form.

13.2.1.3 Literals

Literals are fixed values. Some examples of literals include strings, dates, and num-bers. Literals are related with a data type which allows values to be parsed and

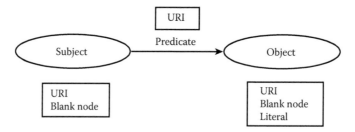

FIGURE 13.2
Structure of RDF graph with subject, object and predicate.

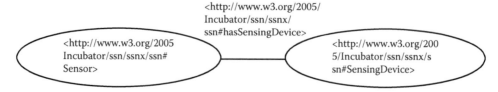

FIGURE 13.3
Example RDF graph.

interpreted appropriately. It may appear only in the **object position** but not in the subject or the predicate positions in the RDF triple. Generally, literals with string data type can be related with a language tag. Consider the following example: "dimanche" can be related with the "fr" language tag and "Sunday" with the "en" language tag.

13.2.1.4 Blank Nodes

When we talk about resources without the consideration of the use of the global identifier, the blank nodes come into picture. Blank nodes are used to denote resources and its naming can be given without an IRI. In an RDF triple, these blank nodes may be used in the **subject and object positions**.

13.2.1.5 Multiple RDF Graphs

Sometimes, it is advantageous to work with multiple RDF graphs combined easily with the data from multiple sources. An RDF dataset is a collection of RDF graphs with zero or more named graphs. In an RDF dataset, the named graphs are the RDF graphs that are associated with IRI or blank node. The blank node or IRI is called the graph name. The other graph which is remaining in the RDF dataset without a related IRI is called the default graph.

13.2.1.6 RDF Vocabulary

An RDF vocabulary is a pre-defined set of predicates that can be used in an application. List of RDF vocabulary is given in Table 13.1.

13.2.1.7 Writing RDF Graphs

For writing RDF graphs, different serialization formats are available such as Turtle, N-Triple, RDF/JSON, RDF/XML, and N3 or Notation3. Nevertheless, using the different formats for writing down the same graph leads to logically equivalent triples.

(a) Turtle

A Turtle serialization format is a textual representation. Single-line comments can be given after a "#" which is not part of the lexical token. Simple triple statement is a chain of subject part, predicate part, and object part, each of which can separated by space and ends with by "."

TABLE 13.1

RDF Classes

Classes	Comment
rdf:Statement	Class of RDF statements
rdf:Property	Class of properties
rdf:nil	Empty list
rdf:Bag	Unordered containers class
rdf:Seq	Ordered containers class
rdf:Alt	Class of containers of alternatives
rdf:List	RDF lists class
rdf:XMLLiteral	XML literal values

<https://w3id.org/saref#EnergyMeter> <https://w3id.org/saref#IsUsedFor>
< https://w3id.org/saref#:Energy>.

Often the same subject part in an RDF triple will be related with more than one predicate and object in RDF triple. Hence, when the predicate and the object vary for the same subject of triples, a ";" can be used to repeat the subject. The following two examples are logically equivalent:

<https://w3id.org/saref#EnergyMeter> <https://w3id.org/saref#IsUsedFor>
 <https://w3id.org/saref#:Energy>
 <https://w3id.org/saref#consistsOf>
 <https://w3id.org/saref#Meter>.

<https://w3id.org/saref#EnergyMeter> <https://w3id.org/saref#IsUsedFor>
 <https://w3id.org/saref#:Energy>.
<https://w3id.org/saref#EnergyMeter> <https://w3id.org/saref#consistsOf>
 <https://w3id.org/saref#Meter>.

When the object varies with the same subject and predicate, a "," can be used to repeat the subject part and predicate part of the triple. The following examples are logically equivalent ways of writing the unit measurement in two different languages.

<https://w3id.org/saref#Meter> <https://w3id.org/saref#hasUnit>
              "meter"@en, "метр"@ru.

<https://w3id.org/saref#Meter> <https://w3id.org/saref#hasUnit>
              "meter"@en.
<https://w3id.org/saref#Meter> <https://w3id.org/saref#hasUnit>
              "метр"@ru.

(b) N-Triples
N-Triples are a sequence of RDF terms representing the subject, predicate, and object. Each RDF term in the sequence of N-Triples may be separated by space. This sequence ends by a ".", and it is optional to the last triple in the document. It is a subset of Turtle and Notation 3.

<https://w3id.org/saref#EnergyMeter> <https://w3id.org/saref#hasFunction>
<https://w3id.org/saref#MeteringFunction>.

<https://w3id.org/saref#EnergyMeter> <https://w3id.org/saref#hasCategory>
<https://w3id.org/saref# Meter>.

(c) RDF/JSON
RDF/JSON serialization format represents a set of RDF triples as a series of nested data structures.

A single JSON object in an RDF/JSON document is called the root object. Subject of the RDF triple will be key in the root object. Key may appear only once in the root object. Each root object's key value is a JSON object, whose keys are the URIs of the properties associated with each subject in a triple. These keys are called predicate keys. The predicate keys may present only once in a JSON object and the value of each predicate key is an array of JSON objects.

An RDF triple can be written in the following way:

{ "Subject" : { "Predicate" : [Object] } }

In a triple, the object is represented as a JSON object as follows:

type (mandatory and must be lower case)
 'uri' or 'literal' or 'bnode'
value (mandatory and full URI should be used)
 its lexical value
datatype (optional)
 the data type of the literal value
lang (optional)
 language of a literal value

Example :
```
{
  " https://w3id.org/saref#Switch" : {
  " https://w3id.org/saref#hasName" : [ { "value" : " "Light switch",
                       "type" : "literal",
                       "lang" : "en" } ]
}
}
```

(d) N3 or Notation3

In N3 notation, the subject, predicate, and object of an RDF triple are separated by spaces and each triple is terminated with a period (.). The syntax of N3 would be:

```
@prefix saref: <https://w3id.org/saref#>.
saref:EnergyMeter saref:IsUsedForr    saref:Energy;
        sasaref:consistsOf   saref:Meter.
```

(e) RDF/\XML

RDF/XML is an XML syntax for RDF in terms of namespaces in XML. With the help of XML syntax, RDF information can easily be exchanged between computers independent of operating systems and application languages. RDF/XML is a syntax defined by the W3C to serialize an RDF graph as an XML document.

<rdf:RDF> is the root element of an RDF document. It defines the XML document to be an RDF document. It also contains a reference to the RDF namespace as follows:

```
<?xml version="1.0"?>
<rdf:RDF
      xmlns:rdf="http://www.w3.org/1999/02/22-rdf-syntax-ns#">
      .............
</rdf:RDF>
```

The <rdf:Description> element contains the description of the resource identified by the rdf:about attribute.

The property elements can also be defined as attributes:

```
<?xml version="1.0"?>
<rdf:RDF
    xmlns:cd=" http:/sensormeasurement.appspot.com/m3#"
    xmlns:rdf="http://www.w3.org/1999/02/22-rdf-syntax-ns#">
<rdf:Description
    rdf:about="http:/sensormeasurement.appspot.com/m3#Measurement"
    cd:name=" Temperature " cd:unit="DegreeCelsius" cd:value="35"/>
</rdf:RDF>
```

The property elements can also be defined as resources:

```
<?xml version="1.0"?>
<rdf:RDF
    xmlns:cd=" http:/sensormeasurement.appspot.com/m3#"
    xmlns:rdf="http://www.w3.org/1999/02/22-rdf-syntax-ns#">
<rdf:Description
  rdf:about="http:/sensormeasurement.appspot.com/m3#Measurement">
<cd:name rdf:resource="http:/sensormeasurement.appspot.com/m3#Temperature"/>
  ...
  ...
</rdf:Description>
</rdf:RDF>
```

RDF/XML allows use of the xml:lang attribute to identify the content language. This attribute can be used in any resource to indicate that the content given in the tag is in the specific language. The xml:lang="" indicates that there is no language specified.

```
<?xml version="1.0" encoding="utf-8"?>
<rdf:RDF   xmlns:saref=" https://w3id.org/saref#"
            xmlns:rdf="http://www.w3.org/1999/02/22-rdf-syntax-ns#">
  <rdf:Description rdf:about="saref:Measurement">
    <saref:hasUnit xml:lang="en">meter</saref:hasUnit>
    <saref:hasUnit xml:lang="ru"> метр </saref:hasUnit>
  </rdf:Description>
</rdf:RDF>
```

In an RDF, typed literals can be used as an object node. It consists of a string and a data-type IRI. These are written in RDF/XML using an attribute rdf:datatype="*datatypeURI*" within the property element.

```
<?xml version="1.0"?>
<rdf:RDF   xmlns:saref=" https://w3id.org/saref#"
            xmlns:rdf="http://www.w3.org/1999/02/22-rdf-syntax-ns#">
<rdf:Description rdf:about="saref:MeteringFunction">
<saref:hasMterReadingType rdf:datatype="http://www.w3.org/2001/XMLSchema#string">
Water
 </saref:hasMterReadingType>
```

```
  </rdf:Description>
</rdf:RDF>
```

Instead of rdf:about, the rdf:ID attribute on a node element can be used, and it gives a relative IRI which is equivalent to "#" symbol concatenated with attribute rdf:ID. The below example shows abbreviating the node IRI of https://w3id.org/saref#TemparatureUnit using an xml:base of http://example.org/case/ and an rdf:ID on the rdf:Description node element.

```
<?xml version="1.0"?>
<rdf:RDF    xml:base="http://example.org/case/"
            xmlns:rdf="http://www.w3.org/1999/02/22-rdf-syntax-ns#"
            xmlns:rx="https://w3id.org/saref#/"gt;
  <rdf:Description rdf:id="Ch13-Temparature">
   <rx:hasUnit rdf:resource="TemparatureUnit"/>
  </rdf:Description>
</rdf:RDF>
```

RDF has container membership properties called rdf:_1, rdf:_2, etc., and these property elements are used with instances of the rdf:Alt, rdf:Seq, and rdf:Bag classes as shown in the following example. The rdf:li is a special property element, and it can be used instead of rdf:_1, rdf:_2, etc.,

```
<?xml version="1.0"?>
<rdf:RDF    xmlns om::"http://www.wurvoc.org/vocabularies/om-1.8/"
            xmlns: saref="https://w3id.org/saref#"
            xmlns:rdf="http://www.w3.org/1999/02/22-rdf-syntax-ns#">
  <rdf:Seq rdf:about=" https://w3id.org/saref#TemperatureUnit ">
   <rdf:_1 rdf:resource="om:degree_Celsius"/>
   <rdf:_2 rdf:resource="om:degree_Fahrenheit"/>
   <rdf:_3 rdf:resource="om:kelvin"/>
  </rdf:Seq>
</rdf:RDF>
```

RDF/XML allows an rdf:parseType="Collection" attribute on a property element which contains multiple node elements. A set of subject nodes of the collection can be given using these nodes. A collection of subject nodes can be represented as given below:

```
<?xml version="1.0"?>
<rdf:RDF    xmlns: sare="https://w3id.org/saref#"
            xmlns om::"http://www.wurvoc.org/vocabularies/om-1.8/"
            xmlns:rdf="http://www.w3.org/1999/02/22-rdf-syntax-ns#">
  <rdf:Description rdf:about="https://w3id.org/saref#TemperatureUnit">
   <sare:hasUnit rdf:parseType="Collection">
     <rdf:Description rdf:about="om:degree_Celsius"/>
     <rdf:Description rdf:about="om:degree_Fahrenheit"/>
   <rdf:Description rdf:about="om:kelvin"/>
   </sare:hasUnitt>
  </rdf:Description>
</rdf:RDF>
```

Note: RDF/XML serialization format is used in all examples throughout this chapter.

13.2.2 RDF Schema

RDF Schema is a semantic extension of RDF. RDF Schema is abbreviated in different ways as RDFS, RDF(S), RDF-S, or RDF/S. RDFS provides a data-modeling vocabulary for RDF data. It provides a way for describing groups of resources and the relationships between these resources. These resources can be used to determine characteristics of other resources such as the domains and ranges of properties.

RDFS uses the class to identify groups, and it can be used to group the resources. The relationship between an individual and its class is represented by using property. RDFS is used to create class hierarchies and of properties and (Table 13.2). Domain and range restrictions can be used to provide restrictions on the subjects and objects of RDF triples. For example, subject of "hasUnit" triples should be of class "Measurement" and object should be "MeasurementUnit". Tables 13.3 and 13.4 provide an overview of the RDFS vocabulary.

TABLE 13.2

RDF Properties

Properties	Comment
rdf:type	Type of an instance
rdf:first	First item in a list
rdf:rest	Rest of a list
rdf:subject	Subject of the subject RDF statement
rdf:object	Object of the subject RDF statement
rdf:value	Idiomatic property used for structured values
rdf:predicate	Predicate of the subject RDF statement

TABLE 13.3

RDFS Classes

Class Name	Comment
rdfs:Resource	The class resource, everything
rdf:Property	Class of RDF properties
rdfs:Class	Class of classes
rdfs:Literal	Class of literal values
rdfs:Datatype	Class of RDF data types

TABLE 13.4

RDFS Properties

Property Name	Comment
rdfs:range	Range of the subject property
rdfs:domain	Domain of the subject property
rdfs:subClassOf	Subject is a subclass of an object class
rdfs:subPropertyOf	Subject is a subproperty of an object
rdfs:comment	Description of the resource
rdfs:label	Human understandable name for the resource
rdfs:isDefinedBy	Definition of the subject resource
rdfs:Container	Class of RDF containers
rdfs:seeAlso	More information about the subject resource

RDF vocabularies used worldwide are given below:

"Friend of a Friend" (FOAF) (Brickley and Miller 2004) vocabulary is used to represent the relations between entities in social networks.

Dublin Core is used to represent the relations between video, image, and other web resources.

schema.org is used to describe event, health, and medical types, Organization, Person, Place, LocalBusiness, Restaurant, etc.

13.2.3 Web Ontology Language

Web Ontology Language (OWL) (Patel-Schneider et al. 2012) is a set of mark-up languages that are designed for use by applications to process the content of information instead of just presenting information to humans. The OWL ontology describes the hierarchical organization of ideas in a domain, in a way that can be parsed and understood by software. OWL has more facilities for expressing meaning and semantics than XML, RDF, and RDFS, and thus OWL goes beyond these languages in its ability to represent machine interpretable content on the web. OWL is part of the W3C recommendations related to the semantic web. It is summarized as follows.

Ontologies are critical for applications that want to search across or merge information from diverse communities. Although XML DTDs and XML Schemas are sufficient for exchanging data between parties who have agreed to definitions beforehand, their lack of semantics prevent machines from reliably performing this task given new XML vocabularies. The same term may be used with different meaning in different contexts, and different terms may be used for items that have the same meaning. RDF and RDF Schema begin to approach this problem by allowing simple semantics to be associated with identifiers. With RDF Schema, one can define classes that may have multiple subclasses and super classes, and can define properties, which may have subproperties, domains, and ranges. In this sense, RDF Schema is a simple ontology language. However, in order to achieve interoperation between numerous, autonomously developed and managed schemas, richer semantics are needed. For example, RDF Schema cannot specify that the Person and Car classes are disjoint, or that a string quartet has exactly four musicians as members.

OWL adds additional vocabulary for describing properties and relations between classes and characteristics of properties as depicted in Figure 13.4.

13.2.3.1 OWL Language Basics

OWL offers three expressive sublanguages intended to be use by ontology developers.

- OWL LITE provides users a classification hierarchy and constraints specifications for classes and properties.
- OWL DL Provides users the most expressive vocabulary including computational completeness (all conclusions are certain to be computed) and decidability (all computations can end in finite time).
- OWL Full has most expressiveness and the syntactic freedom of RDF with no computational guarantees. For example, in OWL Full a class will be treated at the same time as a collection of individuals and as an individual in its own right. OWL Full permits an ontology to boost the meaning of the pre-defined (RDF or OWL) vocabulary.

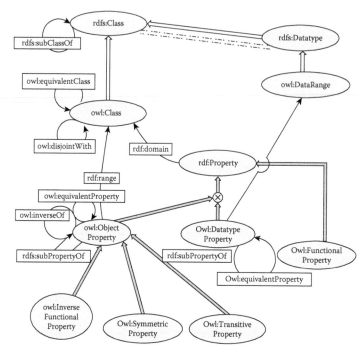

FIGURE 13.4
Properties and relations between classes and characteristics of properties.

13.2.3.2 OWL Lite Features

In OWL Lite, as in RDF, we can outline classes and class/subclass hierarchies in any domain and properties of these classes using RDF constructs such as *Class, rdfs:subClassOf, rdfs:Property, rdfs:subPropertyOf, rdfs:domain,* and *rdfs:range.* For defining the equality/inequality relationships among the ontology components, the following constructs are there in OWL Lite:

- *sameAs*: One individual may be indicated to be the same as another individual.
- *allDifferent*: Many individuals may be identified to be mutually distinct in one allDifferent statement.
- *differentFrom*: An individual may be quantified to be different from other individuals.
- *equivalentClass*: One class is expected to be equivalent to another class.
- *equivalentProperty*: One property may be defined to be equivalent to another property.

In OWL Lite, the following distinctive identifiers offer an influential mechanism for better reasoning about a property to provide information regarding properties and their values:

- *FunctionalProperty*: Properties may be defined to have a single value. If a property is a FunctionalProperty, then it can have no more than one value for each individual. For example, hasdeviceId may be stated to be a FunctionalProperty.

From this, a reasoner may infer that no devices may have more than one deviceId in a sensor network. This does not imply that every device must have at least one device id in a network however.

- *InverseFunctionalProperty*: Properties may be inverse functional. If a property is inverse functional, then the inverse of the property is functional. Therefore, the inverse of the property will have at the most one value for every individual. As an example, hasdeviceId (a distinctive identifier for the device in a network) is also expressed to be inverse functional (or unambiguous). The inverse of this property (which is also mentioned as isThedeviceIdFor) has at the most one value for any device (for a unique network). Therefore, any one device's deviceID is the only value for its isThedeviceIdFor property. From this, a reasoner will deduce that no two totally different individual instances of device have the identical deviceId in an exceedingly unique network. Also, a reasoner will deduce that if two instances of device in a unique network have a similar deviceID, then those two instances confer with a similar device.

- *SymmetricProperty*: Properties may be identified to be symmetric. For example, is_connectedto may be defined as a symmetric property. Then a reasoner can deduce that "Device_2" is_connectedto "Device_2" provided that "Device_1" is_connectedto "Device_2."

- *inverseOf*: Two properties are identified to be inverse to each other. For example, if has_sensor is the inverse of has_network and "Network_A" has_sensor "Sensor_1," then a reasoner can deduce that "Sensor_1" has_network "Network_A."

- *TransitiveProperty*: Properties may be well defined to be transitive. For example, if "part of" property is said to be transitive, and if Device_1 is a partOf SensorA (i.e., (Device_1, SensorA) is an instance of the property partOf and Device_2 is partOf Device_1 (i.e., (Device_2, Device_1) is an instance of the property partOf, then a reasoner can deduce that Device_2 is a part of SensorA. MaxCardinality 1 restriction is not allowed for transitive properties and their super properties.

  OWL Lite lets restrictions on properties that may be utilized by instances of a class. The subsequent two restrictions limit the values that may be used, whereas the cardinality restrictions limit how many values may be used.

- *someValuesFrom:* The restriction someValuesFrom is declared on a property with regard to a class having a restriction that a minimum of one value for that property is of a particular type.

- *allValuesFrom:* The restriction allValuesFrom declared on a property with regard to a class implies that the property incorporates a local range restriction related to it.

  OWL Lite cardinality restrictions on properties allow having value zero or one. They do not permit random values for cardinality, as is the case in OWL dl and OWL Full. This permits the user to specify "at least one," "no more than one," and "exactly one."

- *minCardinality:* This restriction requires that the property should have a value for all instances of the class. That is, if a minCardinality of one is given on a property with regard to a class, then any instance of that class is correlative to a minimum of one individual by that property.

- *maxCardinality:* A maxCardinality one restriction is termed a functional or distinctive property. That is, if a maxCardinality of one is declared on a property with regard to a class, then any instance of that class will be associated with at the most one individual by that property.

In OWL, properties are characterized consistent with individuals, whether or not they relate to individuals (object properties) or to data-type properties. OWL uses most of the intrinsic XML Schema data types.

13.2.3.3 Incremental Language Description of OWL DL and OWL FULL

OWL FULL and OWL DL have the same vocabulary while OWL DL has some restrictions. Furthermore, OWL DL properties can be DatatypeProperties or ObjectProperties. ObjectProperties are relations between instances of two classes while DatatypeProperties are relations between instances of classes, RDF literals, and XML Schema data types. The OWL DL and OWL Full vocabulary that extends the constructions of OWL Lite are described below.

- *oneOf:* (enumerated classes): Classes can be designated by enumeration of a variety of individuals. The class members are exactly the set of enumerated individuals; no more, no less. For example, the class of Sensor_Type for a sensor network can be described by simply enumerating the individuals as UV sensors, touch sensors, humid sensors, pressure sensors, thermal sensors, etc.
- *hasValue:* Certain individual values can be associated for a property, also sometimes referred to as property values. For example, instances of the class of Chemical_sensors can be characterized as those sensors that have gas as a value of their chemical component. (The chemical component value, gas, is an instance of the class of chemical_components.)
- *disjointWith:* Classes can be disjoint. For instance, Analog_sensors and Digital_sensors can be considered as disjoint classes. So a reasoner can infer an inconsistency when an individual is stated to be an instance of both and similarly a reasoner can reason out that if sensor_A is an instance of Analog_sensor, then A cannot be an instance of Digital_sensor.
- *complementOf, unionOf, intersectionOf:* OWL DL allows Boolean combinations of classes and restrictions—complementOf, unionOf, and intersectionOf. For example, using complementOf, we could state that analog sensors are not digital sensors and using unionOf, we can state that a class contains sensors that are either Analog_sensors or Digital_sensors.
- *minCardinality, maxCardinality, cardinality:* FULL OWL allows cardinality statements for arbitrary positive integers, while in OWL Lite, cardinalities are limited to at least, at most, or exactly 1 or 0.

13.3 Existing Standardizations in IoT Semantics

Semantic sensor networks work on the interoperability of physical sensor networks to ease the sensor discovery. Further, they focus on semantically annotating sensor information to unleash it as "Linked Sensor Data." Ontologies and alternative semantic technologies are often key enabling technologies for sensor networks, as they facilitate semantic interoperability and integration, reasoning, classification, different kinds of assurance, and automation not addressed within the OGC standards. A Semantic Sensor Network has to be organized, installed and managed, queried, understood, and controlled through high-level specifications. Further, when reasoning regarding sensors, complicated physical constraints like restricted power availability, restricted memory, variable information quality, and loose connectivity ought to be taken into consideration. Once these constraints are formally depicted in an ontology, reasoning techniques are more readily applied and utilized. They enable classification and reasoning on the capabilities and measurements of sensors, the origin of measurements, and also the association of a variety of sensors as a macro instrument. Following W3C recommendation, OWL 2 dl is the designated language for ontology specification. There are excellent sensor ontologies, few of which are coupled with one another and are obtainable online. The sensor ontologies, to some extent, reflect the OGC standards and may encode sensor descriptions and mapping between the ontologies. Table 13.5 summarizes the details of various published IoT ontologies for adding semantics to sensor networks.

13.3.1 W3C SSN Ontology

The W3C Semantic Sensor Networks (SSN) ontology is a product of all of the aforementioned sensor ontologies. These ontologies mainly focus on the description of physical sensor networks such as sensor capabilities, location of the sensor (latitude and longitude), etc.

TABLE 13.5

IoT Ontology

Ontology Name	Key Concepts	Author	Status	Complexity	Cited
SSN	Stimulus, sensor, and observation	W3C Incubator	Maintained	Complicated	Yes
CSIRO	Sensor and process	Michael Compton	Developing	Simple	None
MMI	Device, capability, and property	Luis Bermudez	Developing	Ordinary	None
CESN	Sensor and physical property	Holger Neuhaus	Cease	Simple	None
SWAMO	Platform, process, and observation	John Graybeal	Developing	Complicated	Yes
A3ME	Device, data, service, and capability	Arthur Herzog	Maintained	Simple	None
OntoSensor	Sensor, capability, and measurand	Danh Le Phuoc	Cease	Complicated	Yes
OBOE	Observation, context, value, and measurement	Kevin Page	Maintained	Simple	None
SeReS	Feature and result and observation	Krzysztof Janowicz	Developing	Complicated	Yes
SemSOS	Observation, process, feature, and phenomenon	Cory Henson	Maintained	Ordinary	None
Sensei O and M	Observation, data, process, and service	Payam Barnaghi	Cease	Simple	None
OOSTethys	Process, system, and observation	Luis Bermudez	Developing	Ordinary	Yes
SERONTO	Abstract, physical, and reference	Laurent Lefort	Cease	Complicated	Yes

They do not address the problem of describing sensor data in an interoperable manner to ease the reasoning task. The development of SSN ontology is based on the following works:

- The CESN ontology (Calder et al. 2010)
- The OntoSensor ontology, a prototype sensor knowledge repository (Russomanno et al. 2005a, 2005b; Goodwin and Russomanno 2006)
- A service-oriented ontology for Wireless Sensor Networks (WSN) (Kim et al. 2008)
- An ontology for sensor networks to describe the topology, network setting, sensor properties, dataflow, and sensor network performance (Jurdak et al. 2004)
- Other ontologies related to sensor networks (Avancha et al. 2004; Eid et al. 2006, 2007)

The W3C Semantic Sensor Networks Incubator Group, the SSN-XG, developed the SSN ontology for defining sensors and the measurement capabilities of sensors, the observations that outcomes from sensing, and deployments in which sensors are used. The ontology is centered on concepts of systems, processes, and observations. It describes the physical and processing structure of sensors. The SSN ontology is a combination of all prevailing sensor ontologies. It outlines the concepts for demonstrating sensors, their observation, and the surrounding settings. The ontology includes large parts of the SensorML and O&M standards from the OGC, omitting regulations as well as process explanations and data types, which were not considered sensor specific and if required can be incorporated from other ontologies. The SSN ontology is presently used in many research projects. Constant use of the SSN ontology by its development community and others will uncover design weaknesses or essential add-ons, and may boost the community to restructure for approved enrichments, either informally or as part of a development or even standards group. The illustration of a sensor in the ontology links together what it measures (the domain phenomena), the physical sensor (the device), and its functions and processing (the models) as depicted in Figure 13.5.

13.3.2 Ontology Structure

The Semantic Sensor Network ontology orbits around the central Stimulus-Sensor-Observation pattern. Several theoretical modules enumerate the pattern to describe key sensor concepts. The relationships between them appear in Figure 13.6 which comprises a summary of the 10 main classes in the ontology modules.

The ontology can be used for a focus on any (or a combination) of a number of perspectives:

- A data or observation perception, with an emphasis on observations and related metadata
- A sensor perception, with an emphasis on what senses, how it senses, and what is sensed
- A system perception, with an emphasis on systems of sensors
- A feature and property perception, with an emphasis on features, their properties, and what can sense those properties

The modules as depicted in the Figure 13.7 allow more refining or combining of these views on sensors and sensing. The report of sensors may be in depth or abstract.

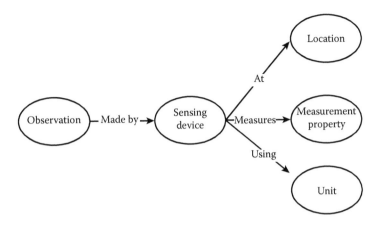

FIGURE 13.5
Sensor activity illustration.

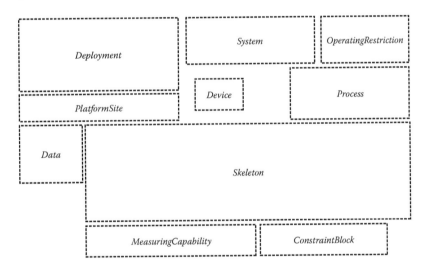

FIGURE 13.6
Overview of the semantic sensor network ontology modules.

The ontology excludes a hierarchy of sensor types; these definitions are left for domain experts to decide, based on the workings of the sensors.

SSN defines high-level concepts for representing sensors, their observation, and the surrounding environment. However, according to the W3C SSN ontology final report, SSN has several limitations:

- Common terms to represent subclasses of ssn:Sensor type, measurement type, unit type, or domain type (ssn:FeatureOfInterest) are not made available in the ontology. This limitation hinders machine automation.
- Domain concepts like time and locations are not included in SSN. But via OWL imports, these concepts can be included from other ontologies. This may result in interoperability issues between domain ontologies, since domain ontologies that are related to IoT are not standardized.

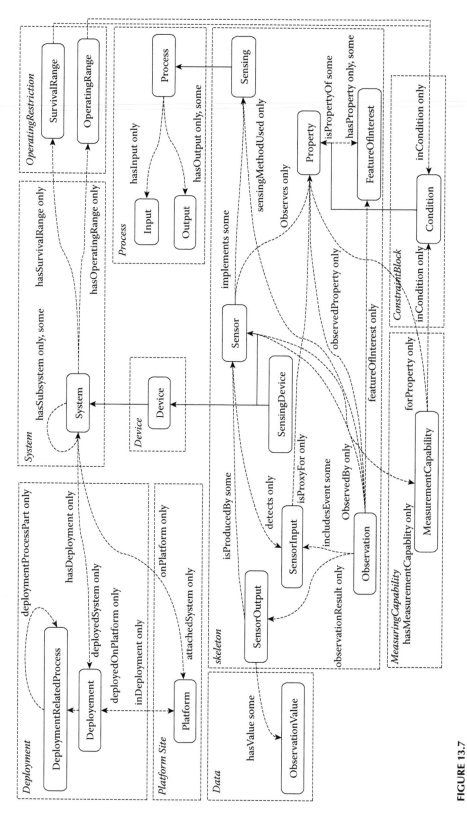

FIGURE 13.7
Overview of the semantic sensor network ontology classes and properties.

In their final report, they have mentioned that they "do not provide a basis for reasoning that can ease the development of advanced applications." This makes it evident that a common description of sensor measurements should be provided.

Future work is to "standardize the SSN ontology to use it in a Linked Sensor Data context" and to "standardize the SSN ontology to bridge the Internet of Things." These facts highlight the need to provide a common description of all related sensor measurements and other details.

13.3.3 IoT-Lite Ontology

Iot-Lite is a lightweight ontology based on SSN to describe IoT concepts and relationships. IoT-Lite is an instantiation of the SSN ontology. The lightweight allows the use of IoT platforms that can be extended in order to represent IoT concepts in a more detailed way in different domains that allows the discovery and interoperability of IoT resources in heterogeneous platforms using a common vocabulary. The ontology includes IoT concepts into three classes—objects, system or resources, and services. The devices are divided into three classes: sensing devices, tag devices, and actuating devices. The services are described with availability or access control and coverage. The coverage discusses about the area covered by the IoT device. To describe the units and QuantityKind that IoT devices can measure, well-known taxonomies qu and qudt are used. Figure 13.8 depicts the concepts of the ontology and the main relationships between them.

13.4 Semantic Web Rule Language

Rule languages are commonly used in most of the business applications. But, there is little interoperability between current rule-based systems. The main goal of the semantic web

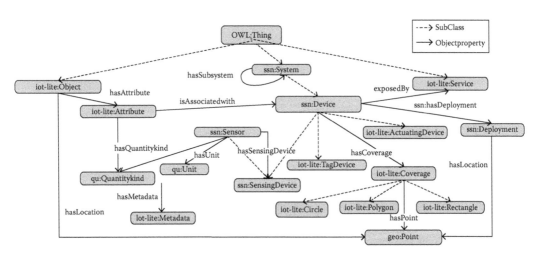

FIGURE 13.8
IoT ontology classes and properties.

is the interoperation. Developing a language for sharing rules is often seen as a key step in reaching this goal.

Expressing all relations using OWL 2 language is very difficult. For example, OWL cannot express the relation *child of married parents*, because there is no way in OWL 2 to express the relation between resources with which a resource has relations. By adding SWRL rules (Horrocks et al. 2004) to an ontology, the expressivity of OWL can be extended. SWRL rules are similar to rules in DATALOG or Prolog languages. SWRL allows users to write rules expressed in terms of OWL concepts to reason about OWL individuals. Using the rules, new knowledge can be inferred from existing knowledge bases. There are many built-ins which will provide an extension mechanism whereby the modeling language can be enhanced with domain-specific built-ins. The predicates can be class expressions, property expressions, data range restrictions, sameIndividual, differentIndividuals, core SWRL built-ins, and user-defined SWRL built-ins.

SWRL rules with class expressions
The following rule uses a class expression:
Observces(?sensor, health), produces(?sensor,?measurement) →
$$BodyTemperature(?measurement)$$

SWRL rules with data range restrictions
The following rule uses a data range restriction:

BodyTemperature(?measurement), hasValue (?measurement,?v), integer[> 36, < 38](?v) →
$$NormalBodyTemperature(?measurement)$$

The following rule uses the core built-ins:
Precipitation(?measurement),hasValue(?measurement,?v),hasUnit(?measurement, Milimet erPerHour),swrlb:greaterThan(?v,20),swrlb:lessThan(?v,50) →
$$HeavyRain((?measurement)$$

Table 13.6 shows the built-Ins which are defined for various comparisons.

TABLE 13.6

Built-Ins for Comparisons

Built-Ins	Syntax with Comment
swrlb:equal	swrlb:equal(?x,?y) If argument1 and argument2 are same
swrlb:notEqual	swrlb:notEqual(?x,?y) If argument1 and argument2 are not same It is the negation of swrlb:equal
swrlb:lessThan	swrlb:lessThan(?x,?y) If argument1 is less than argument2
swrlb:lessThanOrEqual	swrlb:lessThanOrEqual(?x,?y) If argument1 is less than or equal to argument2
swrlb:greaterThan	swrlb:greaterThanOrEqual(?x,?y) If argument1 is greater than argument2
swrlb:greaterThanOrEqual	swrlb:greaterThanOrEqual(?x,?z) if argument1 is greater than or equal to argument2

Table 13.7 shows built-ins which are defined for various arithmetic operations.

TABLE 13.7

Mathematics Built-Ins

Built-Ins	Syntax and Comment
swrlb:add	swrlb:add(?x,?y,?z) - argument1 is addition of argument2 and argument3
swrlb:multiply	swrlb:multiply(?x,?y,?z) - argument1 is multiplication of argument2 and argument3
swrlb:subtract	swrlb:subtract(?x,?y,?z) - argument1 is the difference of argument2 and argument3
swrlb:mod	swrlb:mod(?x,?y,?z) - argument1 is the remainder resulting from dividing argument2(dividend) by argument3(divisor)
swrlb:divide	swrlb:divide(?x,?y.?z) - argument1 is the quotient of argument2 divided by argument3
swrlb:pow	swrlb:pow(?x,?y,?z) - argument1 is the result of argument2 raised to argument3(power)
swrlb:unaryMinus	swrlb:unaryMinus(?x,?y) - argument1 is argument2 with its sign reversed
swrlb:unaryPlus	swrlb:unaryPlus(?x,?y) - argument1 is argument2 with its sign unchanged
swrlb:abs	swrlb:abs (?x,?y) - argument1 is the absolute value of argument2
swrlb:round	swrlb:round(?x,?y) - argument1 is the nearest number to argument2 without fractional part
swrlb:floor	swrlb:floor(?x,?y) - argument1 is the largest number without fractional part
swrlb:ceiling	swrlb:ceiling(?x,?y) - argument1 is the smallest number without fractional part
swrlb:tan	swrlb:tan(?x,?y) - argument1 is the tangent of argument2(in radian)
swrlb:sin	swrlb:sin(?x,?y) - argument1 is the sine of argument2(in radian)
swrlb:cos	swrlb:cos(?x,?y) - argument1 is the cosine of argument2(in radian)

Table 13.8 shows the built-in functions which are defined for strings.

TABLE 13.8

Built-Ins for Strings

Built-Ins	Syntax and Comment
swrlb:stringEqualIgnoreCase	swrlb:stringEqualIgnoreCase(?x,?y) - argument1 is the same as argument2
swrlb:stringConcat	swrlb:stringConcat(?x,?y,?z) - argument1(string) is concatenation of argument2 through the last argument
swrlb:stringLength	swrlb:stringLength(?x,?y) - argument1 is the length of argument2

(Continued)

TABLE 13.8 *(Continued)*

Built-Ins for Strings

Built-Ins	Syntax and Comment
swrlb:substring	swrlb:substring(?x,?y,?z,?n) - argument1 is the substring starting at character offset the argument3 in the string the argument2 with length the argument4(optional)
swrlb:upperCase	swrlb:upperCase(?x,?y) - argument1 is the uppercase of argument2
swrlb:lower-Case	swrlb:lowercase(?x,?y) - argument1 is the lowercase of argument2
swrlb:contains	swrlb:contains(?x,?y) - the argument1 contains argument2
swrlb:containsIgnoreCase	swrlb:containsIgnoreCase(?x,?y) - argument1 contains argument2
swrlb:startsWith	swrlb:startsWith(?x,?y) - argument1 starts with argument2
swrlb:endsWith	swrlb:endsWith(?x,?y) - argument1 ends with argument2
swrlb:substringBefore	swrlb:substringBefore(?x,?y,?z) - argument1 is the characters of argument2 that precede the characters of argument3
swrlb:matches	swrlb:matches(?x,?y) - argument1 matches argument2(in regular expression)
swrlb:substringAfter	swrlb:substringAfter(?x,?y,?z) - argument1 is the characters of argument2 that follow the characters of argument3
swrlb:replace	swrlb:replace(?x,?y,?z,?n) - argument1 is the value of argument2 with every substring matched by argument3(regular expression) replaced by the replacement argument4(string)

Table 13.9 shows the built-in functions which are defined for XML Schema data types such as date, time, and duration.

TABLE 13.9

Built-in Functions for Date, Time, and Duration

Built-Ins	Comment
swrlb:yearMonthDuration	- argument1 consisting of argument2(in year) and argument3 (in month)
swrlb:dayTimeDuration	- argument1 consisting of argument2(in days), argument3(in hours), argument4(in minutes), and argument5(in seconds)
swrlb:dateTime	- argument1 consisting of argument2(in year), argument3(in month), argument4(in day), argument5(in hours), argument6(in minutes), argument7(in seconds), and argument8(in timezone)
swrlb:date	- argument1 consisting of argument2(in year), argument3(in month), argument4(in day), and argument(in timezone)
swrlb:time	- argument1 consisting of argument2 (hours), argument3 (minutes), argument4 (seconds), and argument5 (timezone)
swrlb:addYearMonthDurations	- argument1 is the sum of argument2 (in yearMonthDuration) through last argument (in yearMonthDuration)

(Continued)

TABLE 13.9 *(Continued)*

Built-in Functions for Date, Time, and Duration

Built-Ins	Comment
swrlb:subtractYearMonthDurations	- argument1 is the difference of argument2(in yearMonthDuration) and argument3(in yearMonthDuration)
swrlb:multiplyYearMonthDuration	- argument1 is the product of argument2(in yearMonthDuration) multiplied by argument3
swrlb:divideYearMonthDurations	- argument1 is the remainder of argument2(in yearMonthDuration) divided by the third argument
swrlb:addDayTimeDurations	- argument1 is the sum of argument2(in dayTimeDuration) and through the last argument(in dayTimeDuration)
swrlb:subtractDayTimeDurations	- argument1 is the difference of argument2(in dayTimeDuration) and argument3(in dayTimeDuration)
swrlb:multiplyDayTimeDurations	- argument1 is the product of argument2(in dayTimeDuration) multiplied by argument3
swrlb:divideDayTimeDuration	- argument1 is the remainder of argument2(in dayTimeDuration) divided by argument3(in dayTimeDuration)
swrlb:subtractDates	- argument1 is the difference of argument2(in xsd:date) and argument3(in xsd:date)
swrlb:subtractTimes	- argument1 is the difference of argument2(in xsd:time) and argument3 (in xsd:time)
swrlb:addYearMonthDurationToDateTime	- argument1 is the sum of argument2(in xsd:dateTime) and argument3(in yearMonthDuration)
swrlb:addDayTimeDurationToDateTime	- argument1 is the sum of argument2(in xsd:dateTime) and argument3(in dayTimeDuration)
swrlb:addYearMonthDurationToDate	- argument1 is the sum of argument2 (in xsd:date) and argument3(in yearMonthDuration)
swrlb:addDayTimeDurationToDate	- argument1 is the sum of argument2 (in xsd:date) and argument3 (in dayTimeDuration)
swrlb:subtractYearMonthDurationFromDate	- argument1 is the difference of argument2(n xsd:date) and argument3(in yearMonthDuration)
swrlb:subtractDayTimeDurationFromDate	- argument1 is the difference of argument2(in xsd:date) and argument3(in yearMonthDuration)
swrlb:addDayTimeDurationToTime	- argument1 is the sum of argument2(in xsd:time) and argument3(in dayTimeDuration)
swrlb:subtractDayTimeDurationFromTime	- argument1 is the difference of argument2(in xsd:time) and argument3(in dayTimeDuration)

Table 13.10 shows the built-in functions which are defined for RDF-style lists.

TABLE 13.10

Built-Ins for Lists

Built-Ins	Syntax and Comment
swrlb:listConcat	swrlb:listConcat(?x,?y,?z)concatenation of argument2(list type) through the last - argument(list type) and return the result in argument1
swrlb:listIntersection	swrlb:listIntersection(?x,?y,?z) - argument1 is a list that consists of elements found in both argument2 and argument3
swrlb:listSubtraction	swrlb:listSubtraction(?x,?y,?z) - argument1 is a list that consists of the elements which are in argument2 that are not members of argument3

(Continued)

TABLE 13.10 *(Continued)*

Built-Ins for Lists

Built-Ins	Syntax and Comment
swrlb:member	swrlb:member(?x,?y) - argument1 is a member of argument2
swrlb:length	swrlb:length (?x,?y) - argument1 is the length of argument2
swrlb:rest	swrlb:rest(?x,?y) - argument1 is a list that consists of all members of argument2 except the first member
swrlb:first	swrlb:first(?x,?y) - argument1 is the first member of argument2(list type)
swrlb:empty	swrlb:empty(?x) - return if argument1 is an empty list
swrlb:sublist	swrlb:sublist(?x,?y) - argument1 contains sublist of argument2

13.5 Simple Protocol and RDF Query Language

SPARQL (Prud'Hommeaux and Seaborne 2008) is a standard query language for RDF. ASPARQL query consists of a set of triples where the subject, predicate, and object can consist of variables. The idea is to match the triples in the SPARQL query with the existing RDF triples and find solutions to the variables.

SPARQL 1.0 only allows limited operations on matching results or solutions such as filter or duplicate elimination, ordering, projection, and triple construction. We need to provide more flexible operations such as aggregates, grouping, assignment, and select expressions. SPARQL 1.1 provides all these features.

Like SQL, SPARQL selects data from the query dataset by using a SELECT statement to determine which subset of the selected data is returned. Also, SPARQL uses a WHERE clause to define graph patterns to find a match for in the query data set. A graph pattern in a SPARQL WHERE clause consists of the subject, predicate, and object triple to find a match for in the data.

SPARQL General Form
PREFIX (Namespace Prefixes)
    e.g., *PREFIX iot:* < http://purl.org/IoT/iot#>
SELECT (Result Set)
    e.g., *SELECT ?sensorname*
WHERE (Query Triple Pattern)
    e.g., *WHERE iot:?sensorname iot:location iot:?room1*
ORDER BY(modifier)
    e.g., *ORDER BY ?sensorname*

13.6 Case Study

A number of use cases have been recognized such as web portals, corporate web site management, agents and services, web services, and ubiquitous computing where web ontology language can be applied. We have taken the domain of home automation scenario to provide security to home when occupants are out of home. The actors of the scenario are elucidated below:

- *IoT application consumers and infrastructure owners*: Buy applications and IoT devices for their house such as alarm lights/lamp, air-conditioner, etc.
- *IoT infrastructure and data providers*: Home automation device sellers for devices such as sensors, actuators, gateway boxes, IoT data feeds from the Cloud such as temperature sensor, motion detection sensor, switch actuator, luminosity sensor, and gateway box.
- *IoT application providers*: Software providers for the IoT home automation/building management domain such as application for security monitoring and application for energy monitoring.
- *IoT semantic interoperability providers*: Service providers for the semantic registry and the alignment/matchmaking of IoT entities such as ontology alignment, matchmaking, message translation, and message transformation.

Representing IoT entities and their alignments

The following ontology namespaces were utilized in the design of the required application ontology:

http://purl.oclc.org/NET/ssnx/ssn# for conceptualizations related to sensing and observations of different sensors.
http://www.loa-cnr.it/ontologies/DUL.owl# for high-level generic ones from the DUL ontology
http://purl.org/IoT/iot# for the IoT entities layer of the proposed IoT-ontology namespace.

For the sake of brevity, only the most relevant properties for IoT entities' registration are presented here. A physical object, a lamp can be registered by the end-user in the repository as:

```
<rdf:Description rdf:about="http://www.case_study.in#Lamp">
 <rdf:type
 rdf:resource="http://www.loa-nr.it/ontologies/DUL.owl#DesignedArtifact"/>
  <rdf:type>
   <owl:Class rdf:about="http://www. case_study.in#LampType">
    <rdfs:label xml:lang="en">Light</rdfs:label>
   </owl:Class>
  </rdf:type>
</rdf:Description>
```

The label "Light" is provided as an annotation of the class of the object. The remote-controlled electrical socket is registered as:

```
 <rdf:Descriptionrdf:about="http://www.case_study.in#Switch">
  <rdf:typerdf:resource="http://purl.org/IoT/iot#Actuator"/>
  <rdf:typerdf:resource="http://purl.org/IoT/iot#ActuatingDevice"/>
 </rdf:Description>
```

The lamp that is plugged into and controlled through the socket is expressed as a smart lamp entity:

```
<rdf:Descriptionrdf:about="http://www. case_study.in#SmartLamp">
  <rdf:typerdf:resource="http://purl.org/IoT/iot#SmartEntity"/>
  <ssn:featureOfInterestrdf:resource="http://www.case_study.in#Lamp"/>
  <dul:includesObjectrdf:resource="http://www.case_study.in#Switch"/>
</rdf:Description>
```

The service provided by these classes is also registered in the repository but is omitted here. Another smart entity, that is, a smart room entity(rooms) is described below. Its feature of interest is a room:

```
<rdf:Descriptionrdf:about="http://www.case_studyin#Rooms">
<rdf:typerdf:resource="http://purl.org/IoT/iot#Room"/>
</rdf:Description>
```

Motion detector, a sensing device is associated with this physical entity.

```
<rdf:Descriptionrdf:about="http://www.case_studyin#MotionDetector">
<rdf:typerdf:resource="http://www.w3.org/2005/Incubator/ssn/Sensor"/>
<rdf:typerdf:resource="http://www.w3.org/2005/Incubator/ssn/SensingDevice"/>
</rdf:Description>
```

The smart room entity is registered in the ontology as follows:

```
<rdf:Descriptionrdf:about="http://www.case_studyin#SmartRoom">
  <rdf:typerdf:resource="http://purl.org/IoT/iot#SmartEntity"/>
  <ssn:featureOfInterestrdf:resource="http://www.case_studyin#Rooms"/>
  <dul:includesObjectrdf:resource="http://www.case_studyin#MotionDetector"/>
  <dul:isConceptualizedBy>
    <rdf:Description>
     <rdf:typerdf:resource="http://purl.org/IoT/iot#SoftwareAgent"/>
     <iot:providesServicerdf:resource="http://www.case_studyin#DetectionService"/>
    </rdf:Description>
  </dul:isConceptualizedBy>
</rdf:Description>
```

The service provided by this smart entity through its associated software agent is registered as follows (only the output's description is given here for simplicity):

```
<rdf:Description rdf:about="http://www.case_study.in#DetectionService">
<rdf:type rdf:resource="http://purl.org/IoT/iot##Service"/>
<ssn:hasOutput>

<rdf:Description>
<rdf:type rdf:resource="http://purl.org/IoT/iot#InformMessage"/>
<dul:expresses>
<rdf:Description>
```

```
<rdf:type>
<rdf:Description
    rdf:about="http://www.case_study.in#MotionDetectorSignalType">
<rdf:type rdf:resource=" http://www.w3.org/2002/07/owl#Class"/>
<rdfs:subClassOf rdf:resource="http://purl.oclc.org/NET/ssnx/ssn#Property"/>
<rdfs:label xml:lang="en">Motion</rdfs:label>
</rdf:Description>
</rdf:type>
</rdf:Description>
</dul:expresses>

<dul:isRealizedBy>
<rdf:Description>
<rdf:type rdf:resource="http://purl.org/IoT/iot#CommunicationMessage"/>
<rdfs:label>MotionDetectorSignal</rdfs:label>
<iot:hasTemplate>{'list': [ { 'timeStamp': 1333450241.736228, 'signal': 'PropertiesChanged',
'data': { 'MotionDetected': true }, 'IDeviceId': 23 } ], 'until': 1333450241.741899, 'tobj': 'signals'
}</iot:hasTemplate>
</rdf:Description>
</dul:isRealizedBy>
</rdf:Description>
</ssn:hasOutput>
<iot:hasAccessAddress rdf:resource="http://192.168.50.1/api/signals/listen?ids=23"/>
</rdf:Description>
```

Assume a generic application has been developed to implement the function "Switch a light when a movement is detected in the room." This application has to be registered in the ontology by the IoT service provider and application developer that provides service "LightService" and conceptualizes a control entity.

```
<rdf:Descriptionrdf:about="http://www.case_studyin#Control">
  <rdf:typerdf:resource="http://purl.org/IoT/iot#ControlEntity"/>
  <dul:isConceptualizedBy>
    <rdf:Descriptionrdf:about="http://www.case_studyin#Application">
      <rdf:typerdf:resource="http://purl.org/IoT/iot#Application"/>
      <iot:providesServicerdf:resource="http://www.case_studyin#LightService"/>
    </rdf:Description>
  </dul:isConceptualizedBy>
</rdf:Description>
```

The instantiation of the service provided by the IoT application service provider is described as follows:

```
  <rdf:Descriptionrdf:about="http://www.vit.ac.in#LightService">
  <rdf:typerdf:resource="http://purl.org/IoT/iot#Service"/>
  <ssn:hasOutput>
    <rdf:Description>
     <rdf:typerdf:resource="http://purl.org/IoT/iot#QueryMessage"/>
     <dul:isRealizedBy>
      <rdf:Description>
```

```
      <rdf:typerdf:resource="http://purl.org/IoT/iot#CommunicationMessage"/>
      <rdfs:label>Monitor</rdfs:label>
      <iot:hasTemplate>token=X</iot:hasTemplate>
     </rdf:Description>
    </dul:isRealizedBy>
     <iot:hasTarget>?entity ssn:featureOfInterest [a iot:Room]; dul:isConceptualizedBy
[iot:providesService ?this]. ?this ssn:hasOutput [dul:expresses [ a <http://www.vtt.fi/IoT/
test/app1#HasMovement> ] ]</iot:hasTarget>
    </rdf:Description>
   </ssn:hasOutput>
   <ssn:hasOutput>
    <iot:CommandMessage>
     <dul:isRealizedBy>
      <iot:CommunicationMessage>
       <rdfs:label>Switch</rdfs:label>
       <iot:hasTemplate><command><type>switch</type><value>1</value><token> X</
token></command></iot:hasTemplate>
      </iot:CommunicationMessage>
     </dul:isRealizedBy>
      <iot:hasTarget>?entity ssn:featureOfInterest [http://www.vtt.fi/IoT/test/
app1#LightingDevice]dul:isConceptualizedBy [iot:providesService ?this]</iot:hasTarget>
    </iot:CommandMessage>
   </ssn:hasOutput>
   <ssn:hasInput>
    <iot:InformMessage>
     <dul:isRealizedBy>
      <iot:CommunicationMessage>
       <rdfs:label>MovementDetectorResponse</rdfs:label>

       <iot:hasTemplate><event><type>movement</type><value>false<value></event></
iot:hasTemplate>
      </iot:CommunicationMessage>
     </dul:isRealizedBy>
    </iot:InformMessage>
   </ssn:hasInput>
  </rdf:Description>
```

The custom classes used in iot:hasTarget SPARQL patterns have to be defined along with
the application registration:

```
<owl:Classrdf:about="http://www.case_studyin/app1/#HasMovement">
  <rdfs:labelxml:lang="en">HasMovement</rdfs:label>
</owl:Class>

<owl:Classrdf:about="http://www.case_studyin/app1/#LightingDevice">
  <rdfs:labelxml:lang="en">LightingDevice</rdfs:label>
</owl:Class>
```

Follow the below steps to exploit the Smart Proxy toolset with the abovementioned registrations:

- Ontology alignment task is carried out on the registrations as a whole, aligning classes LampType (label "Light") and MotionDetectorSignalType (label "Motion") with app1:LightingDevice and app1:HasMovement and they are in the ontology using subclass axioms, for example, LampType rdfs:subClassOf app1:LightingDevice.
- Execute patterns of the iot:hasTarget-related SPARQL queries by performing the matchmaking between the registered application and the smart entities.
- The message data format examples given with iot:hasTemplate are processed for both the control entities' services and the smart entity's application by generating an OWL ontology representation for a given XML or JSON so that "MotionDetected" attribute in DetectionService's output will get mapped with the "value" XML tag in the application LightService's MovementDetectorResponse input.

13.7 Concluding Remarks

This chapter introduced key concepts in the semantic technology domain and key business applications in IoT. The chapter also touched upon emerging ontologies such as SSN and IoT and the impact they will have on IoT smart solutions and future collaborative interoperable applications. The chapter examines evolving standards and consortiums that were instigated to advance standardization work and interoperability of IoT. Finally, the chapter presents final remarks using a case study and guidelines to encourage readers to innovate and build differentiated solution that contributes toward refining the quality of lives and smart business practices in the domain of IoT.

References

Avancha, S., Patel, C. and Joshi, A., 2004. Ontology-driven adaptive sensor networks. In *First Annual International Conference on Mobile and Ubiquitous Systems, Networking and Services*, 22–26 August, Boston, MA (pp. 194–202).

Barnaghi, P., Wang, W., Henson, C. and Taylor, K., 2012. Semantics for the Internet of things: Early progress and back to the future. *International Journal on Semantic Web and Information Systems (IJSWIS)*, 8(1), 1–21.

Berners-Lee, T., Hendler, J. and Lassila, O., 2001. The semantic web. *Scientific American*, 284(5), 28–37.

Brickley, D. and Miller, L., 2004. *FOAF vocabulary specification*. Namespace Document 2 Sept 2004, FOAF Project, 2004. Available from: http://xmlns.com/foaf/spec/ updated on 2014.

Calder, M., Morris, R.A. and Peri, F., 2010. Machine reasoning about anomalous sensor data. *Ecological Informatics*, 5(1), 9–18.

Eid, M., Liscano, R. and El Saddik, A., 2006, July. A novel ontology for sensor networks data. In *2006 IEEE International Conference on Computational Intelligence for Measurement Systems and Applications*, 12–14 July, IEEE, Spain (pp. 75–79).

Eid, M., Liscano, R. and El Saddik, A., 2007, June. A universal ontology for sensor networks data. In 2007 IEEE International Conference on Computational Intelligence for Measurement Systems and Applications, IEEE, Ostuni, Italy (pp. 59–62).

Goodwin, C. and Russomanno, D.J., 2006, April. An ontology-based sensor network prototype environment. In Proceedings of the Fifth International Conference on Information Processing in Sensor Networks, Ostuni, Italy (pp. 1–2).

Gruber, T.R., 1995. Toward principles for the design of ontologies used for knowledge sharing?. *International Journal of Human-Computer Studies*, 43(5), 907–928.

Horrocks, I., Patel-Schneider, P.F., Boley, H., Tabet, S., Grosof, B. and Dean, M., 2004. *SWRL: A semantic web rule language combining OWL and RuleML*. Volume 21, W3C Member submission, p. 79. http://www.daml.org/rules/proposal/

Jurdak, R., Lopes, C. and Baldi, P., 2004. A framework for modeling sensor networks. In *Proceedings of the Building Software for Pervasive Computing Workshop at OOPSLA*, Volume 4, (pp. 1–5).

Kim, J.H., Kwon, H., Kim, D.H., Kwak, H.Y. and Lee, S.J., 2008, May. Building a service-oriented ontology for wireless sensor networks. In Seventh IEEE/ACIS International Conference on Computer and Information Science, 2008. ICIS 08, IEEE, Marriott Portland City Center Portland, OR (pp. 649–654).

Klyne, G. and Carroll, J.J., 2004. *Resource description framework (RDF): Concepts and abstract syntax*. W3C Recommendation, 2004. World Wide Web Consortium. http://w3c. org/TR/rdf-concepts (accessed June 20, 2017).

Patel-Schneider, P.F., Hayes, P. and Horrocks, I., 2012. *OWL web ontology language semantics and abstract syntax*. W3C recommendation, p. 10. https://www.w3.org/TR/2012/PER-owl2-overview-20121018/

Prud'Hommeaux, E. and Seaborne, A., 2008. *SPARQL query language for RDF*. W3C Recommendation, p. 15. https://www.w3.org/TR/2013/REC-sparql11-overview-20130321/

Russomanno, D.J., Kothari, C.R. and Thomas, O.A., 2005a, June. Building a sensor ontology: A practical approach leveraging ISO and OGC models. In *IC-AI*, Las Vegas, NV (pp. 637–643).

Russomanno, D.J., Kothari, C.R. and Thomas, O.A., 2005b, March. Sensor ontologies: From shallow to deep models. In *Proceedings of the Thirty-Seventh Southeastern Symposium on System Theory 2005. SSST'05*, IEEE, Tuskegee, AL (pp. 107–112).

Simperl, E., 2009. Reusing ontologies on the semantic web: A feasibility study. *Data & Knowledge Engineering*, 68(10), 905–925.

Suárez-Figueroa, M.C., 2010. NeOn methodology for building ontology networks: Specification, scheduling and reuse (Doctoral dissertation, Informatica). POLYTECHNIC OF MADRID, Spain.

14

Applications of IoT to Address the Issues of Children Affected by Autism Spectrum Disorders

Hrudaya Kumar Tripathy

KIIT University

Bhubaneswar, India

B.K. Tripathy

VIT University

Vellore, India

CONTENTS

14.1 Introduction

The human being is an innovative creature. The progress of the innovation as discussed (Michel 2014) is prompting smart kinds of stuff being evolved for distinguishing, finding, detecting, and associating. Also, the technologies are promoting new types of communication among individuals and things themselves. Innovation constitutes the incredible establishments of our survival, which depends on the inevitable activities in cutting edge of human life. The modern technology is just the down-to-earth utilization of information to a specific region. Researchers always look for and consider better technologies for enhancing the strength of debilitated people and attempt to empower those who have disabilities to lead as near an ordinary life as would be discreet. As discussed by Sula and Spaho (2014), today's technologies undergo a continual enhancement of different tools and utilities varying from low-tech to high-tech which shows man's superiority over other species.

In the course of recent years, it has been observed that there is a relentless increase in disabilities among the children in the general population growth. Hasselbring and Glaser (2000) discussed that children's disabilities range from speech and language weaknesses to mental hindrance, and more than half were portrayed as having a particular learning disability because of a mental disorder. Children who have abilities differ in cause, degree, and the impact they have on the educational and learning progress. Innovation has turned out to be a practical thing for giving chances to take part in fundamental communication, general practice, exploratory, or normal correspondence drills to coordinate their needs and capacities. This chapter focuses on the role of Internet of things (IoT) in promoting the learning/teaching of children with special needs. Presently, the IoT has become an integral part of the radio frequency identification (RFID) tags and future Internet. It allows the things to be individually recognized, to establish the position, capture physical data, and to establish a communication with the corresponding environment. As discussed by Huang and Li (2010), it is known that, in the present situation, there is no medical cure for children with autism. Now the major goal is to create an accommodating environment for their smooth running of life.

This chapter begins with an outline of children with special needs and an initial discourse of how modern information technology with smart environment creates a funnel for providing education and for enhancing the quality of life of children with autism. More detailed sections follow and describe how particular IoT applications and devices make it possible for autistic children to learn in a regular manner together with their normal, healthy companions' environment. It concentrates especially on gadgets and applications that specifically provide an interface between autistic children and the technical methodologies. Emphasis is additionally put on recent improvements, with past technologies serving as cases of the many effective techniques for adapting to an autistic child through technology.

14.1.1 Children with Special Needs

Children with autism have some inabilities that make learning or performing different day-to-day activities difficult for them. They may require additional help because of their medical, emotional, or learning issues. As discussed by Jamdeaf (2015), children with special needs mainly have mental disorder, which causes them to grow more slowly than other normal children. They may have to go for medical treatment or mental therapy to keep them in a normal condition, whereas the normal children do not ordinarily require or just need occasionally. We may have the capacity to recognize a child with special needs; however, we presumably cannot recognize every one of them. Nemours (2016) states that a child could have an issue that isn't noticeable unless you know well that individually. When provided with right instruction and framework, an autistic child individually can make a great success to interact with the normal life. In any case, to oversee such a child, we have to first comprehend what autism spectrum disorders (ASDs) or autism means.

14.1.1.1 What Is ASD?

ASDs are the most widespread neurodevelopmental disorders. Diagnosing children with ASD has become a challenging task for many therapists all over the world. It is a cluster of developmental disorders consisting of different categories such as pervasive developmental disorder, classic autism, and Asperger's syndrome. The intensity of symptoms in ASD ranges from moderate to severe, depending on the behavioral part of the child. Research on the activities of autistic children observed that their less effective involvement in social interaction and imagination, language and communication skill, imagining things and stereotype behavior became more difficult to make them what they did not understand. There are so many symptoms of this pervasive disorder and the ASD covers all the impairments mentioned above.

Researchers have observed that normally autism appears during the initial three years of life of a child and it shows a complex formative inability. It is identified when a child's failure to convey and associate with others is far from an individual interaction. It is normally a brain disorder that affects a child's capability to convey, establish relationships with environments, and counter perfectly to the society. Notbohm (2005) suggested that individuals with autism are not physically disabled and they "look" simply like anyone without disability. A recent study (Mighty 2016) has shown that an autistic child may have a number of behaviors at similar stages. They can be mildly or severely autistic in different ways. Such an individual is therefore said to exhibit behaviors of someone on the range of autistic spectrum and has ASDs.

14.1.1.2 Technology Needs to Assist Children with Autism

The vast majority of today's research is socially assistive research, which focused exclusively on ASDs. It has been observed (Anon 2014) that most of the autistic children are profoundly intrigued and inspired by smart gadgets available around them. By using these sorts of assistive advanced gadgets such as PCs, Tablets, and mobiles, they can be more comfortable and can cooperate more, settle on decisions, react, increase new attitudes, and turn out in to secure zone. Researchers (Sula and Spaho 2014) have suggested that these smart kinds of stuff with assistive technology can be used in primary or secondary schools to help autistic children in a number of ways to overcome communication barriers and build up new dialect abilities.

Nowadays, the progress of technology is important, and researchers and developers are determined on designing a range of smart objects/devices that can help users and make possible identifying, sensing, locating, and connecting things better. Only one type of teaching and learning methods cannot be recommended to the children with autism syndrome. In present market scenario, there are several types of IoT applications available to meet the requirement of these children. Children suffered from ASD may be uncommonly compelled for interpersonal correspondence and association abilities. This leads to the motivation to develop gadgets and IoT devices to support their prerequisites for consolidation in communal life. It is not only for enlightening functions or educational purposes or for entertainment. By using IoT devices/gadgets, people can confer their feelings to others, test/comment on existing substances, and chat among their surroundings.

In this information technology era, the computer-assisted involvement will help autistic children practice on their own time or engage them in daily learning what the regularly trained teacher cannot give them through proper attention. Also, it may happen that the rate of acquisition of skills will be faster than other different manual learning procedures. It has been proven that the IoT became an effective way of learning medium for children with special needs. IoT helps to take them on their basic practice for social communication what they personally need for the potential of abilities in different situations. With the massive advancement of smart technologies and enhancement in digital communication methods, IoT has become a new paradigm for the development and design of associated devices into smart environments. IoT is a vibrant and distributed global network consisting of different objects (Internet connected) including radio-frequency identification (RFID), different sensors, different actuators, mouse-controlled and touch-screen application, and other smart devices that are essential components for the present and future world of Internet.

14.1.2 Background and Functionality of IoT Solutions

In this era, particularly children with ASD may require the help and support of information and communication technologies. As discussed by several researchers (Nicolás 2004; Takeo and Toshitaka 2007), the IoT is a network of networks where the value-added kinds of services are well provided through the proper communication and association of an enormous number of items, things, sensors, or gadgets. As shown in Perera et al. (2013), the term "IoT" was first authored in 1998. The main motivation of the services provided by IoT was that it can be used by anybody and can be connected anywhere, anytime, and with anyone. Also it may avail any kind of services using any network. Woodcock and Woolner (2007) says that IoT makes it feasible and predictable environments to offer multisensory incitement, and cultivate or make it conceivable to work autonomously. Also it helps to build up the limit of capacity for self-control, empowering self-consideration, encouraging attention, and reducing dissatisfaction level of autistic children that may arise from committing mistakes. According to Ingersoll and Wainer (2013), with the help of these new technologies, autistic children can get closer and promote a better understanding to develop their learning skills, which would not be possible without IoT. So IoT helps them to interact with the world that has been separated from them in which they initially lived (Aresti-Bartolome and Garcia-Zapirain 2014).

IoT application directly visible on the computer screen or any display devices creates the actions of interest among autistic children. IoT also creates the infrastructure of information and communications through the different network of networks established by different sensors, devices, and objects or things that became value-added services for autism patients. The human–computer interaction (HCI) in the success of IoT and peer-to-peer (P2P) products and solutions plays a significant role in the development of social communication and language skills among autistic children. Also autistic children are now more involved and being highly encouraged to use smart devices. The usability of IoT among autistic children helps them for more interaction and response to the society, creates new talents and make them self-sufficient.

The significant objectives for IoT applications are the making of smart products, smart cities, smart buildings, smart health application, smart social communication, and many more smart spaces or environments. Today the implementation and deployment of P2P network is the most challenging research (Huang and Li 2010) in IoT because P2P is an excellent approach and a suitable platform for interacting robot control and e-learning for autistic children. Similarly, the parallel research has focused on low-range network such as Wireless Personal Area Network (WPAN) which covers a few meter radiuses. It helps out to connect the devices or gadgets to nearby computers with wireless connectivity. An Assistive Technology (AT) followed by IoT such as Ambient Assisted Living (AAL) also supports autistic children to learn their daily activities and allows them to be self-dependent and secure lifestyle. Also this technology provides learning environment support by the combination of smart technologies and IoT framework.

Researchers have suggested that before recommending the IoT devices it is highly necessary to evaluate the characteristics of different activities of autistic children, which include:

1. Age factor of the child
2. Proper diagnosis of the problem
3. Suitable learning methods
4. Capability to sit and use the system
5. Interesting kind of objects
6. Choosing the kind of environment, and many more

The different analysis proved that smart environment helps autistic children to be more determined on their interest in learning and enhances the ability of their attention toward usual activities. As a result, the performance of these children regarding day-to-day problem-solving task become easier. It is observed that because of this technology a child with autism will be able to increase the interest to learn the new social skills, social languages, and behaviors, and achieve moderate academic performance. This chapter is focused on the different functionalities of IoT solutions. Although at present there is no therapy for autism, the major aim is to care and train autistic children by using such IoT devices that are the key part of a supportive environment to connect and communicate them with the real world.

14.1.3 Strategies for Use of IoT and Media-Based Tools

Children with ASD are confronting a few difficulties that IoT framework can address as discussed by Cafiero (2008). Various general characteristics include:

- *The problem with communication*—The key characteristics of life characterizing attitudes, it is the major problem in ASD. IoT offers specialized devices in the area of perusing, composing, and talking.
- *A problem with the complex sign*—IoT offers the capacity for making a scope of images, from single image to different and more mixed images.
- *The problem with emotional and social learning*—Full of feeling and social learning happens when one is involved with others. Innovation can go about as a support and an extension between correspondence accomplices.

According to Woodcock and Woolner (2007), to support the engagement of kids with ASDs using IoT technology, it is essential to:

1. Comprehend the prerequisites of autistic children and their educators and care takers
2. Develop a versatile, intuitive IoT environment that will support the desire of autistic children
3. Design an IoT framework of broadly useful philosophy empowering such prerequisites to be captured and implemented in programming outline
4. Contribute to the development of user-centered design and learning ergonomics

Assistive technologies are growing at a great extent, from normal innovation to IoT implementation in this present digital stage. It is characterized by Michel (2004) as "a part of an object, or framework, whether procured commercially off the rack, changed, and that is utilized to increment, or enhances the functional abilities of children with disabilities." Today many companies are developing different assistive solutions to help autistic children. These IoT-based assistive solutions are classified (McClanahan and Krantz 2010) as a group of "low," "mid," and "high" technology. It is very difficult to present here every accessible item. However, the following have been discussed for the overview of the different development of helpful assistive technologies that are used to support the requirement of autistic children.

14.1.3.1 Low Technology

Low-tech dealings can be utilized as a part of various approaches to giving the visual data for the child. Normally autistic children are comfortable to handle visual data rather than sound-related data. Particularly pictorial data has been appeared to be very useful. Visual data enormously expands the child's perception of what is anticipated from him or her and is significantly more viable than sound-related data. Henceforth, most low-tech approaches have a cardboard-based or basic paper form. Visual bearings pick up, keep up, and refocus a child's consideration and attention. Visual information is not equipped with any electronic or battery-operated systems. As discussed by Pierce and Schreibman (1994), basically visual data is of low cost and simple to utilize. It also helps children to enhance their independent behaviors. There are many visual information kiosks

FIGURE 14.1
Low-tech methods of aiding comprehension skills: visual schedules, activity schedules, and calendars.

available in the market, such as Clipboard, photo album, normal whiteboard, highlight taps, and many more. Figure 14.1 shows the low tech methods of aiding comprehension skills, using visual schedules, activity schedules, and calendars. Visual activity schedules or calendars help many autistic children to accompany with their daily routines to understand what is going on, what will be the next schedule coming up, and what has already gone. It helps children to engage in a particular activity. McClanahan and Krantz (2010) suggests that these common low-tech approaches focus on enhancing a child's expressive communication skills.

14.1.3.2 Mid Technology

Mid technology deals with simple electronic devices or battery-operated devices that have narrow improvements in technology. Examples of this include Language Master, recorder, calculators, timers, overhead projector, and voice output devices. A self-contained and inexpensive electronic voice output device can be used for the diagnosis of autism. As discussed by Schepis et al. (1998), the voice output communication aids (VOCA) produces a digitized voice when a symbol or picture is selected on its display screen. It has been found that autistic children easily accept and adopt the functionalities of VOCAs.

14.1.3.2.1 Voice Output Communication Aids

According to Harris (2014), VOCA has different shapes and sizes, with different synthesized voices, allowed to record speech. Some VOCAs may use plastic or paper sheets with symbols or words to identify the messages or words that have been recorded, as shown in Figure 14.2. Other VOCAs will display the symbols or words on their display screen.

- Single message VOCAs speak a recorded message.
- Message Sequencer VOCAs allow the user to have a chain of messages.
- Overlay VOCAs permit the child to select from a number of messages, pictures, and words, and symbols are located above the keys to explain which one will say which message.
- Dynamic Screen VOCAs display symbols or graphics on the screen rather than on paper overlays.
- Talk Pad: The gadget can be programmed with simple 1–4 step commands. The child is motivated to hit the buttons and complete the series of steps.

FIGURE 14.2
Different styles of voice output communication aids.

14.1.3.3 High Technology

A number of studies show that information technology provides support for training and therapy for autistic children. It not only provides support to increase the learning ability but also extensively reduces the disturbing behaviors of the autistic children and expands consistency with directions of learning (Moore et al. 2000; Pleinis and Romanczyk 1985). Different software and tools have been developed to expand perusing and general and social relational abilities of children with autism. According to Cheng et al. (2003), computer technology in a teaching environment for autistic children helps increase focused attention, generalization of skills, changing of behavior, etc., whereas in the traditional approaches of learning to autism the child fails in some aspects.

14.1.4 Current Developments and Applications of IoT Technologies

This section discusses a few ultramodern and new applications of IoT technology for providing guidance and help to children suffering from ASDs. Several studies have identified that caution is necessary when taking into consideration the application of IoT made by the large amount of modern research improvements, as indicated in Figure 14.3. Low-tech technology became helpful for autistic children to establish their daily schedule and it would be easy for children to modify the schedule themselves. High-tech technology could support in training additional complicated processing. A noticeable changes has been guaranteed in the study of autism by using the progressive technology. However, most research (Michel 2004) results have not been used to wider testing. The right assures can only be exposed over time, as systematic, restricted scientific testings are approved. This present changing situation, following IoT devices, is coordinated to the particulars of the required theme. It helps the dual target of empowering socialization and encouraging collaboration among children with ASD.

14.1.5 Wearable Devices

As explained by Fletcher et al. (2010), wearable sensors are successfully implanted for a long-term constant physiological monitoring, which is vital for the cure and supervision of many neurological disorders, mental disorder issues, and chronic illnesses. Wearable sensors also have integrated into fashion accessories, garments, wristbands, hats, shoes, socks, headphones, eyeglasses, etc. These gadgets sometimes include the accelerometers

FIGURE 14.3
IoT and smart environment in information technology.

and temperature sensors to classify and monitor an autistic child's physical activities. The IoT-based wearable sensors framework is generally assembled on a three-tier structure. According to Lidwina and Jayanthy (2016), the first is wearable sensors that accumulate the health particulars; the second is communication gadgets that constitute a wireless networking which facilitates the process. The third is a cloud-based online repository for continuing storage of a child's activities to further assist concerned parents or guardians with their monitoring information. A number of control sensors are used for collecting the physical activity signals and uploading. Furthermore, a lot of sensors are operated for detecting the emotional status of the child by comparing the sensor data with the predefined normal scenario values.

The representation of wearable technologies shows the changes from digital replication and separation (simulation) to digital connectivity and responsiveness (augmentation) and follows the three-tier system architecture as explained in Figure 14.4. The wearable gadgets are very sensitive to adapt and recognize their users, identify their place and positions, followed by the specific activities being carried out. In recent years (Viseu 2003), a great increase has been noticed in the variety of wearable technologies available in the

FIGURE 14.4
Three tier system architecture. (From Lidwina Jennifer, J., and Jayanthy, S., *Middle-East Journal of Scientific Research*, 24(S2), 263–267, 2016. With permission.)

open market for autistic children. The report (Bower and Sturman 2015) on 10 April 2015 from Vandrico Wearable Technologies database (http://vandrico.com/ wearables) stated that about 296 types of gadgets are available in the field of entrainment, gaming, lifestyles, and many more areas. As discussed by Barfield and Caudell (2014), all wearable devices are integrated with a large number of various sensors that are capable of gathering the following information:

1. Acoustic information such as volume, frequency, and pitch
2. Mechanical information such as displacement, force, position, and acceleration
3. Environmental information such as humidity and temperature
4. Biological information such as temperature, neural activities, heart rate, and pulse rate
5. Optical information such as light wave frequency, refraction, luminance, and brightness.

14.1.6 Tracking Devices

As we know children with ASD are getting lost due to their inexpressible behavior. Parents are worried to provide the safe environment to their kids when they are outside of the home or away from a safety zone. Healy (2012) has noticed that ASDs are more often found in younger kids and the risk can be major issues. Keeping in view of this problem for an autistic child the tracking devices are developed, and by using such devices autistic children do not go for missing. The use of tracking devices is not compulsory and it is not mandatory for parents to use such devices for their autistic child. However, if parents will feel that the tracking devices are useful for safety of their kids, they may use it without hesitation. Now, the tracking devices can be used in different ways and it depends on the work habits of the individual autistic child. The wearable devices are like anklets, belt loops, clothing, and bracelets. The system will track the child by using global positioning system (GPS) and will send an emergency message to the nearby caretaker of the autistic child. According to the journal *Pediatrics* published in 2012 (Kelly 2014), analysis shows that around 49% of autistic children are gone missing at least once in their childhood time. Around 65% of autistic children reached the close call with traffic accidents.

The benefits of using tracking device for autistic children will provide an additional safety and will also help escape the life-frightening situations. Next couple of years we trust and will see a stream of GPS tracking devices designed particularly for the mental imbalance group or autistic children. Most of the guardians do not know that there are some good GPS tracking devices available to track their autistic child or special disabilities.

Here are the lists of a few awesome gadgets that can help track mentally unbalanced or autistic children and prevent them from getting lost.

14.1.6.1 AngelSense

This device was launched in 2014. It provides the safety solution with voice monitoring system. AngelSense is also GPS enabled and is entirely used by autistic and other special need children as they often wander or dart off. As explained by Lewis (2014), although it helps for GPS tracking and is connected with voice monitoring system, it provides maximum protection and security for autistic children.

14.1.6.2 Amber Alert GPS

The Amber Alert (Figure 14.5) is a very small and pocket-sized (around 7 cm) GPS device and is more child-friendly. It uses the AT&T's 3G network to call the parents by the child or vice versa through a touch button mounted on the device. So it supports to track the autistic children by their parents. It has no display screen and cannot access the Internet, but it helps to protect the child from cyberbullying. It is easier to keep inside the child's pocket and parents will be able to track their child's location using browser, an online portal, or mobile phone applications accompanying iOS and Android. If the child will face any panic attack, he or she can just press the button and a message will be sent to the parent and up to 10 authorized caretakers. It has a rechargeable battery that can last up to 40 hours. This device also sends a text or email alert to the parent or guardian if the child comes within 500 feet of a registered domestic network area.

14.1.6.3 Pocket Finder

It is a very reliable gadget (Figure 14.6) for autistic children that could help them prevent a tragedy in their day-to-day life. In this passionate world, parents may track their child from anywhere and may allow accessing the data from the local website or by using free mobile applications. It gives more peace of mind to the parents as they know that they are well connected with their children.

14.1.6.4 Filip

As a necessity, most of the parents need to reach their autistic children. A majority of parents do not need their children to have a cell phone during their teenage years. All things

FIGURE 14.5
Amber alert GPS. (From Lewis, L., *7 Tracking Devices to Find a Lost Child with Autism-Friendship Circle-Special Needs Blog*, 2014, Available from: http://www.friendshipcircle.org/blog/2014/01/15/7-tracking-devices-to-find-a-lost-child-with-autism/ [Accessed 12 October 2016]. With permission.)

FIGURE 14.6

Pocket Finder. (From Lewis, L., *7 Tracking Devices to Find a Lost Child with Autism-Friendship Circle-Special Needs Blog*, 2014, Available from: http://www.friendshipcircle.org/blog/2014/01/15/7-tracking-devices-to-find-a-lost-child-with-autism/ [Accessed 12 October 2016]. With permission.)

FIGURE 14.7

Filip. (From Lewis, L., *7 Tracking Devices to Find a Lost Child with Autism-Friendship Circle-Special Needs Blog*, 2014, Available from: http://www.friendshipcircle.org/blog/2014/01/15/7-tracking-devices-to-find-a-lost-child-with-autism/ [Accessed 12 October 2016]. With permission.)

considered, cell phones accompany definite risks for children and also the risk of loss of the phone. But to avoid this issue, Filip has made and designed a simple mobile gadget that can be worn on the wrist of the children aged between 5 and 12 years (Figure 14.7). By using this gadget, a child can communicate five reliable contacts and can also get in touch with his/her parents, relatives, and friends.

This gadget has a lot of features with built-in GSM, GPS, and Wi-Fi facilities. It will help parents for tracking their children by using iOS or Android applications. It has facilities to create multiple profiles and restore for more than one child. Its GPS sensor is embedded with geofencing features, and because of this reason Filip is called as "Safe zones."

14.1.6.5 Trax

Trax is a small GPS tracker that can be found suitable for all purposes in anyplace and anywhere. Simply use the mobile application and you will observe the tracker moving continuously on a map. It will make safe zones (GEO-fences) and notice about quick developments and monitor and track the distance. It has waterproof system integrated with lots of modern technology and covers two years of free roaming facilities. The use of Trax (Figure 14.8) is absolutely safe and user-friendly. It runs with Android/iOS application equipped

FIGURE 14.8
Trax. (From Lewis, L., *7 Tracking Devices to Find a Lost Child with Autism-Friendship Circle-Special Needs Blog*, 2014, Available from: http://www.friendshipcircle.org/blog/2014/01/15/7-tracking-devices-to-find-a-lost-child-with-autism/ [Accessed 12 October 2016]. With permission.)

with micro-USB charging port, inbuilt accelerometer, and gyro. It is provided with a free Subscriber Identity Module (SIM) card and roaming facility to 33 countries. It has such features as real-time mapping, expanded proximity searches, geofencing, and speed alarms.

14.1.6.6 Be Luvv Guardian

"Guardian" (Figure 14.9) is the world's first wearable gadget designed to establish a safety network for kids. Parents can effortlessly identify the whereabouts of their autistic children any place at any time through their cell phones. Once you install Guardian application, it becomes the part of its network to enhance the safety for your children around. Guardian has a vast network. If a child is missing, parents can use Guardian application and effectively trace out their missing child within a few minutes. With the support of social network and integration of latest sophisticated technology, Guardian can help out to provide more and more safety to special needs children. It is composed of two segments: the wearable waterproof Watchman gadget itself and an iOS application. The gadget shown in Figure 14.9 can be worn like bracelets, a pendant, or a clasp, and it speaks with the parent's telephone through Bluetooth technology.

Guardian will allow parents to check if their child is with other persons that they have added to their safety contact list, who are additionally running the same application. Furthermore, if the child comes quite close to anybody using the application, the location will be sent to the child's parents through the cloud technology. This gadget is powered by a replaceable battery and the charge will continue 5 to 10 months depending on the use. It covers a maximum range of 70 meters or 230 feet. It will be compatible with iPhones 4S, 5S, 5C, plus iPad.

14.1.6.7 Secure GPS eZoom

Secure GPS offers various benefits for children, drivers, and old age persons. With the help of secure eZoom GPS gadget (Figure 14.10), you can check their area whenever on the web by tracking the device, with your cell phone, or by sending an instant message.

FIGURE 14.9
Be Luvv Guardian. (From Lewis, L., *7 Tracking Devices to Find a Lost Child with Autism-Friendship Circle-Special Needs Blog,* 2014, Available from: http://www.friendshipcircle.org/blog/2014/01/15/7-tracking-devices-to-find-a-lost-child-with-autism/ [Accessed 12 October 2016]. With permission.)

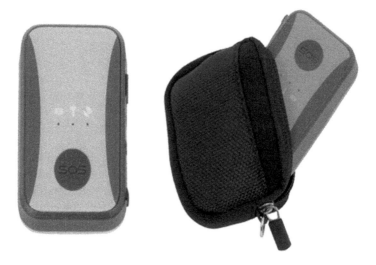

FIGURE 14.10
Secure GPS eZoom. (From Lewis, L., *7 Tracking Devices to Find a Lost Child with Autism-Friendship Circle-Special Needs Blog,* 2014, Available from: http://www.friendshipcircle.org/blog/2014/01/15/7-tracking-devices-to-find-a-lost-child-with-autism/ [Accessed 12 October 2016]. With permission.)

The advantages of this gadget include:

- eZoom's SOS catch rapidly associates children in trouble with their parent or gatekeeper by sending the location of the child's present area.
- Waterproof and strong unit.
- Long-enduring rechargeable battery.
- 1-year guarantee.

14.1.6.8 Bikn

As explained in Coxworth (2011), the Bikn (pronounced Beacon) is a complete solution that works both outside and inside your home or working place within a 250,000 ft^2 range.

FIGURE 14.11
Bikn. (From Coxworth, B., *BiKN uses iPhone to keep track of your stuff…or kids*, 2011, Available from: http://newat-las.com/bikn-iphone-tracking-system/20497/ [Accessed 22 October 2016]. With permission.)

The more features with BiKN is that not only does your iPhone discover your stuff but also your things can help you discover your iPhone even if it is on turned-off mode, muted, or dead. You do not lose your keys or purse; the Bikn may be an incredible arrangement.

This Bikn tracking system shown in Figure 14.11 comprises an iPhone case, an application, and up to eight tags that can be appended to things of your choice. That case can track the radio frequency tags, which can be attached to your or children's belongings, for example, key chains, packs, or child's dress (while the children are wearing them). Both the case and the tags must be charged to work. Now you can monitor or track your stuff/child in three ways. The first is Find mode, in which you have to use the application to find lost things by the help of the display screen and a beep sound tone. The second is Leash mode in which an alert will sound on both the case and the tag if a labeled thing moves to a distance from the case which a user can measure. Finally, there is Page mode in which a client can bring about various tags and additionally different bases to sound a caution or just to find something by sound alone.

14.1.6.9 HereO

HereO is the world's earliest smallest real-time cellular-connected GPS tracking device. It is just like a colorful watch that looks very cute and attractive for children. This is embedded with a clock so that the child can show off and find time. Parents can track the whereabouts of their children through iOS or Android app. This helps to make synchronize with parents' mobile and provide a map popup showing the child's location. HereO shown in Figure 14.12 creates a geofenced safe zone and allows the parent to track their child when the child goes beyond the restricted boundary by sending an alert message to the parent's mobile phone. It is designed in such a way that children above age 3 can use it, and parents will feel free from any type of dangerous situation for their children. HereO GPS gadget is more durable with waterproof system and withstands even the heaviest of play by children. Its battery life is more than 60 hours and depends on the use; however, it will send an alert message if the battery goes down.

FIGURE 14.12
HereO. (From Lamkin, P., *The best kids trackers: Using wearables for child safety*, 2016, Available from: http://www.wareable.com/internet-of-things/the-best-kids-trackers [Accessed 12 October 2016]. With permission.)

14.1.6.10 Lineable

Lineable is noticeable among the most moderate alternatives because of its Bluetooth-powered location gadget. This gadget shown in Figure 14.13 has a silicon-based moderate outline design with dust-proof and waterproof facilities. It holds a no-charging required battery life. This became the choice of parents who actually need a simple gadget without spending a considerable measure of money.

The scope of this gadget is just 6598 feet; yet what makes it emerge is that anybody with the Lineable application can find their missing child by using what the developer calls "Crowdsource GPS."

FIGURE 14.13
Lineable. (From Lamkin, P., *The best kids trackers: Using wearables for child safety*, 2016, Available from: http://www.wareable.com/internet-of-things/the-best-kids-trackers [Accessed 12 October 2016]. With permission.)

14.1.6.11 My Buddy Tag

Another affordable Bluetooth gadget for parents is "My Buddy Tag." This tag uses Bluetooth instead of GPS as a normal child tracker. This device has a panic alarm for the child to their parent. If any types of unwanted situation happen, it supports to send an alert message with personal ID to relocate the child with their parent.

The My Buddy Tag shown in Figure 14.14 is completely waterproof and has an attractive design wristband. You can buy nonreusable silicone wristbands, as well as Velcro or terry fabric wristbands for ordinary use. The range of coverage for this tag is limited to 40 feet indoor area to 80,120 outdoor area. So it is the best Bluetooth gadget available with economy prices.

14.1.6.12 Weenect Kids

It is a very good gadget and handy of 6 cm in size with waterproof, protecting cover with provided USB charger.

The tracking system follows the real and live map followed by a compass. The Weenect Kids gadget shown in Figure 14.15 measures the distance connecting to any smart phone but ensure that the camera of your phone is in the accurate direction. It comes with zone entry and exit option that will alert the parent when their child is back to home. It is available in whole Europe and a few countries, including Madagascar, Korea, Ghana, Malaysia, India, and Australia.

14.1.6.13 Tinitell

This gadget has two features, GPS tracker and live map supervision system, and can be used with any smartphones designed in Sweden. Tinitell is waterproof and has a voice recognition system. The gadget has a great Star Trekstyle communication gizmo for the entertainment of adventurous children.

FIGURE 14.14
My Buddy Tag. (From Lamkin, P., *The best kids trackers: Using wearables for child safety,* 2016, Available from: http://www.wareable.com/internet-of-things/the-best-kids-trackers [Accessed 12 October 2016]. With permission.)

FIGURE 14.15
Weenect Kids. (From Lamkin, P., *The best kids trackers: Using wearables for child safety*, 2016, Available from: http://www.wareable.com/internet-of-things/the-best-kids-trackers [Accessed 12 October 2016]. With permission.)

FIGURE 14.16
FlashMe. (From Lamkin, P., *The best kids trackers: Using wearables for child safety*, 2016, Available from: http://www.wareable.com/internet-of-things/the-best-kids-trackers [Accessed 12 October 2016]. With permission.)

14.1.6.14 FlashMe

This is less like a tracker but more like a method for reaching the parent of a lost child. It is basically a bright silicone wristband with a printed QR code that contains the contact points of interest of the child's parent. FlashMe as shown in Figure 14.16, the printed QR code will help the outsider to recognize the child's identity. As we know children under 4 do not remember or know their home address or their parents' phone number, this gadget can help to save them.

14.1.6.15 C-Way GPS

Well, for one thing, it helps parents monitor their child's location. C-way GPS locator with another idea of Plug & Play! By implanting a GPS sensor in the gadget, parents can open

C-Way's application and see where their child is at all times. The watch-like gadget gives parents a chance to send one-way "instant messages" to their child's watch. C-Way has also a schedule feature for children and parents can enter their child's calendar into the application, and the gadget will send cautions to the child and additionally let the parents know whether the child's present locality coordinates up to where their timetable says they must be. The gadget shown in Figure 14.17 has a 48-hour rechargeable battery and has distinctive connections that can be utilized relying upon how old your child is. Elder children can utilize the device provided there is a little LCD screen.

14.1.6.16 *SkyNanny GPS*

SkyNanny GPS shown in Figure 14.18 is a simple-to-utilize application for your Android cell phone, combined with a little GPS sensor, that can be worn by children. This wonderful item can give parents with significant calmness by letting them know their children's location. The whereabouts of the child can be observed from the parent's cell phone anywhere on the planet.

As discussed by Helga (2015), the Skynanny GPS can help parent for trouble-free monitoring of their children's whereabouts when they go to school, are on excursions, and engage in other activities outside. The gadget utilizes a standard SIM card from your versatile service provider. The GPS tracker gets its directions from GPS satellites and communicates with the SkyNanny GPS application on parent's Android phone by means of SMS. To minimize cost, it is best to get a versatile arrangement that has boundless messaging (for both the tracker and the mobile phone with application).

14.1.6.17 *TIIFA II: Smart and Interactive TeeShirt*

As developed by Bourouis (2015), TIIFA is a wearable gadget fitted inside the Tee Shirt for autistic children that use the Internet Object (IoT), Body Sensor Network (BSN), single

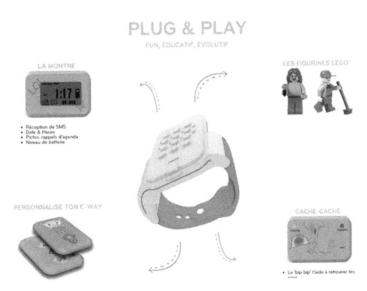

FIGURE 14.17
C-way gps. (From Lamkin, P., *The best kids trackers: Using wearables for child safety*, 2016, Available from: http://www.wareable.com/internet-of-things/the-best-kids-trackers [Accessed 12 October 2016]. With permission.)

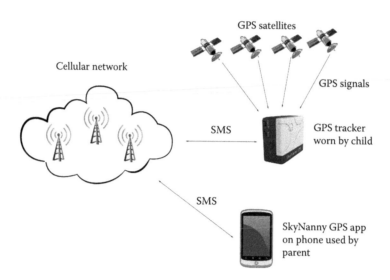

FIGURE 14.18
SkyNanny GPS. (From Helga, *SkyNanny GPS, Top notch child GPS tracker SkyNanny – Review,* 2015, Available from: http://tracking.watch/child-gps-tracker-skynanny-review/ [Accessed 17 October 2016]. With permission.)

controller board, and e-textile. The newest version (TIIFA II) was intended to enhance the accuracy. TIIFA demonstrates that the framework is a brilliant help for autistic children. It also helps autistic children to talk, practice, and do more outside activities and permits the exact checking and control.

TIIFA solution includes:

- Low-cost mobile solution
- Any location and any day in and day out
- Fast reaction and low energy utilization
- Easy to use and no operating cost
- No risk for the child (comfortable, waterproof, and adapted)

14.1.6.18 SensaCalm Weighted Blankets

In view of the comparative calming effect for children with autism, weighted bedspreads/blankets like SensaCalm (Figure 14.19) provide a comfort sleep. You may order the blankets of different size, weight, and texture for your child based on the profound weight and the calming material information their bodies require. In addition, accessible weighted vests and lap pads are also available (Maffei 2014).

14.1.6.19 Sensory Pea Pod

Generally, it has been observed that the cocooning effect of the deep even pressure is found among autistic children. Sensory Pea Pod shown in Figure 14.20 supports such kids and assists them for calm and accomplishes their sensory requirements. It is made up of vinyl and has inflatable and washable surface pad. It is also available in different sizes (Maffei 2014).

FIGURE 14.19
SensaCalm weighted blankets. (From Maffei, M., *9 Must-have gadgets for kids with autism*, 2014, Available from: http://www.sheknows.com/parenting/articles/1034307/9-must-have-gadgets-for-kids-with-autism and http://www.sensacalm.com/ [Accessed 16 January 2017]. With permission.)

FIGURE 14.20
Sensory Pea Pod. (From Maffei, M., *9 Must-have gadgets for kids with autism*, 2014, Available from: http://www.sheknows.com/parenting/articles/1034307/9-must-have-gadgets-for-kids-with-autism and http://www.sensacalm.com/ [Accessed 16 January 2017]. With permission.)

14.1.7 Head-Mounted Display Products: Google Glass

Children with ASDs are facing problems in social communication, facial expression, and eye contacts during face-to-face interaction. So the Stanford specialists created a facial recognition programming, particularly for Glass. The software is a mentor for helping the children to judge the correct emotions on people's face.

According to Smith (2015), Google Glass shown in Figure 14.21 is a comfort gadget that makes possible to detect the facial expression and can record the moments of an interaction of the child. An autistic child has a difficulty to identify and study somebody's look and judge what they are thinking or feeling. Google is utilizing wearable frames, Google Glass, along with the infrastructure designed to bridge the gap. The glass is integrated with facial recognition software synchronized with a front faced camera and enables the technology to get facial information. In the right corner of the glass, children will see icons of feelings like "sad" or "happy" so they can understand the feelings and also learn at their own pace to remember it naturally. The child may wear the gadget every day for short sessions where he/she will collaborate with surrounds face to face, talking and playing. The software keeps running on a cell phone, which records the sessions.

14.1.8 IoT Apps for Children with ASDs

When searching for applications (apps) for children suffered from ASDs, it is essential to take a look at all educational and instructive apps. An autistic child may have a great number of learning needs similar to that of a normal child. The following list was created (Carrington 2015) to give applications in view of normal learning qualities and attributes that are used for children with ASD. Remember that all autistic children learn in a different way and selecting applications should be based on the child's learning behavior. This rundown is just a testing of applications accessible for every aptitude region. This is not,

FIGURE 14.21
Google Glass. (From Smith, S., *The Autism Glass Project at Stanford Medicine*, 2015, Available from: http://www.medicaldaily.com/autism-glass-project-researchers-stanford-are-using-google-glass-help-treat-autism-357844 [Accessed 22 October 2016]. With permission.)

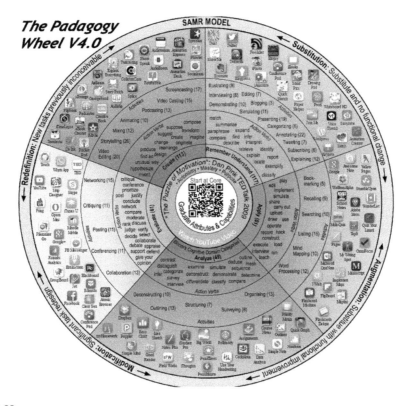

FIGURE 14.22
The Pedagogy Wheel V4. (From Carrington, A., *The Pedagogy Wheel – It's Not About The Apps, It's About The Pedagogy,* 2015, Available from: http://www.wegrowteachers.com/about/facilitators/ [Accessed 13 October 2016]. With permission.)

nor is it intended, to be a conclusive rundown. This list is planned to give you a beginning idea and a method of reasoning for selecting certain applications.

The Pedagogy Wheel shown in Figure 14.22 was developed by Carrington (2015) and is intended to help instructors think—deliberately, reasonably, and with a view to long-term, enormous picture results—about how they utilize apps in their teaching methodology. The Pedagogy Wheel is about attitudes and mindsets. It is a state of mind about advanced digital age training of education which supports inspiration, psychological improvement, learning alteration, and enduring learning objectives.

14.1.8.1 iPrompts® PRO

iPrompts (Figure 14.23) is the first visual support application discussed in Tzvi (2011) to provide the visual structure for autistic children. It collaborated with StoryMaker which is the premier app for making and exhibiting Social Stories, including selective materials from "Carol Gray." Another section of SpeechPrompts, coordinated effort with the Yale Child Study Center, gives activities to practice rate, pitch, stress, and power of discourse. Components of all applications depend on research, granted by the U.S. Branch of Educational program. This application offers the following list of features: Picture Library, Schedules, Countdown, Choices, Stories, Academy, Video Prompts, Voice Match, and Voice Chart. The picture library contains 35 envelopes with many pictures; in addition, children

FIGURE 14.23

iPrompts® PRO. (From Tzvi, iPrompts® PRO on the App Store, Handhold Adaptive, 2011, Available from: http://www.friendshipcircle.org/blog/2011/02/16/seven-scheduling-behavioral-apps-for-children-with-special-needs/ [Accessed 23 October 2016]. With permission.)

can stack more pictures from their picture library or rapidly include photographs utilizing Google, Bing, or Flickr. These pictures can be utilized to make plans, commencements, decisions, and stories.

14.1.8.2 Visual Schedules and Task Managers

As Krull (2015) said that adapting visually is an extraordinary mode for some individuals, including various children and elders with ASD. For a few, utilizing visual timetables and assignment administration frameworks can have all the effect. In spite of the fact that not for everybody, utilizing the cell phones with applications permit us to rapidly include photos, sounds, videos on the spot to help the child in need.

14.1.8.3 Choiceworks Scheduler

The main page of Choiceworks shown in Figure 14.24 is embedded with feature boards like "Schedules," "Waiting," and "Feelings" which can be selected by a child or parent. One can utilize pictures from the Choiceworks program or picture/photographs of their decision and consolidate these with customized sound recordings, which can be printed and shared. The "Schedules Board" provides an assignment with clocks. At the point when a part of the assignment on the left is finished, it can be dragged to a relating square on the right, where it gets a green check mark. At the point when a clock goes off, a general message appears on the screen and the clock reads "Done." The "Waiting Board" provides a clock and two exercises to browse while the individual waits. The "Feelings Board" provides two adaptable techniques when one feels, and two exercises to do after the technique has been finished. For instance, the specimen "disturbed" board says, "When I am vexed, I can get help or enjoy a reprieve."

14.1.8.4 Choiceworks Calendar

This application presents scheduling in a calendar design with pictures and sounds. Schedule events can be smoothly revised and monitored in daily, weekly, or monthly intervals. The scheduled calendar shown in Figure 14.25 can be allowed to print and share. It permits one to insert pictures into right place inside the scheduled section of the calendar. It has a simple button to watch the upcoming activities.

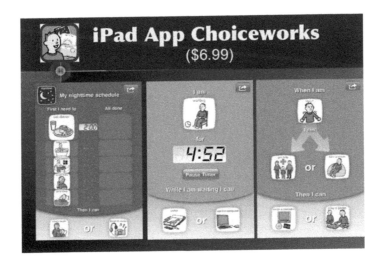

FIGURE 14.24
Choiceworks. (From Krull, J., *Visual Schedules and Task Managers: Apps for Autism and Beyond! Part 1*, IPAT ND Assistive Technology Blog, 2015, Available from: http://ndipat.org/blog/visual-schedules-and-task-managers-apps-for-autism-and-beyond-part-1/ [Accessed 22 October 2016]. With permission.)

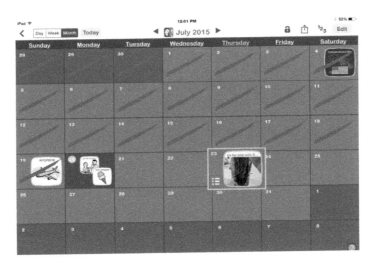

FIGURE 14.25
Choiceworks Calendar. (From Krull, J., *Visual Schedules and Task Managers: Apps for Autism and Beyond! Part 1*, IPAT ND Assistive Technology Blog, 2015, Available from: http://ndipat.org/blog/visual-schedules-and-task-managers-apps-for-autism-and-beyond-part-1/ [Accessed 22 October 2016]. With permission.)

14.1.8.5 *Proloquo2Go*

Proloquo2Go is a popular symbolic communication app. The general symptom of an autistic child includes difficulty to speak, express, and sometimes failing to talk. Proloquo2Go is an Augmentative and Alternative Communication (AAC) application that shows children how to build sentences utilizing images, symbols, and different pictures. Proloquo2Go shown in Figure 14.26 has a text to speech conversation, prediction of the word, and vocabulary system. The application is intended to advance dialect improvement and develop social

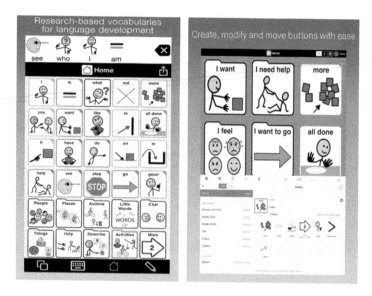

FIGURE 14.26
Proloquo2Go. (From Tzvi, iPrompts® PRO on the App Store, Handhold Adaptive, 2011, Available from: http://www.friendshipcircle.org/blog/2011/02/16/seven-scheduling-behavioral-apps-for-children-with-special-needs/ [Accessed 23 October 2016]. With permission.)

communication abilities. Its creative components support children, parents, trainers, and specialists to quickly customize the vocabulary and settings. Proloquo2Go is used by individuals with ASD, Down's syndrome, and cerebral paralysis, and different other sufferers. It is accessible in English, French, and Spanish, including provincial and bilingual support. It has three levels: Basic Communication, Intermediate Core, and Advanced Core. It helps to enhance the language skills and vocabulary of the autistic or mental disorder children.

14.1.8.6 Pictello: Talking Visual Story Creator

Everybody loves to tell the fun and innovative stories (Goldthwait 2016). Pictello helps make a social story, visual timetable, or a slideshow of different occasion pictures for an autistic child. It provides a breeze to make and share all these stuff with the child. This app allows to record one's own voice or may use natural text to speech voices, and it is an ideal app for visual narration. Every page in a Pictello as presented in Figure 14.27 shows a story that comprises a photograph or video and some content, and it can be read thoroughly. So anyone might hear by a Text to Speech voice or your own recorded sounds. The child can communicate by transforming pictures and recording videos into talking story books with Pictello.

Pictello helps to create the proficiency as follows:

- Support reading with word-by-word content highlighting.
- Develop composing and spelling with implicit word expectation and Speak.
- Create stories to show educational and social attitudes.
- Play stories page by page or as a slideshow.
- Use text to speech voices to pay particular attention to the story.

FIGURE 14.27
Pictello. (From Goldthwait, H., *Pictello—Talking visual story creator*, AssistiveWare B. V., 2016, Available from: https://itunes.apple.com/us /app/pictello/id397858008?mt=8 [Accessed 12 October 2016]. With permission.)

14.1.8.7 iCommunicate

Children may use the iCommunicate app (Figure 14.28) to record their own voice and make a visual schedule planner. This is integrated with text-to-speech including 20 voices of English, Italian, Norwegian, French, Swedish, Spanish, and German. It allows sharing of pictures and communicating with email or iTunes as file sharing.

14.1.8.8 Talking Tom Cat

Talking Tom Cat shown in Figure 14.29 is an entertainment app that reacts to the child's touch and repeats all that he or she says in an amusing voice. By pushing or snatching his tail will make him murmur, fall over, or scratch the screen. The child can share the recordings of the cat's response and reactions over YouTube, Facebook, email, or MMS as a novel video message. So it helps the autistic children to self-learn the responses and reactions.

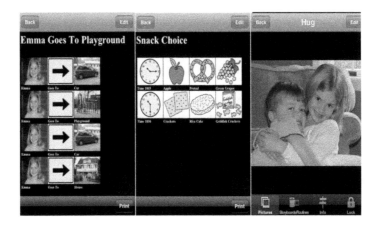

FIGURE 14.28
iCommunicate. (From Prospero, M.A., *13 Best Autism Apps for the iPad*, LAPTOP Reviews Editor, 2014, Available from: http://www.laptopmag.com /articles/best-autism-apps-ipad [Accessed 22 October 2016]. With permission.)

FIGURE 14.29
Talking Tom Cat. (From Prospero, M.A., *13 Best Autism Apps for the iPad*, LAPTOP Reviews Editor, 2014, Available from: http://www.laptopmag.com /articles/best-autism-apps-ipad [Accessed 22 October 2016]. With permission.)

14.1.9 Benefits of Using Assistive IoT Technologies to Support ASD Children

IoT assistive technology is a service that specifically helps (Garretson et al. 1990) a child with an inability in the determination, procurement, or utilization of general assistive gadgets. By handling of individual assistive gadgets that are designed for them as indicated by their capability and inability, children can enhance their quality of learning and increase new aptitudes. The instructors, parents, and experts can also use and suggest such IoT assistive gadgets by knowing the ability and inability of the autistic children for their better learning. This smart environment can help children with ASD to upgrade them, and children will have the capacity to learn new dialect abilities, social attitudes, proper conduct, and scholarly aptitudes to enable them for performing and completing day-to-day activities.

14.1.10 Future Scope

Researchers keep on dedicating themselves to disentangling and finding the causes behind extreme ASD children cases every day. Sometimes, the symptoms of autism are very difficult to differentiate for the parents of such children.

Keeping this in view, in far future, billions of data spouting devices will be connected with the Internet. The real work that can be taken lies in outlining minimal, nonintruding wearable gadgets. Better and more grounded calculations and algorithms can be developed to enhance the productivity of grouping and diminish the idleness. Currently, there is no particular diagnosis for autism; in any case, with proper preparing, training, and instruction by support of today's IoT gadgets, children with other disabilities and children with extreme autistic syndrome can learn and make themselves feel comfortable. With the rapid growth of smart wearable technology in the health sector, industry has leveraged the level of healthcare management for individuals with autism. New technologies can improve communication, help in the skill development, and enhance the ability to learn.

14.1.11 Conclusion

There has been advance in using the IoT structure for assistive technologies in various fields, for example, incorporation and autonomous living for seniors. Such smart environment and related gadgets could have extraordinary potential for conveying and filling in as method for assistive advances for autistic individuals and their families too. Besides, the low expenses for such advancements could make them accessible for locales with low wages, potentially helping enhancing the lives of individuals with a mental imbalance and their families in underserved and financially tested provinces of the world. With the massive advancement in smart technologies and improvement in digital communication methods, the "Internet of Thing" has turned into another standard for the plan and improvement of associated gadgets and smart environments.

The smart device can help to alert and clarify the needs of children, and also they and their parents may feel less pressure as it provides ubiquitous healthcare monitoring system. We analyzed the commitment of every arrangement of IoT toward enhancing the efficiency and effectiveness of autistic children for their way of life and also of society. Unmistakably more and more wearable IoT solutions will advance their life easier over the coming years. We accept that further research that addresses open difficulties will help extend more IoT arrangements to strengthen the present IoT solutions. This chapter is projected to emphasize the prospective benefits of the IoT serving as a means of distributing services and improving the value of living of autistic children and supporting their families.

References

Anon, 2014. Children with Special Educational Needs—Information Booklet for Parents. *National Council for Special Education, 1–2 Mill Street, Trim, Co Meath.* Available from: www.ncse.ie/wp-content/uploads/2014/10/ChildrenWithSpecialEdNeeds1.pdf [Accessed 28 September 2016]

Aresti-Bartolome N, Garcia-Zapirain B, 2014. Technologies as Support Tools for Persons with Autistic Spectrum Disorder: A Systematic Review. *International Journal of Environmental Research and Public Health*, 11, 7767–7802.

Barfield W, Caudell T, (Eds.), 2014. *Fundamentals of Wearable Computers and Augmented Reality. NJ:* Lawrence Erlbaum Associates, pp. 3–26.

Bourouis A, 2015. TIIFA II: Smart and Interactive Tee-Shirt for autistic children, The Young Innovators Competition ITU Telecom World 2015. Available from: https://ideas.itu.int/post/155593 [Accessed 15 October 2016]

Bower M, Sturman D, 2015. What Are the Educational Affordances of Wearable Technologies? *International Journal of Computers & Education*, 88, 343–353.

Cafiero JM, 2008. Technology Supports for Individuals with Autism Spectrum Disorders. *Technology In Action, Technology and Media Division*, 3(3), 1–12.

Carrington A, 2015. The Padagogy Wheel – It's Not about The Apps, It's About The Pedagogy. Available from: http://www.wegrowteachers.com/about/facilitators/ [Accessed 13 October 2016]

Cheng L, Kimberly G, Orlich F, 2003. *Kidtalk: Online Therapy for Asperger's Syndrome* (Tech. Rep.). Social Computing Group, Microsoft Research.

Coxworth B, 2011. *BiKN Uses iPhone to Keep Track of Your Stuff…or Kids.* Available from: http://newatlas.com/bikn-iphone-tracking-system/20497/ [Accessed 22 October 2016]

Fletcher RR, Poh M-Z, Eydgahi H, 2010. Wearable Sensors: Opportunities and Challenges For Low-Cost Health Care. *32nd Annual International Conference of the IEEE EMBS Buenos Aires, Argentina.* pp. 1763–1766.

Garretson HB, Fein D, Waterhouse L, 1990. Sustained Attention in Children with Autism. *Journal of Autism and Developmental Disorders*, 20(1), 101–114.

Goldthwait H, 2016. Pictello—Talking visual story creator. AssistiveWare B. V. Available from: https://itunes.apple.com/us /app/pictello/id397858008?mt=8 [Accessed 12 October 2016]

Harris C, 2014. Communication Matters, More than Just Talking. Available from: http://www. communicationm atters.org.uk/page/vocas [Accessed 12 October 2016]

Hasselbring TS, Glaser CHW, 2000. Use of Computer Technology to Help Students with Special Needs. *The Future of Children and Computer Technology*, 10 (2). Available from: http://www. futureofchildren.org [Accessed 10 October 2016].

Healy M, 2012. Nearly Half of Children with Autism Wander from Safety. USA TODAY. Available from: http://www.usatoday.com/story/news/nation/2012/10/08/autism-wandering-children/ 1612803/ [Accessed 14 January 2017].

Helga, 2015. SkyNanny GPS, Top Notch Child GPS Tracker SkyNanny—Review. Available from: http://tracking.watch/child-gps-tracker-skynanny-review/ [Accessed 17 October 2016].

Huang Y, Li G, 2010. Descriptive Models for Internet of Things. *Proc. of Intelligent Control and Information Processing International Conference (ICICIP-2010)*, pp. 483–486.

Ingersoll B, Wainer A, 2013. Initial Efficacy of Project ImPACT: A Parent-Mediated Social Communication Intervention for Young Children with ASD. *Journal of Autism Development*, 43, 2943–2952.

Jamdeaf, 2015. Who Are The Children With Special Needs? *Jamaica Association for the Deaf*. Available from: http://www.jamdeaf.org.jm/articles/who-are-the-children-with-special-needs [Accessed 28 September 2016].

Kelly M, 2014. GPS Tracking Devices for Children with Autism. Autism Articles. Autism News. Available from: https://www.autismparentingmagazine.com/gps-tracking-devices-children-autism/ [Accessed 15 January 2017].

Krull J, 2015. Visual Schedules and Task Managers: Apps for Autism and Beyond! Part 1. *IPAT ND Assistive Technology Blog*. Available from: http://ndipat.org/blog/visual-schedules-and-task-managers-apps-for-autism-and-beyond-part-1/ [Accessed 22 October 2016].

Lamkin P, 2016. The Best Kids Trackers: Using Wearables for Child Safety. Available from: http:// www.wareable.com/internet-of-things/the-best-kids-trackers [Accessed 12 October 2016].

Lewis L, 2014. 7 Tracking Devices to Find a Lost Child with Autism-Friendship Circle-Special Needs Blog. Available from: http://www.friendshipcircle.org/blog/2014/01/15/7-tracking-devices-to-find-a-lost-child-with-autism/ [Accessed 12 October 2016].

Lidwina J, Jayanthy S, 2016. A Smart Wearable Sensor Device For infants with ASD. *Middle-East Journal of Scientific Research*, 24 (S2), 263–267.

Maffei M, 2014. 9 Must-Have Gadgets for Kids with Autism. Available from: http://www.sheknows. com/parenting/articles/1034307/9-must-have-gadgets-for-kids-with-autism and http://www. sensacalm.com/ [Accessed 16 January 2017].

McClanahan LE, Krantz PJ, 2010. *Activity Schedules for Children with Autism: Teaching Independent Behavior*. Bethesda, MD: Woodbine House.

Michel P, 2004. The Use of Technology in the Study, Diagnosis and Treatment of Autism. *Final term paper for CSC350: Autism and Associated Developmental Disorders, Carnegie Mellon University, Pittsburgh, Pennsylvania*, pp. 1–26.

Mighty Z, 2016. Children with Special Needs. Available from: http://www.childrenwithspecial-needs.com/ disability-info/autism/ [Accessed 14 September 2016]

Moore D, McGrath P, Thorpe J, 2000. Computer-Aided Learning for People with Autism— A Framework for Research and Development. *Innovations in Education and Training International*, 37(3), 218–228.

Nemours, 2016. *Kids with Special Needs*. Available from: https://kidshealth.org/en/kids/special-needs.html [Accessed 14 September 2016].

Nicolás FT, 2004. Tecnologías de Ayuda en Personas con Trastornos del Espectro Autista: Guía Para Docentes. Available from: http://diversidad.murciaeduca.es/tecnoneet/docs/autismo.pdf [Accessed 29 September 2016].

Notbohm E, 2005. *About Autism: What You Need to Know*. Autism Speaks Inc. 2013. Available from: https://www.autismspeaks.org/sites/default/files/afyo_about_autism.pdf [Accessed 12 October 2016].

Perera C, Zaslavsky A, Christen P, Georgakopoulos D, 2013. Context Aware Computing for the Internet of Things: A Survey. *IEEE Commun. Surveys Tuts.*, 16 (1), 414–454.

Pierce K, Schreibman L, 1994. Teaching Daily Living Skills to Children with Autism in Unsupervised Settings through Pictorial Self-Management. *Journal Application Behaviors Anal*, 27(3), 471–481.

Pleinis A, Romanczyk R, 1985. Analyses of Performance, Behavior, and Predictors for Severely Disturbed Children: A Comparison of Adult vs. Computer Instruction. *Analysis & Intervention in Developmental Disabilities*, 5, 345–356.

Prospero MA, 2014. 13 Best Autism Apps for the iPad. *LAPTOP Reviews Editor*. Available from: http://www.laptopmag.com /articles/best-autism-apps-ipad [Accessed 22 October 2016].

Schepis M, Reid D, Behrmann M, Sutton K, 1998. Increasing Communicative Interactions of Young Children with Autism Using a Voice Output Communication Aid and Naturalistic Teaching. *Journal of Applied Behavior Analysis*, 31, 561–578.

Smith S, 2015. The Autism Glass Project at Stanford Medicine. Available from: http://www.medicaldaily.com/autism-glass-project-researchers-stanford-are-using-google-glass-help-treat-autism-357844 [Accessed 22 October 2016].

Sula A, Spaho E, 2014. Using Assistive Technologies in Autism Care Centers to Support Children Develop Communication and Language Skills. A Case Study: Albania. *Academic Journal of Interdisciplinary Studies*, 3(1), 203–212.

Takeo T, Toshitaka N, Daisuke K, 2007. Development application softwares on PDA for autistic disorder children. *IPSJ SIG Technical Report*, 12, 31–38.

Tzvi, 2011. iPrompts® PRO on the App Store. Handhold Adaptive. Available from: http://www.friendshipcircle.org/blog/2011/02/16/seven-scheduling-behavioral-apps-for-children-with-special-needs/ [Accessed 23 October 2016].

Viseu A, 2003. Simulation and Augmentation: Issues of Wearable Computers. *Ethics and Information Technology*, 5(1), 17–26.

Woodcock A, Woolner A, 2007. Facilitating Communication, Teaching and Learning in Children with an ASD: Project Spectrum. *IEEE 6th International Conference on Development and Learning (ICDL)*, pp. 11–13.

Index